BRITAIN AND DISARMAMENT

For my mother, Audrey Mary Brodie Walker, 1930–85, who encouraged me to apply for and then accept the vacant post in the Arms Control and Disarmament Research Unit in the Foreign and Commonwealth Office.

Britain and Disarmament

The UK and Nuclear, Biological and Chemical Weapons Arms Control and Programmes 1956–1975

JOHN R. WALKER
Foreign and Commonwealth Office, UK

Routledge
Taylor & Francis Group
LONDON AND NEW YORK

First published 2012 by Ashgate Publishing

Published 2016 by Routledge
2 Park Square, Milton Park, Abingdon, Oxon OX14 4RN
711 Third Avenue, New York, NY 10017, USA

First issued in paperback 2017

Routledge is an imprint of the Taylor & Francis Group, an informa business

British Library Cataloguing in Publication Data
Walker, John R., 1960–
Britain and disarmament : the UK and nuclear, biological and chemical weapons arms control and programmes 1956–1975.
1. Disarmament--Great Britain--History--20th century.
2. Nuclear weapons--Government policy--Great Britain--History--20th century.
3. Nuclear arms control--Great Britain--History--20th century. 4. Biological weapons--Government policy--Great Britain--History--20th century.
5. Chemical weapons--Government policy--Great Britain--History--20th century.
6. Great Britain--Military policy. 7. Great Britain--Politics and government--1945–1964.
8. Great Britain--Politics and government--1964–1979. 9. Cold War.
I. Title
327.1'74'0941-dc22

Library of Congress Cataloging-in-Publication Data
Walker, John R., 1960–
Britain and disarmament : the UK and nuclear, biological and chemical weapons arms control and programmes, 1956–1975 / John R. Walker.
p. cm.
Includes bibliographical references and index.
ISBN 978-1-4094-3580-8 (hardcover)
1. Disarmament-Great Britain--History--20th century. 2. Arms control--Great Britain--History--20th century.
3. Disarmament--Government policy--Great Britain. 4. Arms control--Government policy--Great Britain.
5. Weapons of mass destruction--Government policy--Great Britain. I. Title.

JZ6009.G7W35 2012
327.1'745094109046--dc23

2011026565

ISBN 13: 978-1-138-11096-0 (pbk)
ISBN 13: 978-1-4094-3580-8 (hbk)

Contents

Acknowledgements

This book is the result of some twenty-five years of research into the dusty recesses of Foreign and Commonwealth Office and other government departmental files on nuclear, chemical and biological arms control and disarmament policies as well as on the weapons programmes themselves. The object of this research was usually to address a specific question on aspects of UK policies in order to find out how an issue had been handled, why a decision had been taken or not taken; or to explain why a treaty had been written the way it had. The past continues to exert an influence on current policy preoccupations and the historical background often helps inform contemporary debates. This is particularly true in the case of Treaties. The files on which these research papers were based have now entered the public domain at The National Archives, Kew, so there was a chance to flesh out the original papers referencing the original sources. In more recent times I was on a part-time secondment to the University of Southampton's Mountbatten Centre to participate in the Arts and Humanities Research Council-funded project that ran from 2004 to 2006 under Professor John Simpson on the history of the UK's nuclear weapons programme. I volunteered to address the relationships between the weapons programme and arms control. In addition to the material that led to my book *British Nuclear Weapons and the Test Ban 1954–1973* published by Ashgate in 2010, I wrote papers on Britain and the Non-Proliferation Treaty (NPT), the cut-off in the production of fissile material, and the Strategic Arms Limitation Talks and the Anti-Ballistic Missile Treaty – one of these, on Britain and the Atlantic Nuclear Force and the Nuclear Non-Proliferation Treaty – was published as UK Nuclear History Working Paper No. 4 on the Mountbatten Centre's website.[1] All of these addressed the relationships between weapons programmes and arms control and disarmament. When looking at the various papers on nuclear, chemical and biological arms control and the UK's own programmes it struck me that here was a collection that told an interesting story about UK approaches and attitudes to disarmament during the Cold War from the 1950s up until the 1970s that has not really been addressed in the academic literature. It is moreover rare to find a study that looks at nuclear, chemical and biological weapons programmes and

1 *British Nuclear Weapons, the ANF and the NPT 1965–1968*, UK Nuclear History Working Paper No. 4, at http://www.mcis.soton.ac.uk/Site_Files/pdf/nuclear_history/Working_Paper_No_4.pdf, last accessed 14 March 2011.

arms control in the same volume – hence the idea for this book, which attempts to pull together all the diverse 'single issue' papers and chapters with new linking material and analysis. I hope the joins are not too obvious. The hope is that this effort might help other scholars working in the field by identifying relevant issues and gaps that could benefit from further research into the voluminous archival sources.

Over the years many colleagues and former officials who were involved in some of the original events discussed in the following pages have offered comments on the drafts – some of whom are alas no longer with us such as David Summerhayes, Ronald Hope-Jones and Ivor Porter. David was particularly proud of his contribution to the negotiation of the Biological and Toxin Weapons Convention (BTWC). I would also like to thank Lorna Arnold, Andrew Barlow, Gradon Carter, Peter Hammond, William Hillier-Fry, Ken Johnston, Frank Panton, Kate Pyne and Clive Rose for their comments, advice and help over the years. My colleagues in the academic world have also offered their encouragement and I have certainly benefited from many hours of discussions with them – Milton Leitenberg who urged that I try to publish some of my historical papers on the origins of the BTWC; Brian Balmer, Matthew Jones, Caitriona McLeish, Richard Moore, Julian Perry Robinson, John Simpson, Nicholas Sims and Kristan Stoddard.

This is also the place to acknowledge gratefully the agreement of Peter Herby at the International Committee of the Red Cross (ICRC) to include a version of the paper that appears in Chapter 3 originally prepared for an ICRC conference held in Montreux in March 2010 on incapacitating chemical weapons – the report of that conference can be found on the ICRC website.[2] Thanks are due too to Julian Perry Robinson at the Harvard-Sussex Program for allowing me to include the article (appearing here in Chapter 4) that I wrote for the February 2010 issue of *The CBW Conventions Bulletin*. And finally, thanks to Professor John Simpson for permission to include the material that appears in Chapter 9 – as noted above this was originally UK Nuclear History Working Paper No 4.

Thanks are due to all at Ashgate – Tom Gray, Gemma Hayman, Emma Gallon, Beatrice Beaup, Nicole Norman and Jonathan Hoare – for their commendable efficiency and courtesy in the preparation of this book.

As ever responsibility for errors of fact or interpretation in the subsequent pages are my responsibility alone; this is also the place to note that the views expressed in this volume are mine and not necessarily those of the Foreign and Commonwealth Office. This is not an official history.

---

[2] See *Expert Meeting Incapacitating Chemical Agents Implications for International Law*, Montreux, Switzerland 24 to 26 March 2010, ICRC; see also http://www.icrc.org/eng/resources/documents/publication/p4051.htm, last accessed 14 March 2011.

Glossary

ABM	Anti-Ballistic Missile
ABMT	Anti-Ballistic Missile Treaty
ACDA	Arms Control and Disarmament Agency (US)
ACDRU	Arms Control and Disarmament Research Unit (Foreign Office/Foreign and Commonwealth Office)
ACDS (Pol)	Assistant Chief of the Defence Staff (Policy), MOD
ACSA (N)	Assistant Chief Scientific Adviser (Nuclear), MOD
AEDD	Atomic Energy and Disarmament Department (Foreign Office/Foreign and Commonwealth Office)
AE (O) C	Atomic Energy (Official) Committee, Cabinet Office
ANF	Atlantic Nuclear Force
ARTE	Admiralty Reactor and Test Establishment, Dounreay
ASR	Air Staff Requirement
AST	Air Staff Target
AWRE	Atomic Weapons Research Establishment, Aldermaston
BTWC	Biological and Toxin Weapons Convention
BW	Biological Weapons/Warfare
BZ	3-Quinuclidinyl benzilate (incapacitating chemical agent)
CBW	Chemical and Biological Weapons/Warfare
CCD	Conference of the Committee on Disarmament, Geneva
CDAB	Chemical Defence Advisory Board, MOD
CDE	Chemical Defence Establishment, Porton Down
CDEE	Chemical Defence Experimental Establishment, Porton Down/Nancekuke
CDS	Chief of the Defence Staff, MOD
CEGB	Central Electricity Generating Board
CN	ω-Chloroacetophenone (riot control agent)
CR	Dibenzoxazepine (riot control agent)
CS	2-Chlorobenzalmalononitrile (riot control agent)
CTBT	Comprehensive Nuclear Test Ban Treaty
CW	Chemical Weapons/Warfare
CWC	Chemical Weapons Convention
DBO	Dibenzoxazepine (riot control agent – initial name for CR)
DCA (PN)	Deputy Chief Adviser (Policy and Nuclear), MOD

DM	10-Chloro-5,10-dihydrophenarsazinine (Adamsite) (riot control agent)
DOPC	Defence and Overseas Policy Committee (Cabinet)
DRPC	Defence Research Policy Committee, MOD
DS	Defence Secretariat, MOD
ENDC	Eighteen Nation Disarmament Committee, Geneva
EURATOM	European Atomic Energy Agency
FBS	Forward Based Systems
FCO	Foreign and Commonwealth Office
FO	Foreign Office
G 12	Group of 12 Non-aligned Countries in the CCD, Geneva
GB	O-Isopropyl methylphosphonofluoridate (sarin nerve agent – chemical warfare agent)
GCD	General and Complete Disarmament
HEU	Highly Enriched Uranium
HMG	Her Majesty's Government
IAEA	International Atomic Energy Agency
ICBMs	Intercontinental Ballistic Missiles
ICRC	International Committee of the Red Cross
INFCIRC	Information Circular (IAEA documentation notation)
JCAE	Joint Congressional Committee on Atomic Energy (US)
JIC	Joint Intelligence Committee, Cabinet Office
LEU	Low Enriched Uranium
LSD	Lysergic Acid Diethylamide (incapacitating chemical agent)
LTC	Long Term Costings (MOD)
MIRVs	Multiple Targetable Re-entry Vehicles
MLF	Multilateral Force
MOD	Ministry of Defence
MRBMs	Medium Range Ballistic Missiles
MRC	Medical Research Council (UK)
MRE	Microbiological Research Establishment, Porton Down
NAC	North Atlantic Council
NASR	Naval Air Staff Requirement
NATO	North Atlantic Treaty Organisation
NCA	National Command Authority
NGST	Naval Ground Staff Requirement
NIO	Northern Ireland Office
NNWS	Non-Nuclear Weapons States
NP (O) C	Nuclear Policy (Official) Committee, Cabinet Office
NPT	Nuclear Non-Proliferation Treaty

NRDC	Nuclear Requirements for Defence Committee, Cabinet Office
NSA	Negative Security Assurance
NSC	National Security Council (US)
PAL	Permissive Action Links
PNE	Peaceful Nuclear Explosions
PPNN	Programme for Promoting Nuclear Non-Proliferation
PTBT	Partial Test Ban Treaty
RCA	Riot Control Agent(s)
ROE	Rules of Engagement
SACEUR	Supreme Allied Commander Europe
SALT	Strategic Arms Limitation Talks/Treaty
SIPRI	Stockholm International Peace Research Unit
SNDVs	Strategic Nuclear Delivery Vehicles
SPAD	Self-Protection Aid Device
SSEB	South of Scotland Electricity Generating Board
TNA	The National Archives, Kew (UK)
UKAEA	United Kingdom Atomic Energy Authority
UN	United Nations
UNDC	United Nations Disarmament Commission, New York
UNSCR	United Nations Security Council Resolution
UNSG	United Nations Secretary-General
USAEC	United States Atomic Energy Commission
VX	O-Ethyl S-2-diisopropylaminoethyl methyl phosphonothiolate (chemical warfare agent – nerve agent)
WHO	World Health Organisation, Geneva
WOCD	Western Organisations Coordination Department, Foreign Office
WOD	Western Organisations Department, Foreign Office

Chronology of Key Events: Britain and Disarmament 1956–1975

10 July 1956	Cabinet Defence Committee decides to abandon UK's offensive chemical weapons programme.
27 June 1962	UK announces cessation of production of highly enriched uranium at Capenhurst for military purposes.
3 May 1963	Cabinet Defence Committee decides that the UK should re-acquire an offensive CW capability for both lethal and incapacitating chemical agents.
21 April 1964	Prime Minister Alec Douglas-Home announces that military plutonium production in the UK is being gradually terminated.
17 August 1965	USA tables draft of a Nuclear Non-Proliferation Treaty (NPT) in the Eighteen Nation Disarmament Committee negotiations in Geneva.
20 June 1966	UK concludes agreement with the IAEA for safeguarding Bradwell for a period of five years to expire on 1 September 1971.
5 December 1967	During the NPT negotiations the UK offers that at such time as international safeguards on civil nuclear facilities were put into effect in implementation of a provision of the NPT, it would accept similar safeguards subject to exclusions of national security reasons only.
1 July 1968	NPT opened for signature.

6 August 1968	UK tables working paper in the ENDC in Geneva on Microbiological Warfare.
10 July 1969	UK tables draft in the ENDC of a Convention for the Prohibition of Biological Methods of Warfare.
26 August 1969	UK tables revised draft Convention for the Prohibition of Biological Methods of Warfare.
17 November 1969	USA and USSR open first round of Strategic Arms Limitation Talks in Helsinki (subsequent rounds rotate between Helsinki and Vienna).
18 December 1969	Cabinet decides to change UK interpretation of 1925 Geneva Protocol to exclude CS from its scope.
18 August 1970	At the CCD the UK tables further revised Convention for the Prohibition of Biological Methods of Warfare.
5 March 1970	NPT enters into force.
10 March 1971	IAEA's Model NPT safeguards agreement (INFCIRC/153) agreed at Vienna.
30 March 1971	USSR and its allies table draft Biological Weapons Convention in the Conference of the Committee on Disarmament (ENDC successor).
5 August 1971	USA and USSR table separate but identical draft Biological Weapons Convention texts in the CCD.
28 September 1971	Text of Biological and Toxin Weapons Convention tabled in the CCD co-sponsored by UK, USA, USSR, Bulgaria, Canada, Czechoslovakia, Hungary, Italy, Mongolia, Netherlands, Poland and Romania.
10 April 1972	Biological and Toxin Weapons Convention open for signature in London, Washington and Moscow.

26 May 1972	USA and USSR sign at Moscow the Anti-Ballistic Missile Treaty and Interim Agreement with Respect to the Limitation of Strategic Offensive Arms.
4 October 1973	Ministers agree on revised rules of engagement for use of CR in Northern Ireland on 4 October 1973.
15 October 1974	Authority given to use CR at the Maze Prison, but not used in the event.
26 March 1975	Biological and Toxin Weapons Convention enters into force.

Chapter 1

Introduction

Britain and Disarmament: Weapons Programmes and Arms Control: Balancing Conflicting Requirements

Little has been written about Britain's arms control and disarmament and how it has impacted on the UK's own nuclear, chemical and biological weapons programmes.[1] There is a wealth of material in The National Archives (TNA) at Kew covering not only the UK's role in multilateral and bilateral arms control negotiations since the late 1950s, but also on the UK's own weapons programmes. Now whilst the nuclear weapons programme was avowed and the centre of public and Ministerial concerns, the UK's own chemical and biological weapons programmes lurked in the shadows in comparison, there being no clear statement on whether the UK even possessed such weapons or had active programmes for much of the period covered in this book.[2] The reality was that essentially economic pressures and the nuclear programme itself had led to the abandonment of offensive CB programmes and the redirection of effort onto defensive work – protective clothing, agent detectors, medical countermeasures and decontamination. However, in the chemical area there was a renewed interest in reacquiring an offensive programme for both lethal and incapacitating weapons in the early 1960s. Although the nuclear programme faced potential or actual constraints from arms control measures, there was nothing comparable facing the CW programme in the way that a Comprehensive Nuclear Test Ban Treaty or a Cut-off in the Production of Fissile Material for weapons purposes

1 See author's own *British Nuclear Weapons and the Test Ban 1954–1973* (Farnham, 2010).

2 In March 1960 the Minister of Defence Harold Watkinson told the House of Commons in response to a question from Mr A. Henderson that the UK was 'not stockpiling these weapons' (Hansard Oral Answers 23 March 1960 column 497). The issue of UK capabilities would emerge again in 1968. See also Brian Balmer, 'Keeping Nothing Secret: United Kingdom Chemical Warfare Policy in the 1960s', *The Journal of Strategic Studies* vol. 33, no. 6 (2010), pp. 871–93.

could have hobbled the UK's nuclear weapons programme. The only constraint in place in the 1950s and 1960s was the 1925 Geneva Protocol, which was effectively only a no-first use agreement (covering both chemical and biological weapons) as most parties had reserved the right of retaliation if attacked first.

There are several key questions that can be addressed by a detailed review of a wide range of National Archive papers on UK nuclear, chemical and biological programmes and the formulation and conduct of arms control and disarmament policy. What were the histories of the weapons programmes? What pressures impacted on them domestically in terms of economic and doctrinal as well as tactical considerations? What was the impact of arms control and disarmament? Was the UK an enthusiastic or reluctant player on the arms control stage and did arms control adversely affect the directions and content of weapons programmes? Hopefully the following chapters will provide some initial answers to these and other questions. There are several possible areas in the time period 1956 through to 1975 that provide good case studies and enable us to look in detail at the nuclear, chemical and biological programmes, which in turn allows some useful comparisons to be made and, hopefully, useful conclusions to be drawn. We can also explore UK attitudes to effective verification in these areas, which for the entire period and beyond was an ever present requirement.

Inevitably since nuclear weapons were so prominent in UK defence policy, much of the material in this study reflects preoccupations in this area – something that can be seen from the volume of original departmental files covering nuclear issues compared to biological or chemical and hence the level of detail in those chapters covering nuclear topics. There is an intentionally strong narrative element in this study; this, in part, reflects the origins of some of the chapters, which were originally internal Foreign and Commonwealth Office (FCO) research papers designed to address specific questions of interest to policy makers, such as why was the Biological and Toxin Convention (BTWC) written this way? Or why did the UK abandon its offensive chemical weapons (CW) programme? Other topics have a contemporary resonance with issues facing the 1997 Chemical Weapons Convention (CWC), in this case incapacitating chemical agents, riot control agents and law enforcement. Understanding the past can help make sense of the present and future. This is no less true in arms control than in any other subject. The objectives in the research papers were often to explain how an issue evolved, to identify the key factors explaining the decisions and to characterise the key turning points. Although the chapters in this volume therefore started off as separate ventures, and the nuclear ones were contributions to an Arts and Humanities Research Council-funded project at Southampton University on the UK nuclear weapons programme 1953 to 1973, when pulled together it seems that they do present a coherent narrative and illuminate common cross-cutting issues. It is also a first attempt to look at UK nuclear, chemical and biological arms control policy and programmes in the Cold War in a single place based on extensive use of primary sources. The period under review here (1956–75) starts with the landmark decision to abandon the offensive CW programme and ends with the internal travails over the use of CR (a potent riot control agent to emerge from the UK's incapacitating chemical programme) *in extremis* to deal with a mass break-out of Republican prisoners from the Maze prison in Northern Ireland during the 'Troubles'. Just why this is relevant to a history of the UK and disarmament will hopefully become clear.

Chapter 2 begins with a brief overview of chemical warfare in the UK from the decision in 1956 to abandon its offensive programme through the reversal of that decision in 1963 and the subsequent abortive attempts to implement that rearmament requirement. It also reviews the UK CW programme and any impact of arms control considerations in the years to 1975. There was, however, one more promising area of chemical warfare research in this period and that concerns identification and development of incapacitating chemical agents for UK military use; the UK programme from the late 1950s through to the early 1970s is considered in Chapter 3 along with the impact of the 1925 Geneva Protocol on concepts of use and policy. We then look at another aspect of chemical warfare, riot control agents: Chapter 4 considers UK policy on the interpretation of the 1925 Geneva Protocol and the internal sensitivities over the use of the potent riot control agent CR in Northern Ireland and in Hong Kong.

The next three chapters switch to biological warfare with Chapter 5 describing the BTWC's origins in the late 1960s and more particularly the UK's role in it. Negotiations leading to the Convention ran from 1969 through to 1971, but we focus in Chapter 6 on the decisive months from March to September 1971 when the Convention was finalised during which period the UK sought to maintain its original vision of a Convention. Chapter 7 looks at a specific issue within the negotiations; namely whether the Convention in the end did extend to a prohibition on offensive research as originally envisaged by the UK in 1968. This helps illustrate one of the problems of treaty development and interpretation where its negotiating history can play a key part in determining the nature and extent of its basic obligations, in this case the scope of the prohibitions.

Chapters 8, 9 and 10 respectively address the UK and three key nuclear arms control treaties. The first of these, cut-off in the production of fissile material for weapons purposes and safeguards on civil nuclear power programmes, are reviewed in Chapter 8. A cut-off treaty could have imposed severe restrictions on the UK nuclear weapons programme given its perennial shortage of fissile material without which nuclear weapons could not be produced. It is thus most instructive to see how the UK managed to balance the need to be seen supporting arms control whilst protecting its defence interests at the same time. Is such a balancing act achievable? Hopefully the discussions here will show that this is possible. This chapter will also discuss the UK role in the evolution of safeguards for civil nuclear facilities, which sheds light on British attitudes to verification or, as it used to be known, 'control', without which acceptance of any disarmament treaty would not happen. Chapter 9 looks at the proposals for an Atlantic Nuclear Force raised by the Labour Government of Harold Wilson as an alternative to the US inspired Multilateral Nuclear Force – both were variations of nuclear sharing arrangements within NATO – and the extent to which these constrained or risked proliferation. It then goes on to see how the UK's nuclear programme and cooperation with the US could have been affected by the 1968 Nuclear Non-Proliferation Treaty as the negotiations grappled with the difficult issue of nuclear sharing arrangements in NATO. The final nuclear chapter (10) reviews UK attitudes to the bilateral US and USSR strategic nuclear arms control talks as well as the negotiations for an Anti-Ballistic Missile Treaty and the extent to which the UK was able to protect the Polaris Improvement Programme that had begun at roughly the same time and was aimed at maintaining the UK nuclear deterrent's effectiveness in the face of feared Soviet ballistic missile defence developments in the 1970s.

Chapter 11 offers some concluding thoughts on three key questions: To what extent has the UK's need to protect its own weapons programmes constrained its role in arms control? What, if any, impact have the UK's own programme experiences such as overcoming production and weaponisation problems had on shaping its policies? Does the UK experience tell us anything more generally about the arms control process? We end with some thoughts on areas for further research.

This might appear at first sight a somewhat eclectic selection of topics, but they do show the breadth of UK engagement and record for the first time in detail how nuclear, biological and chemical programmes (offensive and defensive) developed in the UK through a period of gradual retrenchment where even the nuclear weapons programme was not immune to economic pressures. These topics provide insights into UK thinking about arms control and disarmament negotiations, agreements and regimes as well as on their implementation; they also open a window on the tortuous and protracted process of agreeing treaty language in a multilateral setting.

A brief word on some of the technical aspects of nuclear, chemical and biological weapons will help explain a little of the context. Nuclear weapons cannot be produced without access to the fissile materials highly enriched uranium or plutonium, which are produced in large and expensive facilities;

the exact quantities required for individual weapons depends on a number of variables such as the sophistication of the warhead design and the isotopic ratios of the fissile material involved. Generally speaking the more advanced the design, the smaller the quantity required. This is why the prospects of halting production of these materials for weapons purposes posed problems for the UK. Further explanations of nuclear weapons design issues are in Chuck Hansen's *US Nuclear Weapons: The Secret History*.[3] Chemical weapons on the other hand can be, and have been, made from a wide range of materials whose effects range from lethal in very small doses to the merely incapacitating or irritating. Dose is important here too as lethal chemicals can be merely incapacitating in effect at very low levels. The chemicals needed to manufacture chemical warfare agents moreover are produced in civil plants and often have many extensive legitimate peaceful uses from pesticides through to pharmaceuticals. Any ban or controls on their production would require extensive verification measures – declarations of facilities and on-site inspections – if they were to be at all credible. Finding an ideal incapacitating agent – one that acts quickly with minimal adverse side effects on the target population – presented significant technical problems in the 1960s and these have not gone away even today in the early twenty-first century. Biological weapons present a unique set of problems in terms of disarmament: they are comparatively easy to produce, certainly when compared to nuclear weapons – although this is by no means a trivial exercise as the UK found out in the early 1950s. The agents exist in nature as microorganisms that cause disease, death and incapacitation in humans, animals and plants. Microorganisms (bacteria, viruses, rickettsia) and toxins (poisonous substances produced by bacteria, plants, scorpions and algae for example) exist in nature, some of which have legitimate and extensive uses in medical research, industry, agriculture, vaccines and in the pharmaceutical industry, which make the concept of control especially challenging or even impossible in the view of some observers. A more detailed discussion of the technical aspects of chemical and biological weapons agents can be found in a World Health Organisation (WHO) publication *Public Health Response to Biological and Chemical Weapons*.[4]

A word about the distinction between arms control and disarmament: until the late 1950s, when negotiations started between the UK, US and USSR on a Comprehensive Nuclear Test Ban Treaty in 1958, diplomatic efforts focused

---

3 Chuck Hansen, *U.S. Nuclear Weapons: The Secret History* (Arlington, TX, 1988), Chapter 1: 'Weapons Physics', pp. 11–29.

4 World Health Organization, *Public Health Response to Biological and Chemical Weapons. WHO Guidance, Second edition of Health Aspects of Chemical and Biological Weapons: Report of a WHO Group of Consultants, Geneva World Health Organization, 1970* (Geneva, 2004).

on disarmament measures – the elimination of entire categories of weapons – and this was to be achieved in stages, usually three in the various plans of the period. However, by this time and into the early 1960s many academic and strategic thinkers such as Thomas Schelling, Morton Halperin and Hedley Bull were developing a new concept of arms control for the modern era, which was not necessarily about disarmament *per se*, but more about ensuring stability and predictability in the strategic arms race between the superpowers. Force levels might in fact end up being higher, but the net result would be greater security for all with deterrence strengthened. For Schelling and Halperin arms control, 'can be as formal as a multilateral treaty or as informal as a shared recognition that certain forms of self-control will be reciprocated. It may involve "cops and robbers" activities like cheating and detection, but may also involve many of the continuing regulatory and negotiatory processes that we associate with bureaucracy and diplomacy.'[5] As Hedley Bull, who went on to become the first head of the Foreign Office's Arms Control and Disarmament Research Unit in 1965, noted, arms control is 'concerned with discriminating between those kinds or quantities of forces and weapons that promote the stability of the balance of power, and those which do not'.[6] Therefore in this study international agreements such as the Nuclear Non-Proliferation Treaty of 1968 and the Anti-Ballistic Missile Treaty 1972 are examples of arms control, whilst the Biological and Toxin Weapons Convention (BTWC) 1972 is a disarmament treaty. Throughout the period of this book pursuit of arms control and disarmament measures was an activity that no British government – whether Conservative or Labour – could be seen to be avoiding or obstructing. Exactly why they felt this way will hopefully become clearer in the chapters that follow.

[5] Thomas C. Schelling and Morton H. Halperin, *Strategy and Arms Control* (Washington, DC: 1985 edition) (first published 1961), p. 141. One of the other key works from this era is Donald G. Brennan (ed.), *Arms Control and Disarmament and National Security* (New York, 1961).

[6] Hedley Bull, *The Control of the Arms Race* (New York, 1961), p. 61.

Chapter 2

The UK's Offensive CW Programmes: Abandonment and Aspirations 1956–1969

Introduction

The negotiations that were to lead to the Convention on the Prohibition of the Development, Production, Stockpiling and Use of Chemical Weapons and on their Destruction did not to start until February 1984. Negotiations concluded in September 1992 and the Convention opened for signature in January 1993 and entered into force in April 1997 – well beyond the timeframe of this volume. By that time, however, the UK had long since ceased to possess an offensive chemical weapons programme, but what is much less clear is exactly why the UK abandoned this capability and whether there were any contemporaneous arms control or disarmament considerations. For the answers to these questions we must look back to the 1950s. At that stage in international discussions on disarmament the focus was on staged approaches, with chemical and biological weapons generally featuring in the second phase in many proposals. In September 1955, for instance, the UK's memorandum to the UN Disarmament Subcommittee on the Control Organ of a future disarmament agreement, noted that inspections would begin in Stage 2 at plants making chemical and biological weapons.[1] The only extant agreement on chemical and biological warfare in this period was the 1925 Geneva Protocol, which was in practice a prohibition on the first use of these weapons. Many of its parties had entered reservations upon ratification, which in essence stated that they would cease to be bound by the non-use prohibition if any other state attacked them with CB weapons. Did this have any bearing on UK decisions? What was the status of UK chemical warfare programmes from the 1950s until the end of our period? This chapter seeks to address these issues.

1 British Memorandum Concerning the Control Organ, submitted to the Disarmament Subcommittee, 13 September 1955 in Disarmament Commission Official Records: Supplement for April to December 1955, pp. 43–4.

The Cabinet Defence Committee Decision to Abandon the UK CW Programme: July 1956

The background and reasons for the UK decision to abandon its offensive CW programme can be found in a Memorandum submitted to the Cabinet Defence Committee by the Minister of Defence on 4 July 1956.[2] Following an earlier instruction from the Minister, the Defence Research Policy Committee (DRPC) had reviewed all R&D expenditures in the face of planned significant reductions in the defence budget. One of the conclusions to emerge from the DRPC's review was a recommendation that the UK abandon the means of large-scale production of nerve agent (GB) and of the development of weapons and ammunition to use such agent.[3] Four years previously the Cabinet Defence Committee had authorised construction of a 50-ton-a-week GB nerve agent plant in April 1952 – this was to be built at Nancekuke in Cornwall, but this was not acted upon at the time on account of the then financial stringency and of the possibility of obtaining aid from the US. The Minister also thought that it would be reasonable in the current circumstances as they applied in the summer of 1956 to abandon the UK's reserve capacity for making mustard gas currently held by the Ministry of Supply as well as the stocks of weapons (artillery projectiles and gravity bombs) held by the Ministry of Supply and the Service Departments. The retention of this capability was increasingly expensive and new storage facilities were required at RAF stations if the mustard-filled bombs were to be kept. UK nuclear weapons as well as the massive US nuclear stockpile would justify the UK, in its then economic circumstances, giving up its own chemical weapons. It is worth noting that the RAF's first squadron of Valiant bombers equipped with the Blue Danube bomb went operational in July 1956.[4] Furthermore, it was also noted that the US was devoting considerable resources to developing nerve agents for its own armed forces; this, it was thought, would provide a 'powerful deterrent' against any Soviet use.

The full Cabinet Defence Committee considered the Minister of Defence's Memorandum at its 10 July 1956 meeting.[5] The Prime Minister, Defence

---

2 TNA CAB 131/17, Cabinet Defence Committee, D.C. (56) 13 (Revise), 4 July 1956. For further historical background on the UK's CW programme see G.B. Carter and Graham S. Pearson, 'Past British Chemical Warfare Capabilities', *RUSI Journal* vol. 141, no. 1 (February 1996), pp. 59–68.

3 GB was one of the Second World War German nerve agents: O-isopropylmethlyphosphonofluoridate.

4 Humphrey Wynn, *RAF Nuclear Deterrent Forces* (London, 1994), pp. 117–18.

5 TNA CAB 131/17, Cabinet Defence Committee, D.C. (56) 6th Meeting Item 3, 10 July 1956.

Minister, Chancellor, Foreign Secretary, Secretary for War, Secretary for Air, the First Lord of the Admiralty, the Commonwealth Relations Secretary, First Sea Lord and Chief of the Naval Staff, the Minister of Supply, Chief of the Air Staff and the Vice Chief of the Imperial General Staff were in attendance.

In his opening remarks as he spoke to his Memorandum the Minister of Defence added that substantial resources were occupied in maintaining large-scale production of nerve agent and delivery systems. The UK would never be the first to prepare to initiate the use of chemical weapons in war unless such weapons had been used against it – the UK's role as a State Party to the Geneva Protocol was a key consideration here. (It is also worth noting in this context that there was no reference to Soviet CW capabilities as intelligence was extremely limited.[6]) In the subsequent Committee discussion Ministers noted that the UK's general policy was to rely on 'the deterrent power of thermonuclear and nuclear weapons' and as such it would be difficult to reconcile this policy to maintain productive capacity and stocks for chemical warfare. There was no argument or difference of opinion; the Committee decided to abandon the offensive programme on economic grounds. As one official noted, 'the present economic situation rather than the military value of CW ... is really dictating the change.'[7] This meant both the mustard gas capability and the means of large-scale nerve agent manufacture. However, the Minister of Defence, Chancellor and Minister of Supply would need to discuss the scale on which research work in the CW field and development work for defence against chemical weapons should be maintained in future. There were to be no public announcements or any use of this decision in the UN Disarmament Commission. The UK did not seek any political benefits from its unilateral disarmament plans.

Implementation of the 1956 Cabinet Defence Committee Decision to Abandon the UK's Offensive CW Programme

As we have seen, the Cabinet Defence Committee decided to abandon the UK's offensive CW programme, but members also agreed that it was desirable that the United States should be informed of what the UK had in mind. The Committee thought that before any action was taken the UK should inform the US of its decision. Consequently the Minister of Defence was invited to prepare

6 TNA CAB 131/16, Memorandum by the Secretary of State for Foreign Affairs, Likely Soviet Courses of Action up to 1 January 1957 Joint Intelligence Committee D.C. (55), 25 October 1955.

7 TNA WO 286/77, Chemical Warfare Policy Reference Ministry of Defence D.C. (56) 13 (revise) of 4 July A.E. Childs D.C.D.R.D to C.M., 9 July 1956.

a draft letter that the Prime Minister Anthony Eden might send to President Eisenhower explaining the intended UK action. However, the War Office noted in October 1956 that no action could be taken on the July decision until it was clear to it that the US had been so informed.[8] The Defence Secretary had submitted a draft letter for the Prime Minister to send to President Eisenhower on 3 September to advise him of the UK decision on its CW programme. However, Eden suggested that because of the UK's other pre-occupations (Suez) it was not a good time to approach the President and proposed to delay sending the message for a month.

MOD reminded No.10 when the month was up, but as of 17 October 1956 there was no response.[9] By the end of February 1957 there were still no further developments. The matter was by then becoming urgent for the War Office since for economic reasons the ammunition supply depot storing the CW stockpile was scheduled to close in March 1958. Given the stockpile's nature, the War Office had to dump it at sea and as a dumping programme was only practicable between April and September, early arrangements had to be made if the War Office was to meet this schedule. Therefore War Office officials needed to know whether the US had in fact been notified of UK policy intentions; the Defence Committee had made this an essential precondition.[10] However, by August 1958 inquiries in the MOD and Foreign Office had failed to discover whether the US had in fact ever been informed of the July 1956 decision. Nevertheless, it was clear that the Suez crisis had resulted in deferral of any message to Eisenhower.[11] It appears too that it was subsequently decided at an official level not to pursue the matter at the Prime Ministerial–Presidential level.[12]

Officially the US was never told of the July 1956 decision and of the UK's intention to rely on the US for the provision of CW agents and munitions. Nevertheless, the Americans were made aware of the changes as they affected the UK's CW R&D at the technical level through the Tripartite Conferences

8 TNA CAB 21/4505, Brigadier R.H.G.C. Spence, War Office to J.M. Wilson, Defence Secretariat Cabinet Office, 16 October 1956.

9 TNA CAB 21/4505, Ministry of Defence Minute to Mr J.M. Wilson, Cabinet Office, 17 October 1956.

10 TNA CAB 21/4505, Mr D.M. Evans, War Office to J.M. Wilson, Defence Secretariat, Cabinet Office, 27 February 1957.

11 TNA CAB 21/405, J.G. Boyd, Foreign Office, Biological and Chemical Warfare, 20 August 1958; TNA WO 286/77, Scale of Future Programme for Chemical Warfare Research and Development for Defence against Chemical Warfare, 23 October 1956.

12 That said, it seems that the Private Secretary at No.10 was still thinking in terms of the PM sending the letter in 'a week or two' on 12 November 1956. TNA WO 286/77, P.de Zulueta to J.M. Gibbon, MOD, 12 November 1956.

– joint UK, US and Canada CBW collaborative programmes on the offensive and defensive aspects of CB warfare that had been underway since the end of the Second World War.[13] US scientists expressed the view that if at any time the UK was unable to continue to make worthwhile contributions at the Tripartite Conferences, they would find it difficult to justify either participation in the Conference or to continue the exchange of detailed information with the UK.[14] (This was remarkably similar to US attitudes to continued cooperation with the UK on nuclear weapons R&D in the mid 1960s when it became clear that there was to be a replacement of the Polaris nuclear deterrent.[15]) For this reason the Defence Research Policy Committee could not accept in January 1959 any proposals to further curtail UK R&D on chemical warfare.[16] Meanwhile, all the UK's old CW weapons agents were destroyed by the end of the decade.[17]

UK CBW Offensive Programme 1963–68: Deferral of Decision to Reacquire CW Capability

In the wake of various reviews of CBW policy, the Cabinet Defence Committee decided in May 1963 that the UK should re-acquire an offensive CW capability for both lethal and incapacitating chemical agents.[18] The Chiefs of Staff reviewed UK CBW policy at the end of September 1965.[19] They concluded that developments since 1962 had reinforced the opinions that they had then expressed – namely that:

---

13 As Margaret Gowing observed, the UK and US CBW programmes at the end of the Second World War remained so closely in step as to be virtually integrated. Margaret Gowing, *Independence and Deterrence: Britain and Atomic Energy 1945–1952*, Vol. 1: *Policy Making* (London, 1974), p. 94. See also Graham Pearson and Gradon Carter, 'North Atlantic Chemical and Biological Research Collaboration: 1916–1995', *The Journal of Strategic Studies* vol. 19, no. 1 (March 1996), pp. 74–103.

14 TNA CAB 21/4505, DRP/P (59) 6, 28 January 1959, Ministry of Defence, Defence Research Policy Committee Biological and Chemical Warfare, Note by the D.R.P. Staff, p. 3.

15 John R. Walker, *British Nuclear Weapons and the Test Ban* (Farnham, 2010), Chapter 11.

16 TNA CAB 21/4505, DRP/P (59) 6 28 January 1959, Ministry of Defence, Defence Research Policy Committee Biological and Chemical Warfare, Note by the D.R.P. Staff.

17 N.J. McCamley, *Secret History of Chemical Warfare* (Barnsley, 2006), p. 169.

18 TNA CAB 131/28, Cabinet Defence Committee, 6th Meeting, 3 May 1963.

19 TNA AIR 8/2391, Annex A Report of Chemical and Biological Warfare to COS 169/65 MOD COSC Chemical and Biological Warfare, Note by the Secretary J.H. Lapsley, AVM Secretary to COSC, 28 September 1965.

- in view of the Soviet CW threat to the armed forces, improvements were required in their defensive equipment and training;
- for defensive needs against BW the UK should implement measures to assess the BW threat, to develop an early warning system, to investigate immunisation problems, and to acquire intelligence of enemy BW potentialities;
- continued research into the offensive possibilities of BW was required in order to ensure that defensive means were adequate, and so as to be aware of any likely breakthrough which might increase the potential effectiveness of BW;
- the UK should not attempt to provide a strategic CW or BW capability;
- the UK did not require a tactical offensive BW capability; and that,
- Britain required a lethal and an incapacitating tactical CW capability in the context of limited war.

In approving the report the Chiefs of Staff took note that the Chief of the Defence Staff would forward it to the Defence Secretary as an expression of their views, inviting him to seek Ministerial approval for the continuance of the present CBW policy. This approval would also extend to the plans for the production of an aircraft spray system to deliver CW.[20] This now envisaged a simplified system using expendable instead of re-fillable spray tanks and was required to be operational by June 1968, the earliest date at which the stipulated lethal agent (VX nerve agent) would have been available. The Army was also considering buying 22,000 105 mm GB nerve agent filled artillery projectiles from the US.[21] However, nothing came of this either – wider pressure on the defence budget and economic retrenchment being the likely explanation. Although delivery of the 105 mm shells could be expected within 18 months of placing an order, special facilities were needed for their storage and completion of these storage facilities would have taken about two years. Therefore the UK could not have possessed any offensive CW retaliatory capability before at least 1968–69 and the very limited capability currently envisaged could not, in the MOD view, be regarded as satisfactory without the inclusion of the aircraft spray system.[22] Original plans called for 320 aircraft spray tanks to be fitted to

[20] TNA AIR 8/2391, R.A. Hull, Chief of the Defence Staff to Secretary of State, Chemical and Biological Warfare, 28 October 1965.

[21] TNA AIR 8/2391, Annex A to COS 1972/2/7/65, Staff Requirement for an Aircraft Spray System for Retaliatory CW, 2 July 1965.

[22] TNA AIR 8/2391, DCGS to CGS Chemical Warfare Weapons, 13 June 1966.

Naval and RAF aircraft and filled with VX. The VX was to be manufactured in the UK from basic materials purchased from the US.[23]

On 8 November 1965 the Defence Secretary minuted the Prime Minister seeking his agreement to develop 'a small, lethal and incapacitating tactical CW capability for retaliatory use (and therefore deterrence) in the context of limited war' outside Europe. Harold Wilson deferred consideration of this proposal. His Private Secretary wrote to Pat Nairne in the MOD on 10 November to note that the Prime Minister would like to discuss the issue with the Defence Secretary at 'some time when things are a little quieter';[24] a view the Foreign Secretary endorsed.[25] The Chancellor also replied to note that he had some views on this matter, which he preferred to raise in a meeting rather than in further correspondence.[26] We cannot tell what these were, but given the state of the economy and the ongoing defence review, we can guess the probable tenor of the Chancellor's views.

By early May 1966 the issue was still in abeyance; nothing was being done about the operational requirements set out in departmental documents NASR 1199 or NGAST 1203 because neither CDEE nor the Ministry of Aviation had a mandate to proceed.[27] Moreover, negotiations for the procurement of VX agent for NASR 1199, probably from the US, were still at a standstill.[28] The Ministry of Aviation thought that it would take three years from the date of contract to delivery of the first operational system to meet the NASR 1199 requirement; allowing for completion of the paper work 1970 now seemed to be the earliest target date, which was five years late on the original plan. The time scale for NGAST 1203 could not be assessed until the feasibility studies were completed, but 1975 seemed probable.[29]

---

23 TNA AIR 8/2391, DCGS to CGS Chemical Warfare Weapons, 13 June 1966.

24 TNA AIR 8/2391, J.O. Wright, No.10 to P.D. Nairn, MOD, 10 November 1965.

25 TNA AIR 8/2391, C.M. Maclehose, Foreign Office to J.O. Wright, No. 10, 26 November 1965.

26 TNA AIR 8/2391, R.J. Painter, Treasury Chambers to P.D. Nairn, MOD, 11 November 1965.

27 NASR – Naval Air Staff Requirement, i.e. a joint service project. NGAST – Naval Ground and Air Staff Target.

28 TNA AIR 8/2391, W.D. Hodgkinson, AVM ACAS (OR) to DCAS, Chemical Warfare Weapons, 11 May 1966.

29 NGAST 1203 called for a study of the additional precautions needed to apply the concept of 'two-man' control to a single as well as multi-crew aircraft. TNA AIR 8/2391, Chemical and Biological Warfare 1963–67, P. de L. LeCheminant, Air Commodore DASB to PS CAS, 7 July 1965, Staff Requirement for An Aircraft Spray System for Retaliatory Chemical Warfare (COS 1972/2/7/65). NGAST 1203 was also a feasibility study of more sophisticated and flexible means of delivering either incapacitating or lethal agents in the

In May 1966 the Assistant Chief of the Air Staff (Operational Requirements) was keen to obtain a clear decision in favour of the programme. He argued that in a limited war scenario – outwith NATO – the UK's adversaries could be expected to have superior numbers as well easier logistics and knowledge of the terrain; conventional firepower would not be enough to counter these advantages, but CW would, especially incapacitating agents, which would increase the odds in the UK's favour.[30] The Deputy Chief of the Air Staff agreed that there was now some urgency in this matter and that the wheels should be set in motion once more.[31] Senior Air Staff personnel agreed that the requirement still remained valid and consequently the Deputy Chief of the Air Staff asked the Chief of the Air Staffs' Private Secretary to consult the Secretary of State's Private Office with a view to re-opening the subject.[32] However, after these consultations in June 1966 the Secretary of State's Private Office was of the view that Healey would not be prepared to resurrect the subject simply on the basis that his minute of the 8 November to the Prime Minister was still formally unanswered, despite the fact that its contents were still correct. In the circumstances, there was nothing for it but to prepare a submission from the Chief of the Defence Staff to the Secretary of State for Defence that would need to demonstrate that circumstances were now such that MOD would need to press for a decision on this issue.[33] It is not clear from available files at The National Archives whether this was in fact done. However, it seems that MOD officials felt by February 1967 that Ministers could not be expected to deal with the issues raised in Healey's 1965 minute until the Defence Research Committee (DRC) had completed its review of UK CBW research.[34] This study in turn appears to have been to some extent overtaken, or at least accelerated, by the need to conduct an R&D study as part of the Defence Expenditure studies then taking place in the late 1960s.[35]

Sir Burke Trend, the Cabinet Secretary, wrote on 2 August 1967 to the Defence Secretary's Private Secretary to ask whether it would be timely to pick

1970s. See A.C.L. Mackie, Air Commodore DASB to PS CAS Research Programme on Chemical and Biological Warfare (AHC/72/76), 22 February 1967.

30 TNA AIR 8/2391, W.D. Hodgkinson, AVM ACAS (OR) to DCAS, Chemical Warfare Weapons, 11 May 1966.

31 TNA AIR 8/2391, D.J. Pearson, PS to DCAS to ACAS (OR), Chemical Warfare Weapons, 1 June 1966.

32 TNA AIR 8/2391, M. Ealey, APS to DCAS to PS to CAS, Chemical Warfare Weapons, 21 June 1968.

33 TNA AIR 8/2391, M. Gainsborough, PS to CAS to PS DCAS, Chemical Warfare Weapons, 27 June 1966.

34 The DRC had replaced the DPRC in 1964.

35 TNA AIR 8/2391, PS to CS, 24 February 1967.

up this subject again in the autumn of that year. MOD replied that the Chiefs of Staff had recently set in train an examination of the UK's CBW warfare capability. Although it would indeed be appropriate for both BW and CW to be considered by Ministers after the summer recess, it was not yet clear when the Defence Secretary would be ready to put a paper to his Cabinet colleagues.

In the Chiefs of Staff paper 124/67 dated 15 December 1967, the Chiefs concluded that UK forces required, 'lethal and incapacitating CW agents and delivery systems for deterrence and retaliatory use outside Europe, with particular emphasis given to nominal-risk incapacitants. Lethal CW agents and delivery systems for retaliatory use inside Europe which, in the absence of firm assurances from the US, would have to be provided either by purchase from abroad or by British development and provision.'

In forwarding this report to the Defence Secretary, the Chief of the Defence Staff (CDS) noted that although there were strong military reasons for settling this matter, the Chiefs recognised that Ministers might prefer to defer a decision for the present. In advising the Secretary of State, the Minister for Defence (Equipment) Roy Mason suggested that the government should defer for at least two years consideration that the UK should develop a CW retaliatory capability, unless NATO allies raised the subject during that time; it would not have been appropriate then to take a positive decision against the development of such a strategy. If this was indeed the case, there was no reason to re-open the issues of BW or CW, unless as a result of the work of the House of Commons Select Committee on Science and Technology, which was reviewing work at the Chemical Defence Establishment, it became necessary to formulate a definite Ministerial doctrine on these CB matters.

On 24 January 1968 the Secretary of State for Defence minuted both Mr Mason and the CDS that on the question of developing an offensive CW capability, for use either inside or outside Europe, he agreed that HMG should, for the present, at least defer a decision. Nothing had changed by early summer 1968 when, as a result of the interest in Porton Down and the leaks about UK CBW activities in *The Observer* newspaper, it became public knowledge that the UK had no CW or BW stockpiles, and that it would take a good deal of time and effort to acquire the means of delivery. A Parliamentary Question to this effect was answered on 24 June 1968.[36] This also made clear that the UK did not manufacture CW, or had any plans to do so.[37] There was at that stage no point in making any further change in policy. This remained the case for the remainder of the 1966–70 Labour government. Nor does it appear that these decisions

---

36 Hansard Column 26, Written Answers.

37 TNA DEFE 11/672, G. Leitch, DUS (E), MOD to PS/S of S, 9 July 1968.

were taken with reference to the emerging UK policy on taking a disarmament initiative on biological warfare – see the discussion in Chapter 5.

Conclusions

So despite the intentions the UK never re-acquired an offensive programme, although the Chiefs of Staff continued to see a role in out-of-area operations – those outside the Central European Front well into the 1970s. However, this remained an aspiration. Arms control and disarmament considerations had no impact whatsoever on UK policies here; financial imperatives, or more accurately pressures on the defence budget, were the main constraints on chemical warfare although there may have been a reluctance to embark on new programmes given the adverse international reaction to US use of riot control agents in the Vietnam War. However, there is no direct documentary evidence that this was a factor. There were no meaningful arms control negotiations underway on chemical weapons throughout the period 1956 through to the mid 1970s; chemical disarmament, if it figured at all, was in Phase II of the General and Complete Disarmament plans. Now there was one area where the UK maintained an interest in chemical warfare and made some progress too, and that was in the case of new incapacitating chemical agents, where ironically an agreement, albeit with limited provisions, from the 1920s (the 1925 Geneva Protocol) appeared to have some impact on UK deliberations and on a planned programme and concepts of use. It is to this issue that we now turn.

Chapter 3

'Inappropriately Hilarious': The UK and Incapacitating Chemical Agents[1]

Introduction

Over the years vast numbers of toxic materials – whether man-made or of natural origin – have been considered as potential chemical warfare agents.[2] The intent was most often to find a lethal chemical, but a variety of factors meant that whilst many were screened few were ever selected and adopted into the munitions stockpiles of nations as they lacked all the properties to make them useable as weapons. Although many toxic chemicals were tested and used in the First World War – some hundreds – the effects of most of them were far from lethal. Indeed the majority – the irritants, lachrymators, blistering agents, sternutators, emetics – were incapacitating in effect. Even mustard, the 'king of the war gases', was in most cases only incapacitating. A Porton Down study in 1967 noted that in the First World War one-third of all US casualties were caused by gas, but of these only two per cent died.[3] By the mid to late 1950s this realisation had led the UK, US and Canada to look into the whole question of designing chemical agents that were specifically intended to be incapacitating in effect. A whole new field of CW research and development then began. Considerable thought was given to the requirements for such agents and concepts of use – including definitional issues. Work quickly uncovered that looking for the ideal agent was not going to be easy. Much depended on the intended scenario and factors such as speed of action required, safety margins, the duration of incapacitation and

1 'Inappropriately hilarious' was a phrase used to describe the condition of some UK soldiers after a field trial of an incapacitating agent when they were under observation in a hospital. Scc TNA WO 195/16694, Applied Biology Committee: SMALL CHANGE (field trials with and incapacitant), 1968.

2 In the First World War alone some tens of thousands of agents were examined as candidate agents – see SIPRI, *The Problem of Chemical and Biological Warfare, Volume I: The Rise of CB Weapons* (Stockholm, 1971), p. 38.

3 TNA WO 195/16429, Applied Biology Committee, 26 April 1967, Item 4: A Review of Some Concepts of Incapacitation, R.W. Brimblecombe, F.W. Beswick and D.F. Downing.

method of delivery all complicated the researchers' task. Questions of dose and variable and lack of predictable effects were major issues that emerged in the research programmes.

This chapter provides a brief historical overview of the developments in incapacitating agents from the First World War through to the mid 1970s – including what might be called the main period (1957–67) when there were active research programmes into the development of incapacitating chemical agents and which is of course of direct relevance to the period covered by this book. This overview concentrates on UK efforts, drawing on primary sources in the main and therefore focusing on the late 1950s and 1960s. It will be clear from discussion of some of the programmes and their problems that these have a clear resonance for early twenty-first-century concerns and can hopefully inform current discussions on incapacitating chemicals and international law. And on legal matters we can, for instance, see that selection of potential agents in the UK programme was shaped by interpretations of the 1925 Geneva Protocol.[4] It is striking that this arms control agreement from the 1920s should still continue to exert an influence on UK policies well into the 1950s and 1960s. Definitions are key to this discussion and it is therefore important to begin with a review of what experts working at the Chemical Defence Experimental Establishment (CDEE) at Porton Down understood by the term 'incapacitant' in our period and how it related to other contemporaneously used terms such as 'ant-riot agent', 'harassing agent' and 'irritant'.

Definitions, Understandings and Developments

In the late 1950s and 1960s when work began in earnest on R&D on new incapacitating agents and especially from 1963 onwards when the UK Chiefs of Staff had decided that the UK needed to acquire both a lethal and incapacitating chemical warfare capability, terminology seems to have been somewhat fluid. It was only when attempts were made to clarify the operational requirements and uses for agents that clearer definitions started to appear; and for this reason it is difficult to separate definitions from the then emerging concepts of use. In the UK the main work was done by the Ministry of Defence and Porton Down. In August 1963 we see that CDEE Porton Down was defining an incapacitating agent as, 'a substance which can render hostile persons incapable of either

4 See, for instance, a note from the Equipment Policy Branch and other policy and operations branches in the War Office entitled 'Incapacitating Agents', 11 September 1963 in TNA WO 32/20163, Development of Defensive Chemical Warfare Capability with Incapacitating Agents, 1962–66.

aggression or resistance for a limited time without permanently injuring them'.[5] This report went on to note that the range of effects that might be produced by incapacitating agents was very wide and that although it was possible to divide them into classes, this was somewhat arbitrary and that some of the potential materials in practice would overlap the divisions.

Porton identified three categories: harassing agents, physical incapacitating agents and mental incapacitating agents. In the middle category there was a greater risk of causing death or permanent injury in a small percentage of cases. The chance of producing death or injuries in a target population attacked with an incapacitating agent depended on the intrinsic toxicity of the agent; the likelihood of its effects producing some permanent damage and the characteristics of the weapon systems used to deliver it; and the density and nature of the target population. There was, moreover, a further important point to bear in mind – the very great difficulty of predicting the secondary effects produced amongst a non-combatant population after exposure.

CDEE concluded that the prospect of finding an incapacitating agent exactly suited to all possible operational needs was remote and that the final choice of agent would inevitably represent a compromise between conflicting requirements. In order to obtain a clear sense of what was needed, service requirements had to be spelled out to guide CDEE's further research and development work. Three considerations were crucial. The first of these concerned the upper limit of primary lethal casualties that would be acceptable in different types of operation. As a starting point CDEE suggested that primary lethal casualties might be five per cent in limited war, two per cent in counter-insurgency and nil in civil riots. Secondly, advice was needed on the maximum and minimum periods of incapacitation demanded by operational considerations and the best compromise if only a single agent could be accepted for Service use. And thirdly, Porton needed to hear views on whether a delay of some hours between exposure and onset of symptoms would materially restrict the usefulness of an incapacitating agent.

MOD reaction makes for interesting reading. Whilst broadly agreeing with the assessments in the CDEE paper, there is one key response on the question of acceptability of fatalities by percentages by category.[6] Some MOD officials noted that the UK was a party to the 1925 Geneva Convention and if it were to use an agent inflicting two per cent casualties by direct effects in general or limited war or in counter-insurgency, then Britain would be accused of using

5 TNA WO 32/20163, The Operational Use of Chemical Incapacitating Agents by M.A.P. Hogg, CDEE Porton Down, 29 August 1963.

6 TNA WO 32/20163, Director for Equipment Policy, War Office to DBCD, Incapacitating Agents, 28 October 1963.

CW and would be likely to meet retaliation with the full range of lethal agents.[7] For this reason the MOD's Equipment Policy Branch felt that there should be only two types of agent: safe incapacitating agents, such as CS, to be used without fear of contravening the Geneva Protocol; and incapacitating agents with possible direct lethal effects to be used only after someone else had used them first – something only likely to occur in limited or general war.

The general view at CDEE by 1967 appears in a paper for the Applied Biology Committee. In this R.W. Brimblecombe, F.W. Beswick and D.F. Downing note in a review on some concepts of incapacitation that the object of incapacitating a subject was to produce in man certain changes which effectively prevented him carrying out his duties without endangering his life. By late 1973 there were attempts to agree definitions of incapacitants for a NATO terminology standardisation project.[8] The MOD at the stage wanted to find agreed language to define 'incapacitants', 'riot control agents' and 'irritants'. A major concern seems to have been that it was most important to ensure that the use of riot control agents was not affected by connotations of chemical warfare or by possible international agreements banning the use of chemical agents. This was important in view of recent UK policies on the place of CS under the Geneva Protocol – see discussion in the following chapter for the context. It is well to recall too that the Biological and Toxin Weapons Convention (BTWC) had only recently been opened for signature and this called for its States Parties to reach early agreement on effective measures for the prohibition of chemical weapons.

Such concerns meant that, for the MOD, the correct definition of incapacitating and riot control agents was most important. However, and echoing the CDEE and MOD understandings from the early 1960s, officials recognised that both incapacitating agents and RCAs had many similarities. For this reason it was essential to consider both definitions together to ensure that they highlighted the differences. The preferred terms, which apparently built on US proposals, were:

---

[7] TNA WO 32/20163, Col GS EP 4 to MO1, CD 2, AMD 8 CSR 2, Incapacitating Agents, 11 September 1963.

[8] TNA FCO 66/500, ACGS (OR) to ASD1b 31 December 1973, NATO Terminology (this refers to standardisation proposals within NATO – differences between US and UK definitions for these terms, thus need for agreed UK terminology to be recommended to UK Joint Services Terminology Committee – see folio 91 ACGS (OR) to DS22, MO1, AD RCB, Defence NBC School. CDE and FCO also invited to meeting to address problem on 12 December 1973, date of letter 4 December).

> Incapacitating Agent: A chemical agent which produces temporary disabling conditions which unlike those caused by RCAs can be physical or mental and persist for hours or days after exposure to the agent has ceased. Medical treatment, while not usually required, facilitates a more rapid recovery. See also chemical agent, RCA.
>
> RCA: A substance which produces temporary irritating or disabling physical effects that disappear within minutes of removal from exposure. There is no significant risk of permanent injury and medical treatment is very rarely required.

For the MOD there were four areas common to both definitions in which differences needed to be highlighted:

- Classification: an incapacitating agent is a 'chemical agent' by definition. Similarly an RCA is not a chemical agent (a view that was to change with the Chemical Weapons Convention (CWC) in 1993). It is therefore described as a substance.
- Type of effects: both riot control and incapacitating agents produce effects which are 'temporary' and 'physically disabling'; incapacitants can also disable mentally; RCAs can be 'irritating' as well as 'disabling'.
- Duration of effects: it was difficult to be too specific, but 'minutes' for RCAs and 'hours' for incapacitants drew a clear enough distinction.
- Medical aspects: with incapacitating agents there was some risk of permanent injury and it was incorrect for US definitions to suggest that there were none. With RCAs there can be no significant risk of permanent injury.

The definitions were adopted and remain in the NATO terminology guide to this day.

In their 1967 review of concepts of incapacitation the Porton scientists Brimblecombe, Beswick and Downing noted that if the anti-riot control agents were excepted, the concept of incapacitation really only went back to the late 1950s. Before then chemical warfare agents had been developed for their lethal potential, but even with these lethal agents records showed that success on the battlefield – defined as military effectiveness – was measured by the large number of non-lethal casualties, who subsequently made complete, or almost complete, recoveries and the lowered state of enemy morale whenever the use of these chemicals was threatened rather than by the large numbers of fatal casualties they might be capable of producing. Indeed some recent research has suggested that the actual numbers of fatalities from the very first use of CW on the Western Front in April 1915 was in actual fact much lower than later claims.[9] Gordon Corrigan argues that at most there were some 1,500 casualties of which only 200 died. Evidence indicates therefore that lethal agents could be effective weapons of war if the net effect in battle was actually to incapacitate rather than to kill the enemy.

As SIPRI noted in 1971 in the course of the First World War just about every known chemical had been investigated for its offensive potential and in this process a very large number of new compounds were prepared for the first time as CW agents. Between October 1917 and June 1919 the UK investigated the likes of oil from poison ivy and the pepper-derivative compound capsicine, which was rejected on account of supply problems.[10] In the post-war years new agent research had four main elements:

- a re-examination of compounds that had been inadequately examined in the war;
- a search for the active ingredients in highly toxic naturally-occurring substances;
- a surveillance of new compounds emerging from academic and industrial labs; and,
- further programmes of synthesis of candidate CW agents guided by existing leads or new theories of molecular toxicology.[11]

For much of the inter-war period, however, offensive research programmes in general and in the UK in particular staggered on with little investment. When the First World War began on 3 September 1939 the UK had virtually no chemical munitions of any sort.[12] In the UK from about 1937 onwards the whole trend of Porton's work was to demand a more exacting definition of key terms such as 'casualty production' and 'harassment'. This along with more 'realistic' trials together with what were seen as the increasing bitterness of modern warfare as witnessed in the Second World War led to an outlook at Porton and presumably the War Office and Ministry of Supply that totally discounted even severe sensory irritation as a military objective. In fact even weapons used

[9] See Gordon Corrigan, *Mud, Blood and Poppycock* (London, 2003), pp. 164–5.

[10] Albert Palazzo, *Seeking Victory on the Western Front: The British Army and Chemical Warfare in World War I* (Lincoln, NE, 2000), p. 163.

[11] SIPRI, *The Problem of Chemical and Biological Warfare*, pp. 58–9.

[12] Gradon B. Carter, *Chemical and Biological Defence at Porton Down 1916–2000* (London, 2000), p. 47.

primarily for harassment purposes had to carry at least the threat of lethality or total incapacitation for a long period.[13]

In the immediate post-war years and into the early 1950s the main interest at places such as Porton and Edgewood in the US was in characterising the new series of organophosphorus nerve agents that had been developed and produced by Germany in the 1930s and 1940s. Plans were afoot to develop UK and US production facilities for such agents and to look for new lethal ones – the discovery of the phosphonothiolates in the early 1950s by the UK was eventually to lead to adoption and production of VX by the United States in 1961 and for which the UK had helped develop a stabilising agent.[14] Interest in incapacitating agents appears to have been minimal in this period. One of the earliest indications of post-Second World War thinking on the 'incapacitation concept' can be found in 1947 when Professor Corwin from Johns Hopkins University wrote to the US Army Chemical Corps including a reference to the possibility of attacking enzyme systems and of producing mass hallucinations and uncontrollable hysteria.[15]

An active interest in incapacitating agents was spurred on by consideration in the Tripartite Conferences on Toxicological Warfare – the UK, US and Canadian CBW offensive and defensive CBW collaboration that started in the 1940s. The Thirteenth Conference in Canada held in September 1958 decided that the participating countries should concentrate efforts on searching for new agents, both incapacitating and lethal.[16] A CDEE Note from February of that year describes the work of a recently established – September 1957 – 'search party' which would be looking for incapacitating agents and psychochemicals as well as for lethal agents.[17] This was not driven by any formal UK military requirement for a new CW capability – that was to come later – as the UK had decided to abandon its offensive programme in 1956. By 1959 the Tripartite

13 TNA WO 195/16429, Applied Biology Committee, 26 April 1967, Item 4: A Review of Some Concepts of Incapacitation, R.W. Brimblecombe, F.W. Beswick and D.F. Downing.

14 TNA WO 195/15019, War Office Advisory Council on Scientific Research and Technical Development, Chemistry Committee Conclusions and Recommendations of the C.W. Agents Group at the 15th Tripartite Conference.

15 TNA WO 195/16429, Applied Biology Committee, 26 April 1967, Item 4: A Review of Some Concepts of Incapacitation, R.W. Brimblecombe, F.W. Beswick and D.F. Downing.

16 J.P. Robinson, *Disabling Chemical Weapons: A Documented Chronology of Events, 1945–2003* (HSP, November 2003), p. 15 citing US Army Chemical Corps Historical Office document January 1960.

17 TNA WO 195/14272, Chemistry Committee: The Search for New Agents, 1958 PN 5 of 27 February 1958 cited in Robinson, *Disabling Chemical Weapons*, p. 13.

Conferences agreed in its new agent work to place an emphasis on incapacitating agents.[18] Work on the search for incapacitants now expanded at Porton.

Various bodies in the UK started to review policy and research on CW in 1960. This includes the Chiefs of Staff and the Defence Policy Research Committee. The outcome of various studies up until 1962 was that incapacitating agents were given higher prominence following recognition that they could have significant military utility.[19] The varied approaches taken by the UK, US and Canada in the search for incapacitating agents had contributed to the discovery of significant numbers of compounds of potential interest.[20] By the end of 1962 the Chiefs of Staff decided to recommend that the UK should now possess both a lethal and incapacitating tactical CW capability and that Ministers should be invited to endorse this new requirement.[21] This was then put to the Cabinet Defence Committee in April 1963 by Peter Thorneycroft (Defence Secretary) and then discussed in the Committee in May 1963.[22]

The policy agreed by the Cabinet Defence Committee in May 1963 called for the production of an incapacitating agent when one was developed.[23] The search for such an agent was already well underway and to assist in this CDEE had prepared a paper setting out the operational use of incapacitating agents as seen from a technical point of view and based on the trends of research to date. This was the report referred to above. Before the programme could be taken any further, guidance was needed on the type and scale of operations envisaged; the weapons best used; the upper limit of primary lethal casualties that would be acceptable in different categories of operations; the maximum and minimum periods of incapacitation demanded by operational considerations and the best compromise if only a single agent could be accepted and whether a delay of some hours between exposure and onset of symptoms would materially restrict the usefulness of incapacitating agents and if so the particular operational categories affected.

---

18 Robinson, *Disabling Chemical Weapons*, p. 22 citing US Army Chemical Corps Historical Office document from April 1961.

19 Robinson, *Disabling Chemical Weapons*, pp. 31–5.

20 TNA WO 195/15386, War Office Advisory Council on Scientific Research and Technical Development Chemistry Committee 16th Tripartite Conference, Conclusions and Recommendations of the C.W. Agents Discussion Group, June 1962.

21 TNA DEFE 5/131, Chemical and Biological Warfare COS (62) 432. See Chapter 2 for detailed discussion of the policy-making aspects of this period.

22 TNA CAB 131/28, Cabinet Defence Committee, D (63) 14, 16 April 1963.

23 TNA WO 32/20163, DCDRD to DEP, Incapacitating Agents, 9 September 1963; TNA CAB 131/28, Cabinet Defence Committee, D (63) 6th Meeting, 3 May 1963.

UK requirements were thus further reviewed and fine tuned and this resulted in agreement that the programme should aim for two categories of incapacitating agents. The first of these should be a safer 'intensely harassing' incapacitating agent for use in riot control and counter-insurgency effective in one or two minutes; and the second should be an incapacitating agent with possible direct lethal effects to be used only in retaliation. Susceptibility of agents to treatment was also an important consideration. Agents should be selected where there were more difficulties in developing prophylactic or therapeutic defence. This would be an important factor in deciding research priorities.[24] However, there were problems. Finding a completely safe incapacitating agent effective up to six hours was going to be extremely difficult, if not impossible to find. That was the view of Gerry Heath – from the MOD's Directorate for Biological and Chemical Defence in late 1963.[25] The requirement for an 'intensely harassing' agent was ultimately to be met by dibenzoxazepine – later known as CR, which had been first identified as a promising agent in 1961 as a result of CDEE's industrial and academic liaison programme.[26]

A year later and after much discussion the first draft of a new Naval General Staff Target (NGST) on incapacitating agents was circulated to relevant MOD branches for comment on 18 September 1964. This covered the requirement for a group of agents with differing degrees of incapacitating effect, and included agents with minimal lethal effect. Detailed specifications for two types of agent were set out along with the operational requirements ranging from internal security and counter-insurgency to limited and general war. It is worth setting out in detail the specific requirement as this provides a clear insight into contemporary thinking on what constituted incapacitating agents and what they would be used for.

NGST No. 3333 Chemical Warfare Agents with Incapacitating Effects called for two categories of agent:

> a. Safe Incapacitating Agents – for use in anti-riot and counter-insurgency operations. Agents within this category should be usable without contravening the Geneva Gas Protocol. These should cause no deaths from toxicity of the agents, but minimal deaths may be expected from secondary effects due to a man's incapacity to react to a situation endangering his life. They should be of two types:

---

24 TNA WO 32/20163, Director Equipment Policy War Office to DBCD Incapacitating Agents, 28 October 1963.

25 TNA WO 32/20163, G.D. Heath for DBCD to EP4, Incapacitating Agents, October 1964.

26 TNA WO 188/2704, Dibenzoxazepine (DBO).

> (i) an improved anti-riot agent (CS type, known to the press as tear gas) is required primarily for ant-riot control and counter-insurgency operations. It should be at least as discomfiting as CS and should act immediately, with its effects lasting from 15 to 30 minutes, a compromise being 20 minutes.
>
> (ii) another safe non-lethal incapacitating agent is required which should act in one or two minutes and which should have effects lasting from one to six hours, preferably adjustable by the concentration administered, the compromise being four hours.
>
> b. Incapacitating Agents with Possible Direct Lethal Effects
>
> (1) Such agents could only be used after contravention of the Geneva Gas Protocol, which would be unlikely to occur only in General or perhaps Limited War, and would be used in preference to lethal agents in circumstances such as against mixed military and civilian targets. Speed of effect is not so necessary as in anti-riot or counter-insurgency operations and onset of symptoms after say, one hour would be acceptable. The effects should last from 6 to 24 hours, with 15 hours as the compromise. A compromise figure between 12 to 18 hours would be acceptable, preferably nearer 18 hours than 12 hours. The aim should be to limit direct lethal effects to 2% maximum.
>
> (2) In exceptional circumstances, if no safe non-lethal incapacitating agent were available, such agents might be used against insurgents despite a risk of accusation of contravening the Geneva Gas Protocol, for example to rescue prisoners in imminent danger of assassination by their captors. It should be possible to reduce risk of minimal lethality by adjustment of the concentrations administered.[27]

This was the first time that a specific research requirement for an agent had been set forth by the MOD or any of its predecessors and the Director of Biological and Chemical Defence asked for its withdrawal.[28] In his view agents on their own had no military properties and only acquired them when they were built into a weapon system. Agents could not be designed to a specification. This did not go down too well in the MOD and elicited the comment that, 'I am somewhat surprised at the apparent unhelpful attitude he is adopting. It is time, I feel, that the CW boys got down to a bit of practical application of all the research they

[27] TNA WO 32/20163, NGST No. 3333 First Draft Chemical Warfare Agents with Incapacitating Effects, 18 September 1964.

[28] TNA WO 32/20163, DBCD to DAEP, GSTs 3332 and 3333, 1 October 1964.

have been doing.'[29] It is not immediately apparent from the available record exactly what happened next and the chronology of events is unclear. There were some further exchanges within MOD branches on this topic in the autumn of 1964 and early 1965. However, it does seem that elements of NGST No. 3333 were later incorporated into what was to become Naval General Air Staff Target (NGAST) No. 1203, which called for an incapacitating agent with minimal lethal effects for use in war where it was desired to keep casualties low and a safer incapacitating agent for use primarily in anti-riot and counter-insurgency operations.[30] There was, it seems, recognition that the characteristics of both the agent and its delivery system had to be specified.[31]

Agent Research

So what about the chemicals that were being investigated as candidate incapacitant agents from 1957 onwards? CDEE's effort was initially conducted under the Chief Scientist's Research Programme. During the period August 1961 to December 1963 240 new compounds were tested as potential agents for example.[32] Some of these came from industrial liaison where research for new civil products often found particularly toxic materials that could have defence interests. There was an active Porton programme to consult industry. These tests involved those for toxicity, physiological effects, physical incapacitation and behaviour.[33] Some 13 types of chemicals were investigated: indole derivatives, pyrrole derivatives; aminoxyalkanoic acids and esters, benzimodazoles, pthalimides, tremorine, tremoram and analogues, thebaine or oripavine derivatives; sterically hindered amines and amides, cyanoalkylamines,

---

29 TNA WO 32/20163, Development of Defensive Chemical Warfare Capability with Incapacitating Agents 1962–1966, DMGO to DCS (A), 22 October 1964.

30 TNA WO 32/20163, 2nd Draft Naval General Air Staff Target No. 1202, Chemical Warfare (CW) Weapon Systems, Group Captain A.J. Peart, Deputy Director Op Reqs 10 (RAF), October 1964.

31 TNA WO 32/20163, Master General of the Ordnance (Dev) to SO/CS (A), SO/DCS (A) and AEP 4, 19 February 1965.

32 TNA WO 189/422, Biological Testing of Incapacitating Agents Part IV, 1964, Further Results of Screening Tests on New Compounds, R.W. Brimblecombe Porton Technical Paper No. 902, 2 October 1964.

33 TNA WO 195/15033, Fifteenth Tripartite Conference on Toxicological Warfare Screening Methods for Incapacitating Agents, UK Discussion Paper Medical Aspects, 5 August 1960.

sulpholanes, hydrazines, natural products and miscellaneous compounds not falling into these classes.[34]

By 1965 CDEE had narrowed down the scope of potential incapacitating agents to some nine materials: glycolates – BZ, T3021, T3126; oripavine derivatives – TL2636, TL2696 and TL2833; and other compounds – LSD-25, CS and T2806 (CR).[35] Porton's New Agent Applications Working Party further reduced this list. LSD-25 was not recommended at the end of the day despite work having started in the autumn of 1961.[36] LSD had a tendency to produce unpredictable behaviour, not a very desirable characteristic for an incapacitating agent.[37] However, in some circumstances, as shown in a Porton trial, it could eliminate units as a fighting force – as happened during one trial, an event which suitably impressed the military umpire.[38] The UK was aware as early as 1963 that a US Army study had shown that BZ had certain deficiencies; it could, for instance, well take more than one hour before effects became manifest.[39] Instead the Working Party came down in favour of T2806 as a short-term incapacitant and T3021 (a cyclopentyl derivative of BZ) as a medium-term incapacitant. These should, in its view, become candidate agents after further evaluation. As noted earlier T2806 was taken into service as CR. CS and BZ would be held in reserve in case the potential of these agents was not realised. T3021 was not in the end taken any further.

Porton recognised that there were problems in finding suitable incapacitating agents. A review in 1967 noted that up until that time vigorous efforts to find an ideal CW incapacitant had only qualified success. Sensory irritants had limited utility against determined and well trained troops as they were able to tolerate the level of personal harassment caused by such substances. High morale could be overcome by using agents that caused physical incapacitation; however, since such agents usually exerted their effect by severe depression of either the cardiovascular system or the central nervous system there was a considerable

---

[34] TNA WO 189/422, Further Results of Screening Tests on New Compounds, R.W. Brimblecombe, Porton Technical Paper No. 902, 2 October 1964.

[35] T numbers were allocated by CDEE to agents under investigation – each one was given a unique number. TNA WO 188/2705, New Agent Applications Working Party Selection of an Incapacitating Agent, 23 August 1965.

[36] TNA WO 195/16213, Applied Biology Committee Summary of Work on LSD at Porton in the Last Five Years, R.J. Moylan-Jones, 29 March 1966.

[37] Alexander Kelle, Kathryn Nixdorff and Malcolm Dando, *Controlling Biochemical Weapons* (Basingstoke, 2006), p. 104.

[38] TNA WO 195/16213, Applied Biology Committee of Work on LSD at Porton in the Last Five Years, R.J.Moylan-Jones, 29 March 1966.

[39] TNA WO 32/20163, R. Holmes Munitions/TA British Defence Staff, Washington to DCDRD, MOD Tactical Incapacitating Agents, 26 July 1963.

risk that they would cause death either by their own pharmacological action, or because they exposed the helpless victim to natural and man-made dangers of the environment.[40] At the MOD policy end of the debate A.H. Cottrell noted in early 1967 that,

> On straight military grounds, the (Defence Research) Committee felt that a successful development of incapacitating agents could provide a capability of real value, particularly in the counter-insurgency and limited war situations which, may perhaps, be the most likely forms of action in which the UK will be involved in the future. An incapacitating chemical weapon could be more effective, and cheaper, than existing weapons, and in situations where friendly civilian populations are involved it could offer the unique advantage that its effects are short-term and reversible. Against these points, however, we have to remember that there are always likely to be some fatal casualties, especially amongst elderly and infirm people and infants, if these happen to be in the area of operations; that our use of such weapons would invite retaliation against ourselves; and that warfare with incapacitating agents is not different in kind from that with lethal agents so that, once begun, a slide into the most severe forms of CW becomes, perhaps, more likely.[41]

Despite the research no decision was taken at Ministerial level to implement the 1963 plan in full on either lethal or incapacitating agents. As Neil Davison observes, UK research continued until at least the early 1970s, but the ideal incapacitating agent still remained elusive.[42]

Conclusions

So what does the British experience tell us about incapacitating agents? The UK programme between the late 1950s and later 1960s highlights themes that are general in their applicability. As the UK, US and Canadian CB programmes were closely integrated in this period the same problems and research results would have been shared – and the challenges posed by the need to find the

---

40 TNA WO 195/16429, Applied Biology Committee, 26 April 1967, Item 4: A Review of Some Concepts of Incapacitation, R.W. Brimblecombe, F.W. Beswick and D.F. Downing.

41 TNA AIR 8/2391, A.H. Cottrell, CA (Studies) to CDS Research Programme on Chemical and Biological Warfare, 8 February 1967.

42 Neil Davison, *'Non-Lethal' Weapons* (London, 2009), p. 110.

ideal incapacitating agent were doubtless the same on both sides of the Atlantic. Perhaps the following points have a contemporary resonance:

- Definitions of what constitutes the ideal incapacitant can depend on the intended use; and the operational requirements.
- In UK thinking there were distinctions to be made between different types of incapacitant and this did blur the edge between what were called as anti-riot control, irritant and harassing agents; operational requirements and scenarios were critical factors here; although as the specifications show 'incapacitating agent' was used in this period as more as an umbrella term – certainly in the 1960s.
- Some compounds of potential interest came from industry and academe, including CR.
- It was, and is, extraordinarily difficult to develop an agent that can produce almost instant effects, last for hours and have no health risks. UK research clearly shows that whilst many compounds might have some attractive features, virtually all of them could not be taken further. After all the effort only one new agent was brought into service (CR) and controls on its use were considerable as we will see in the next chapter. And of course this is not an incapacitating agent as this term would be understood today and even in the minds of contemporaries it was seen as a more effective advance on CS.

There were, however, considerable Ministerial anxieties on the use of the one new agent to emerge from the incapacitants programme. All this was to become apparent in perhaps, given the original scenarios envisaged by Porton and the MOD, the most unusual of circumstances – in the prisons of Northern Ireland. Whilst the 1925 Geneva Protocol had placed some constraints on how such agents might be used, Ministers and officials by the late 1960s were coming to a different view of how materials such as CS should be regarded under the Protocol. The next chapter looks at these two separate but closely related issues in policy and legal terms and how these factors impacted on policy execution. This also helps put the research work on incapacitants into context.

Chapter 4
A Tale of Two Riot Control Agents: UK Attitudes to CS and the 1925 Geneva Protocol and CR in Warfare and Law Enforcement 1969–1975

Introduction

In December 1969 the then Labour government grappled with the problem of the use of riot control agents in war and how they were to be handled under the 1925 Geneva Protocol, which prohibited the use of chemicals and bacteriological weapons in war.[1] The immediate issue at hand was how the riot control agent CS should be treated: was it covered by the Protocol's ban on 'asphyxiating, poisonous or other gases?' This issue was given added piquancy by extensive US use of CS in the Vietnam War. Harold Wilson's government was coming under pressure to state its views clearly on this matter.

A few years later and the then Conservative government and its Labour successor grappled with the problems over rules of engagement governing the use of CR as a last means of defence in the event of prison break-outs in Northern Ireland. As we saw in the previous chapter CR is a more potent riot control agent than CS. In this chapter we look at the main issues and arguments set out in Cabinet Conclusions and Memoranda, Ministry of Defence and Northern Ireland Office (NIO) papers.[2] What emerges clearly from both episodes is the very careful weighing of the scientific, political and legal arguments and, in the case of possible domestic use, the need for strict political control that led in practice to exceedingly tightly drawn rules of engagement (ROEs) for

---

1 At the 1925 Geneva Conference for the Supervision of the International Traffic in Arms France proposed a protocol on the non-use of poisonous gases. Poland suggested that this should be extended to cover bacteriological weapons. Such a protocol was signed on 17 June 1925 and entered into force on 8 February 1928.

2 TNA CAB 128/44, Part 2 C (69) 168, 16 December 1969, The Geneva Protocol and the Use of Riot Control Agents in War. Note by the Prime Minister; and TNA CAB 129/146, Part 2 CC (69) 61st Conclusions Item 3, 18 December 1969.

British troops deployed in Northern Ireland. It will become clear that for the UK international legal norms and domestic sensitivities were considerable constraints on the extent to which even the use of riot control agents could be contemplated.

The CS Case: The Prime Minister's Memorandum

During 1969 the Cabinet Defence and Overseas Policy Committee reviewed the legal, foreign policy and defence implications of the emergence of CS and its place under the 1925 Geneva Protocol. Ministers concluded that there were in practice two possible courses of action:

> (a) to reaffirm the 1930 interpretation of the 1925 Protocol without qualification (namely, that whilst smokes were excluded, tear gases were prohibited) while making it clear that the UK would not object if any other state expressed the view (as the US had) that the Protocol did not prohibit the use in war of riot control agents; or
>
> (b) to reaffirm the UK's 1930 statement, but explain that the use in war of recently developed RCAs such as CS, which the UK did not regard as significantly harmful, were not in the UK view covered by the Protocol.

The majority of the Committee favoured option (b), but felt that the full Cabinet ought to discuss the issue before endorsing any conclusion. Harold Wilson as Prime Minister therefore submitted a note to his colleagues on 16 December 1969 setting out the arguments for and against the proposed courses of action. He included memoranda from the Foreign and Commonwealth Office (FCO) and Ministry of Defence (MOD), which set out contrary views as well as those of the Attorney General on the legal aspects. The note said significantly that, 'it can be maintained that chemical agents, whether regarded as lethal or incapacitating (including harassing materials) are all toxic agents. The term "lethal" and "incapacitating" agents are not absolute terms but imply statistical probabilities of response. Toxicity of these gases varies in different animal species and in different conditions. Depending on circumstances, not all individuals will die from an attack with a given lethal agent, whereas, again depending on circumstances, some might succumb to an attack with an incapacitating agent.'

The FCO case emphasised that HMG had taken the lead in 1930 in trying to establish that the Geneva Protocol's prohibition applied to the use of tear gas in war. If the UK were to change its position, then it would lay itself open to charges that it had done so in order to please the US and because the UK approved of the use of RCAs in Vietnam. The note went on to state that any such change in policy would 'gravely prejudice our standing ... and the position which Britain now justly enjoys as the country which has taken the lead in trying to deal with the growing menace of chemical and biological weapons'. Although the FCO recognised that RCAs could in fact be used in war in order to save life, the manner in which RCAs had been used by the US in the Vietnam War – spray tanks and long range artillery – showed that the agents were not being used with this sole intention. Moreover, in the FCO view RCAs such as CS clearly were caught by the phrase 'or other gases', which appears in the Geneva Protocol prohibition. When used for domestic purposes RCAs such as CS were intended as an alternative to lethal force; in war the risk of it being used to increase the effect of lethal weapons, or its use resulting in escalation was so great that the wisest course was to prohibit it altogether.

MOD insisted that in scientific terms CS was a smoke and not a gas and could only be brought within the terms of the Geneva Protocol on the grounds that it was significantly harmful or deleterious to man – an argument which it rejected. If it were so, then its use for law enforcement, or to deal with riots would be difficult to justify. In fact use of CS had saved innocent lives and gave the police and army a much more humane option than batons, bayonets and bullets or bombs. Science had provided a humane and harmless alternative to older methods and had helped to keep down the number of injuries and deaths in Northern Ireland during the 'Troubles'. (The proximate cause of the 'Troubles' had been the 'battle' of Burntollet Bridge on 4 January 1969 between police and Orangemen on one side and nationalist Catholic civil rights marchers on the other. Northern Ireland descended into sectarian conflict with the British Army deployed on the streets to keep the peace. Events soon spiralled out of control with an escalation of violence and the re-emergence of the Irish Republican Army (IRA).) If the UK were to stick with the 1930 interpretation this would focus attention on differences with the US and in addition, would cause endless difficulties over definitions of substances and the circumstances in which they could be used.

The Attorney-General thought that tear gases were significantly harmful and deleterious to man. The development of CS, which was lower in toxicity than tear gases used in the First World War, did not place CS outside the Protocol. The interpretation of the Protocol as prohibiting all gases that were significantly harmful or deleterious to man applied to such gases in 1925 or developed later. CS despite its sophistication was still significantly harmful or deleterious, as witnessed by its use in Vietnam and in a recent Home Office warning to Chief Constables. He went on to argue that interpretation of the Protocol had to take account of HMG's 1930 interpretation that tear gases were covered. To take a different view would have required very convincing evidence to justify a change and this did not exist. Although the matter was not beyond some doubt, it was in law very much the better view that tear gases, CS especially, were covered by the Protocol.

Cabinet Conclusions

This topic came to the full Cabinet at its 61st meeting in 1969 on 18 December. The Foreign Secretary, Michael Stewart, remained convinced of the arguments for sticking with the traditional interpretation, which was option (b) above. Stewart went on to rehearse the argument in the FCO note. Denis Healey, Defence Secretary, then laid out the views expressed in the MOD paper. He emphasised that to announce that the UK regarded itself as prohibited from using CS in war would strengthen the hands of those who argued that it should not be used in peacetime. The UK's 1930 statement had excluded screening smoke and since CS was less toxic than such smokes it made no sense to include it in the scope of the Protocol.

In Cabinet discussions some support was expressed for the view that the UK should reaffirm the 1930 statement without qualification. The alternative of a qualified statement would unite critics of the government's Northern Ireland policy with those who criticised its disarmament policy. It would lead to charges that the UK was again 'seeking to whitewash the US'. However, the majority of Ministers thought that the conclusion in option (b) from the Defence and Overseas Policy Committee should be supported. In summing up Wilson said that it would be necessary for the qualification statement to be carefully drafted to make it clear that in referring to gases such as CS, the UK had in mind those of similar or lower toxicity. In addition, he thought that an accompanying statement, explaining the reasons leading to the change in the UK's position should be prepared and held ready to use in international or domestic discussion. The Foreign Secretary, in consultation with the Defence Secretary and the Attorney-General, was tasked to prepare statements on the lines indicated in the Prime Minister's summing up. It should be noted that this was a UK decision and no pressure had been brought to bear by the US.[3]

The UK change in position – namely that CS was not in fact covered by the Geneva Protocol was announced in Parliament by Michael Stewart in February 1970. He said,

[3] TNA WO 188/2704, Lord Chalfont/Arthur Gavshon, 4 February 1970.

> ... modern technology has developed CS smoke, which unlike the tear gases available in 1930, is considered to be not significantly harmful to man in other than wholly exceptional circumstances; and we regard CS and other such gases accordingly as being outside the scope of the Geneva Protocol. CS is in fact less toxic than screening smokes, which the 1930 statement specifically excluded.

Interestingly enough this review also concluded that two other types of harassing agents – CN and DM – were within the scope of the Geneva Protocol.[4] In terms of how CS might be used by UK armed forces, Lord Chalfont, Minister of State at the FCO, in an interview on 4 February with Arthur Gavshon, the London Correspondent of Associated Press, noted that CS would only be used, 'with the object of saving life and taking prisoners'.[5] Later that summer the MOD gave the FCO a clear statement on the circumstances in which CS might be used by UK armed forces:

> British troops will never use substances such as CS or for that matter any other weapon in a manner inconsistent with the generally accepted rules of war as embodied, for example, in the 1948 Geneva Convention for the 'Amelioration of the Condition of the Wounded and Sick in Armed Forces in the Field'. Accordingly, an enemy rendered hors de combat would not deliberately be subjected to further acts of violence. Furthermore, CS provide British troops with the means of rendering an enemy temporarily hors de combat without resorting to lethal weapons, thus saving lives.

Instructions to British soldiers on the use of CS were drafted in accordance with these principles.[6]

Domestic and International Reactions

Ronald Hope-Jones, head of the FCO's Atomic Energy and Disarmament Department (AEDD – the lead policy department at that time) was so appalled by this *volte face* that he resigned. In fact this decision to reinterpret the Geneva Protocol was not at all popular in Parliament, with non-governmental organisations, the public and in Geneva at the Conference of the Committee

---

4 TNA WO 188/2704, R. Holmes, AD/BD to D/CDE, The Toxicities of World War I Lachrymators, Screening Smoke and Riot Control Agents, 22 January 1970.

5 TNA WO 188/2704, Lord Chalfont/Arthur Gavshon, 4 February 1970.

6 TNA WO 188/2704, D.A. Nicholls, AUS (GS), MOD to D.V. Bendall, AUS, FCO, 20 July 1970.

on Disarmament (CCD). Between February and April some 30 Parliamentary Questions were tabled on this specific issue. In the four months following the announcement, the FCO's Disarmament Department (successor to AEDD) had received more than 200 letters of protest from MPs and the public. The Quakers and the United Nations Association were vocal too in their protests. In Geneva, Sweden and India were scathing: they argued that the decision weakened the Geneva Protocol and they also questioned the sincerity of the UK's approach to CB disarmament notwithstanding the fact that the UK had tabled a draft Biological Weapons Convention in 1969.[7] Alva Myrdal, Swedish Ambassador to the CCD, said that the, 'UK statement is most regrettable. It is particularly so since it comes from a government which in the past as well as in the present has shown such positive interest in getting ahead with further arms regulation measures in the field of biological and chemical disarmament.'[8]

At the change in government in June 1970, Edward Heath's Conservatives replaced Harold Wilson's Labour Party in office. The new Foreign Secretary, Sir Alec Douglas-Home and his Minister of State, Lord Lothian were not fully convinced by the change in UK policy. The Disarmament Department in the FCO advised that, although it remained strongly of the view that the policy was mistaken, there was little point in re-opening the decision given the strength of MOD opposition to any such move, which its initial soundings had revealed. Ministers were not so sure. Home was worried that some 'more harmful gas could be used under cover of tear gas'. He decided that MOD should be asked to reconsider. A Private Secretary letter was duly despatched on 11 September 1970. Lord Carrington's Private Secretary responded in negative terms on 6 October and made clear that Carrington thought that the Labour government's decision had been the right one. The MOD argued too that the then government had managed to ride out the storm of protest over its change in policy. One of the other reasons cited for maintaining the new policy was that from the point of view of UK–US relations, it was surely better to stick to announced policy, which was in line with Washington's.[9] Not upsetting the Americans was a recurring MOD preoccupation at this time (see the next chapter). Although FCO and MOD officials agreed a joint letter to the Attorney-General's office in December to seek a further formal legal view, at the end of the day there

---

7 TNA FCO 66/221, D.M. Summerhayes, Disarmament Dept to Mr Bendall, Mr Godden, Private Secretary, CS Gas and the 1925 Geneva Protocol, 8 July 1970.

8 CCD/PV.457, 12 March 1970.

9 TNA FCO 66/221, D. Benest, Disarmament Dept to Sir T. Brimelow, Mr Duff and Mr Barrington, CS Gas in the 1925 Geneva Protocol, 12 August 1970; J.A.N. Graham to A.R.M. Jaffray, MOD, 11 September 1970; A.R.M. Jaffray, MOD to J.A.N. Graham, 6 October 1970.

was no change in the policy as laid out by Stewart on 2 February 1969. In fact Home and Carrington agreed that their decision to stick with the Labour line on CS would not be promulgated until, 'the force of circumstances requires it'.[10] There were no further changes in UK attitudes to the use of CS in war until the Chemical Weapons Convention (CWC) was signed in 1993 and entered into force in 1997. The CWC prohibits the use of RCAs as a method of warfare.

The CR Case: Northern Ireland

The problem here was largely domestic and unrelated to international treaties and their interpretation: how could the authorities cope with an attempted mass break-out of Republican prisoners from HM Prisons in Northern Ireland held under the highly controversial policy of internment during the 'Troubles'? The policy of swooping into Catholic areas and summarily arresting and interning supposedly known Republicans in prison camps on hearsay evidence began in August 1971. And in January 1972 the British Army fired on and killed 14 unarmed demonstrators in what was known as 'Bloody Sunday'.[11] Both of these events inflamed passions and only served as a powerful recruiting agent for the IRA. Control of Republican prisoners became a pressing problem for the authorities as the internees did not take their confinement lying down. This question was all the more acute as the physical defences of the prisons concerned – the Maze and Magilligan – were weak and were in the process of being strengthened. A new riot control agent, CR (dibenzoxazepine), was available and it offered an effective solution. Safety and political considerations were, however, significant countervailing pressures. Just how did the then Conservative and Labour governments resolve these dilemmas?

By early 1973 CR had been under investigation at the Chemical Defence Establishment (CDE) since 1961 and both its effectiveness and its safety had been extensively tested on animals and human volunteers.[12] During these tests DBO, as it was initially known, was found to be several times more effective as an irritant than CS. Tests had suggested that it was safer than CS, although examination of the carcinogenic and other long-term medical effects had yet to be completed and were expected to be undertaken by early 1974. CR was

---

[10] TNA WO 188/2704, G.D. Heath, DBCD, MOD to D/CDE, CS and the Geneva Protocol, 17 June 1971.

[11] Robert Fitzroy Foster, *Luck and the Irish: A Brief History of Change 1970–2000* (London, 2007), pp. 114–15; see also Richard English, *Irish Freedom The History of Nationalism in Ireland* (London, 2006), pp. 370–71.

[12] TNA WO 188/2704, Dibenzoxazepine (DBO).

available in a liquid form, and had been studied at a strength of 0.0025 per cent by the Medical Committee of the Chemical Defence Advisory Board (CDAB), which operated under the aegis of the Defence Scientific Advisory Council.[13] It was in this concentration that CR would be used in any agent delivery system and in the first instance this meant SPAD (Self-Protection Aid Device) – a small aerosol container that fitted into a pouch and was designed to be used at very close range of about three to four feet.

The Army thought that CR in solution was the more effective and less physically risky choice of the possible alternatives such as baton rounds for dealing with rioting prisoners or attempted break-outs. CR use in smoke rounds was considered too hazardous in case it drifted off and affected civilians living and working nearby. The prison guard force itself would need to wear protection, which would have operational implications for training and use of CR. HQ Northern Ireland therefore recommended that SPAD should be authorised for use as a last resort at the Maze in the two eventualities of either a riot beginning to run out of control, or of preventing escapes and arresting escapers.[14] This request was referred to a Cabinet sub-committee (known as GEN 79) following agreement between the Defence and Northern Ireland Secretaries of State that a joint paper explaining the issues to Ministerial colleagues would be prepared by the departments.[15]

William Whitelaw, Northern Ireland Secretary of State, was particularly concerned by a subsequent MOD proposal that the Commanding Officer of the military stand-by force at the Maze should be authorised to use SPAD as a last resort in the interim period until the question was considered by GEN 79. Although Whitelaw fully recognised the potentially dangerous situation at the Maze, he took the view that any authorisation could place in jeopardy a political solution to the troubles in Northern Ireland. He felt that it would be a gift for Republican propaganda. Whitelaw was no doubt sensitive to all of this in case events damaged prospects in the parallel negotiations on a new Northern Assembly based upon power sharing between the Ulster Unionists and the nationalist Social Democratic Labour Party.[16] The Prime Minister agreed that

13 TNA CJ 4/456, D.R.J. Stephen AUS (GS) to PS/Minister of State, CR, 31 July 1973.

14 TNA CJ 4/456, DMO to Secretary of State, Security of HM Prison the Maze, 7 August 1973.

15 TNA CJ 4/456, Frank Cooper, MOD to Mr Howard-Drake, Use of CR, 9 August 1977.

16 Foster, *Luck and the Irish*, pp. 117–18.

any decision on possible use of CR should await consideration and decision by a GEN 79 meeting.[17]

A GEN 79 meeting held on 13 September was inconclusive. Whitelaw remained concerned about the potential adverse political implications following any CR use and there were anxieties too over the agent's safety. The Himsworth (Medical Research Council) Report Part II in 1971 into the toxicological aspects of CS and its use for civil purposes had made two recommendations for the introduction of new chemical riot control agents. The first of these noted that the effects of any chemical agents intended for use in internal security operations should be studied in a manner 'more akin to that in which we regard the effects of a new drug'.[18] MOD believed that this requirement was met by the CDAB's Medical Committee in its study of CR liquid. The second recommendation noted that, 'if the competent authorities feel it justifiable to release a chemical agent for use in civil circumstances, the medical and scientific research relevant to this decision should straightaway be published in the appropriate scientific journals so that informed medical and scientific opinion may assess the situation itself'. Ministers and officials believed that this requirement would be addressed when a suitable report was published in the July 1973 edition of the journal *Medicine, Science and Law*. This was delayed, however, which resulted in some GEN 79 members fearing that the second Himsworth recommendation had not been fully implemented.

MOD, however, were quite sure that the Himsworth recommendations had been adopted. CDE had conducted a very extensive series of tests, the results of which were assessed by independent medical advisers who confirmed MOD's conclusions that CR was safe under the conditions of proposed use. Although a rigorous long-term study to determine if CR could have carcinogenic effects was not yet quite completed, the evidence to date from this and from an earlier pilot study indicated that there was no cause for concern about possible long-term effects. After examining the matter afresh, Ian Gilmour (Under-Secretary of State, MOD) saw no reason to depart from the statements made by him in previous minutes to the Prime Minister and by Lord Carrington (Secretary of State for Defence) at the GEN 79 September meeting that the requirements laid down by the Himsworth committee for introducing a new chemical riot control agent had been fully met, and he did not consider that those statements in any way misrepresented the position.[19]

---

17 TNA CJ 4/456, S. Bampton, NIO London to A.W. Stephens DS 10, MOD, 31 August 1973.

18 Sir Harold Himsworth had been Secretary of the Medical Research Council from 1949 until 1968 and had been tasked to lead a review on the safety of CS.

19 TNA CJ 4/1338, D.E. Young, PS to M. Alexander, FCO, 2 October 1973.

In light of this it appears that Whitelaw was prepared, pending a further GEN 75 meeting, to reach an understanding with the General Officer Commanding Northern Ireland (GOC NI) about the circumstances in which SPAD might be used.[20] GEN 79 gave provisional approval for CR, but under strict conditions: authority for deployment would only be extended until the completion of new security fences; it would be a last resort for mass break out – at least 100; and would not be aimed at non-prisoners aiding and abetting any escape.[21] Whitelaw also told GOC NI that he should be consulted first whenever possible. MOD, although not disagreeing with the terms of the decision, argued that it would not be easy to translate these into workable rules of engagement, especially the point about 'at least 100'. It could well be difficult to count in some circumstances. Whitelaw finally agreed to revised rules of engagement on 4 October 1973.[22]

Carrington felt that as long as the possibility existed and so long as the results of a successful mass escape remained as daunting as they undoubtedly were at that time, it would be quite wrong to deny the guard force the use of CR in the last resort – in the form of SPAD – until better systems became available.[23] Moreover, the MOD subsequently argued that SPAD should still be available even after completion of the new security fences. This did not go down too well with Whitelaw and the Northern Ireland Office. They continued to see considerable dangers in the proposal. Whitelaw thought he had to oppose it believing that any disclosure that the use of CR was authorised at the Maze would be extremely damaging to the political progress towards some sort of settlement that was then being made.[24]

By December 1973 Whitelaw, however, was no longer Secretary of State at the Northern Ireland Office having become Employment Secretary instead. His successor Francis Pym noted shortly after taking up office that the present authority on CR use should be continued under the existing rules; but no public announcement specifically relating to the Maze or to Northern Ireland should be made. In addition, further efforts should be undertaken to dispel the scientific doubt about the effects of CR.[25] The Prime Minister commented that in light of Pym's comments the way was now clear for the present authority for the use of

20 TNA CJ 4/456, Robert Stevens, PS to Christopher Roberts, No. 10, Use of SPAD at the Maze, 5 October 1973.

21 TNA CJ 4/1338, F. Cooper to Mr Platt, 18 September 1973.

22 TNA CJ 4/1338, D.E. Young, PS to M. Alexander, FCO, 2 October 1973.

23 TNA CJ 4/456, Carrington to Prime Minister, 31 October 1973.

24 TNA CJ 4/456, Whitelaw to Carrington, 28 November 1973.

25 TNA CJ 4/1338, Martin Reid, NIO to N.H. Nicholls, MOD, 14 December 1973.

CR to be continued under the existing rules, subject to the further points made by the Secretary of State.[26]

Labour's Approach: 1974 and 1975

The key problem for the new Labour government under Harold Wilson was what to do about rules of engagement governing the use of CR in Northern Ireland: just how strict should these be? Non-MOD Labour Ministers displayed the same degree of sensitivity on this matter as their predecessors. However, by spring 1974 Ministers were much more confident on the safety case. Roy Mason, Secretary of State for Defence stated in a written parliamentary question answer that an intensive testing programme into CR over several years was largely completed and gave no cause for concern. The results were being made available in reports to appropriate scientific journals, some of which had already been published with others awaiting publication or in the course of submission.[27] Intensive research had been conducted at CDE into the possible biological and environmental effects of CR including its toxicology, pharmacology, carcinogenicity, teratology (effect on unborn children) and its possible contamination of food and the environment. The results of these tests were submitted to the independent scrutiny of the CDAB's Medical Committee. No long-term or adverse effects were discovered, and although some of the test programme elements were too long-term to have been completed as of summer 1974, the results thus far had given no cause for concern.[28]

The issue, as it had been for the previous government, boiled down to the scope of the ROEs for the use of SPAD. Other delivery systems such as water canon were also coming into the picture, but for the time being the focus was on SPAD. It was very clear, notwithstanding the safety advice received, that Ministers were still keen to control any use of CR at the prisons, the Maze in particular. We can see these concerns in exchanges between the Secretary of State for Defence and the Northern Ireland Secretary in May 1974. Mervyn Rees, the Secretary of State at the Northern Ireland Office, insisted that the ROEs make clear at the outset that the use of SPAD was permitted only on the authority of the GOC or, in his absence, the Commander Land Forces and if

---

26 TNA CJ 4/1338, R. Butler, 10 Downing Street to N.H. Nicholls, 18 December 1973.

27 Hansard Written Answers, 18 March 1974, Roy Mason, CR Gas.

28 TNA CJ 4/1338, Annex to Secretary of State for Defence and Secretary of State NI to PM, Use of CR at the Maze and Magilligan Prisons, Authority for Other CR Devices, 29 July 1974.

possible after prior consultation with the Secretary of State. Moreover, it had to be fully understood by all involved that SPAD was only to be used as a last resort and for the sole purpose of stopping a mass escape. Knowledge of the authority for the use of SPAD was to be limited to those with an absolute need to know.[29]

SPAD was to be deployed from secure storage only on the authority of the Commanding Officer of the Guard Unit at the Maze and could only be discharged on the authority of the Commissioned Officer responsible for the action in the area in which it was required. Use could only be authorised when escaping prisoners reached at least the line of the inner weld mesh fence and either were attempting to cross that fence in such numbers that the available security forces would otherwise be unable to prevent escape if the prisoners succeeded in breaching or crossing the outer fence; or they had to have gone beyond the outer fence and there was no other means of preventing them from making good their escape.[30]

As noted above, thought was being given to using additional CR delivery systems for the guard forces. In July 1974 the Defence Secretary and the Secretary of State for Northern Ireland minuted the Prime Minister asking for his agreement for deployment of additional CR delivery devices at both the Maze and Magilligan prisons: these were water cannon and an enlarged SPAD or vehicle mounted device. These were still subject to approval by CDAB's Medical Committee.[31] However, the Prime Minister took the view that it was important that the Army should not be authorised to use these devices at the prisons until the Medical Committee had given its approval to the enlarged SPAD and trolley or vehicle-mounted device.[32] In December 1974 Harold Wilson was informed that the Medical Committee had concluded that the devices carried no risk of serious injury, but as with the hand-held SPAD, all devices were to be subject to stringent rules of engagement agreed between the two departments – MOD and NIO.

---

[29] TNA CJ 4/1338, Proposed ROEs for SPAD at HMP Maze I Annex A to Roy Mason Secretary of State for Defence to Secretary of State Northern Ireland, Use of CR at the Maze and Magilligan Prisons, 3 May 1974.

[30] TNA CJ 4/1338, Proposed ROEs for SPAD at HMP Maze I Annex A to Roy Mason Secretary of State for Defence to Secretary of State Northern Ireland, Use of CR at the Maze and Magilligan Prisons , 3 May 1974.

[31] TNA DEFE 24/966, Draft Submission PS/Secretary of State, Use of CR at Northern Ireland Prisons, Attached to S.S. Brampton, NIO to N. Bevan, DS10 MOD, 7 December 1976.

[32] TNA CJ 4/1338, Tom Bridges, No. 10 Downing Street to Martin Reid, NIO, Use of CR, 1 August 1974.

The ROEs were too constraining for Lt. Gen. Sir Frank King (GOC Northern Ireland). He felt strongly that delegated authority was needed for a fast moving situation. In such circumstances it would not always be possible to get in touch with Ministers quickly. Recent disturbances at the Maze had highlighted this problem. Authority for CR to be deployed for the first time had in fact been given to the Commanding Officer of the Maze guard unit on the night of 15–16 October 1974. However, as Brigadier Garret told the Northern Ireland Office on 1 November CR was not used.[33] Nevertheless, given the current prison threat the GOC was extremely keen that the limits on CR use by the guard unit Commanding Officers on their own initiative in an extreme situation should be removed as a matter of urgency.[34] Roy Mason urged Mervyn Rees to agree that they should accept the GOC's request to modify existing ROEs to the extent that if in the last resort the Commanding Officer at the prisons was unable to obtain the requisite superior authority for the use of CR in time, he should be permitted to do so on his own authority.[35]

The up-shot of these exchanges was that Ministers acceded to the military request. This meant a revision to the ROEs, whose salient features were as follows:

- the Commanding Officer should not use CR without obtaining on any occasion prior approval of GOC; or Commander Land Forces or CO 3 or 8 Brigade if neither previous two available;
- whichever of the aforementioned officers were consulted, they should seek personal approval from the Secretary of State for Northern Ireland (anywhere in the UK), only in the unlikely eventuality that he was not readily available should authority be sought from the Duty Minister; if neither available then GOC, CLF or COs 3 or 8 Brigade may do so;
- only if the relevant CO was unable to contact either, or *in extremis* a Minister direct, they may then approve CR use;
- only if the situation at the prison is desperate and every second counts, may a unit commander himself authorise the use of CR without first seeking higher authority.

---

33 TNA CJ 4/1338, Brig. H. Garrett to J.B. Bourn, NIO Belfast, Delegation of Authority to Deploy/Use CR, 1 November 1974.

34 TNA CJ 4/1338, Brig. H. Garrett to J.B. Bourn, NIO Belfast, Delegation of Authority to Deploy/Use CR, 1 November 1974.

35 TNA CJ 4/1338, Roy Mason to Secretary of State Northern Ireland, Use of CR Devices at HMP Maze and Magilligan, 11 November 1974.

This last point was emphasised by Rees as he wanted it to be made unmistakably clear to the Commanding Officer concerned that by a desperate situation he meant one in which, for example, an uncontrollable mob of prisoners was using violence and was about to escape.[36] The Prime Minister agreed to this arrangement on 2 January 1975.[37] However, this did not mean that soldiers could train how to use CR; Ministers placed an absolute prohibition on any training since they feared that this might leak and cause political trouble.[38] HQ NI found that this imposed severe constraints but there the matter appears to have rested.

Further Reservations: Exercises in Hong Kong

Sensitivities over the use of CR were not just confined to the tinder box of Northern Ireland, even the proposed use of CR in training exercises in Hong Kong in 1973 fell foul of London's anxieties.[39] Military authorities in Hong Kong, initially supported by the Governor, saw the value of CR in an exercise intended to simulate massive influxes of Chinese refugees into the colony. These exercises would only have involved the British forces and Hong Kong police and with volunteers willing to play the part of refugees and to subject themselves to exposure to CR – Exercise Gentle Rain.[40] There appears to have been no sense of irony in the choice of title even though the objective was to test the utility of CR in controlling refugees. However, once the MOD and FCO got wind of this plan they quickly feared that this would cause unwanted controversy and embarrassment for HMG given that the exercise was to take place so close to the Chinese frontier and that its taking place was bound to leak to the press and local population. If it did it was bound to be misinterpreted by the Chinese military. The recent UK reinterpretation of the scope of the 1925 Geneva Protocol as discussed earlier in this chapter was still a live issue and one moreover that had been controversial given the differences between the MOD and FCO; use of

---

[36] TNA CJ 4/1338, Merlyn Rees to Roy Mason, Secretary of State for Defence, 16 December 1974.

[37] TNA CJ 4/1338, PM agreed on 2 January 1975. Patrick Wright, 10 Downing Street to M.F. Mumford, MOD, Use of CR at the Maze and Magilligan Prisons, 2 January 1975.

[38] TNA DEFE 24/966, J.R.A. MacMillan, Col GS MO4 to DS10, Training with SPAD/MAXI SPAD, 31 December 1975.

[39] TNA FCO 66/514; FCO 66/513, Trials of CR in Hong Kong, 1973, G.F. Clark, FED to Mr Evans, 16 August 1973.

[40] TNA FCO 66/513, GSO 1 GS (OR) 4 to DMG, 18 July 1973 and Head DS6, MOD Exercise Gentle Rain, 10 August 1973.

CR in an exercise was bound to stir things up again. Giving the impression that the UK was prepared to use such material in an international conflict when the majority of states at that time did not share the UK view on the Geneva Protocol was something to be avoided.[41] The up-shot of all of this was that the exercise was to be deferred for another six months, but in the event it appears to have been postponed indefinitely as there is no further record currently available at TNA of such an exercise taking place, certainly up to the end of 1975.[42]

Political sensitivities coupled with the legacy of the UK's reinterpretation of the Geneva Protocol scuppered even a training exercise; the UK was thus constraining its own programmes and tactical requirements for the use of riot control agents through its fear of adverse international reaction.

Conclusions

So what are we to make of UK approaches to CS and CR? What do these two cases tell us about UK thinking and do the events of 1969 to 1975 have any resonance for the contemporary handling and understanding of the riot control, incapacitating chemicals, and the law enforcement chemical issue in the Chemical Weapons Convention (CWC)? And what do they tell us about UK approaches to the issues of this aspect of chemical warfare? We must be careful of course not to assume that mindsets in one period are necessarily going to remain the same in later periods separated by some 30 to 40 years. Science and technology have changed dramatically too since 1969, but perhaps some of the core concerns are the same and still relevant today.

Arguments in 1969 over lethal and non-lethal agents, their use in situations other than war, the humanitarian benefits, the potential threat posed to existing treaties, attempts to strengthen the CBW disarmament regime and the possible impact on UK–US relations all have echoes in the early twenty-first century. We should recall, however, that extensive US CS use in the war in Vietnam gave the debate a much higher profile in the late 1960s and in 1970. CS had been used as a force-multiplier to attack the Vietcong, most notably by flushing them out of their networks of tunnels, so that they could be attacked with artillery and aircraft. Matt Meselson, Professor of Biology at Harvard University, had noted in testimony to the US Senate Defense Committee in May 1970 that the US had by that stage used 6,803,850 kg of CS in Vietnam and as such there was risk that such use would break the psychological barrier to the use of lethal chemical

41 TNA FCO 66/513, N. Marshall, ACDD to Crowson, HKIOD, 8 August 1973.

42 TNA FCO 66/513, D ES Baltherwick, ACDD to Miss Kelly, HKIOD and Mr Davies (FED), Exercise Gentle Rain, 7 September 1973.

agents in combat.[43] Contemporary debates over incapacitating agents and the CWC by way of contrast have a much lower impact than CS and the Geneva Protocol did in its day. That said, given scientific understanding and the pace of technological change today in chemistry, genomics, proteomics, synthetic biology and incapacitant chemicals, the threat posed to the long-term health of the CWC is infinitely greater than the one posed to the Geneva Protocol by CS.

As for CR, it is very clear that Ministers – both Conservative and Labour – were acutely sensitive to both the safety aspects and what both saw as the inevitable adverse political consequences of any CR use in the prisons. The insistence on exceptionally tight ROEs shows that it was only ever to be used as the absolute last resort – political control too was a key requirement with Ministerial authorisation to be sought in almost every case and it should only ever be used for escaping prisoners. General use in Northern Ireland was never contemplated. Even training to use CR dispersal devices was prohibited. And as we saw in the Hong Kong context, this nervousness extended to even training exercises. The connection with CS was also evident following from the Himsworth Committee's recommendation that any new riot control agent should be subject to the same sorts of safety testing that would apply to a new drug. Proposals to deploy CR to Northern Ireland therefore could only be agreed once the various safety studies had been completed and published and not before. Such testing did not apply just for the agent, but also had to take into account the different delivery systems that were becoming available. One cannot help but conclude that despite the very clear and conclusive safety studies, Ministers perhaps only subconsciously, may still have harboured lingering doubts over safety.

One might infer from this that in today's climate, where health and safety and liability concerns are very much greater than they were in the early 1970s, Ministers and officials would be even more sensitive to safety and ROE issues for any chemical that might be considered for law enforcement or riot control purposes. Although the political sensitivities surrounding the use of a new riot control agent at the height of the troubles in Northern Ireland were a special case, we can still conclude that similar concerns as applied in 1973–75 would still in all probability obtain today. Indeed they may well be even greater given much greater preoccupations with safety. It is clear that there were very significant constraints on UK plans to use CR in the 1970s, but these were self-imposed and arose from acute political sensitivities surrounding the situation in which CR might be used; arms control *per se* had little if any impact. So in our period arms control generally had little impact on UK chemical warfare programmes

43 Mathew Meselson, 'Tear Gas in Vietnam and the Return of Poison Gas', *Bulletin of Atomic Scientists* vol. 27, no. 3 (March 1971), pp. 17–19.

apart from the impact of the Geneva Protocol on moderating potential uses of any new incapacitating chemical agent; and, as we have seen, the UK decided that the Protocol did not apply to any use of CS. But what of UK policy making on biological weapons? This is the subject of the next three chapters.

Chapter 5

The Biological and Toxin Weapons Convention: Origins of the UK Proposal to Separate BW from CW 1968

Introduction

Although the UK had an offensive BW programme during the Second World War and in the immediate post-war years, the status of that programme was shifting increasingly towards a defensive posture by the 1950s.[1] Several decision points were critical here. The UK policy on BW adopted in 1952 required that NATO allies should not take up a position that would deprive them of the ability to use BW in retaliation, if this were to their advantage. In 1953 in the interests of economy and efficiency the UK programme of R&D was directed to be co-ordinated with that of the US.[2] However, the existing directive on biological warfare policy that had dated from 1945 was revised and a new policy proposal submitted by the Chiefs of Staff and Ministry of Supply to the Prime Minister in October 1953 for approval. This essentially saw no change in the policy to concentrate efforts primarily on defence against biological attack, but noted that the UK should be prepared to retaliate in kind should the government of the day decide to do so.[3] However, the services decided simultaneously that they could no longer support a BW manufacturing capacity.[4] In any case the UK had struggled to develop workable requirements for an anti-personnel weapon and it had been hoped in 1947, when an Air Staff Requirement had been set out,

---

1 See Brian Balmer, *Britain and Biological Warfare Expert Advice and Science Policy, 1930–65* (London, 2001) and 'The UK Biological Weapons Program', in Mark Wheelis, Lajos Rozsa and Malcolm Dando (eds), *Deadly Cultures Biological Weapons Since 1945* (Cambridge, MA, 2006).

2 TNA CAB 21/4505, D.C. (52) 8th Meeting, Minute 1, C.O.S. (58) 196, 19 August 1958.

3 TNA CAB 21/4505, Minute to PM, Biological Warfare Research and Development Policy D (53) 44, 1953.

4 TNA CAB 21/4505, DRPC Biological and Chemical Warfare, Note by the DRP Staff, 28 January 1959.

to have this in service by 1955. However, by July 1954 it had become apparent to the Air Staff that because of the magnitude or the technical problems more research was required in both the agent and weapon fields before a satisfactory weapon could be recommended to the RAF for service use. Key issues were storage, transportation, testing, preparation for use and the biological agent itself.[5] The 1947 Air Staff Requirement was cancelled in July 1954 and replaced with a new target that required research to determine effective agents and suitable weapons for waging biological warfare. Things were to change by the end of the decade. The Chiefs of Staff agreed in 1958 that the UK did not need either a CW or BW retaliatory capability, provided that the US continued to possess one.[6] So therefore by the mid 1960s, when thoughts were turning to CBW disarmament, the UK had long since ceased to have any offensive programme, but it had continued with significant biological defence work at the Microbiological Research Establishment (MRE) at Porton Down.[7] Here the main focus was on hazard assessment. There was thus no offensive programme, even a dormant one, that could be adversely impacted by any arms control or disarmament initiative launched by others or indeed the UK itself; care would be needed to ensure that defensive work was not impeded or otherwise unduly constrained. The UK had never publicly offered any confirmation that it possessed any biological weapons and generally remained coy about the status of its CB programmes. It was against such a background that the UK began to consider an initiative on biological weapons in the aftermath of the conclusion of the Nuclear Non-Proliferation Treaty in 1968, which was one of the major international arms control agreements of the 1960s. Just why the UK decided to light upon biological weapons as the main preoccupation of its multilateral arms control policy and the process in which that decision was reached has remained somewhat opaque. This chapter explores the background and explains the evolution of UK thinking and the events that led to the tabling in 1968 of one of the landmark proposals in the history of the main multilateral disarmament forum – the Eighteen Nation Disarmament Committee (ENDC), which was based in Geneva in the UN Palais des Nations. This was to lead to the Biological and Toxin Weapons Convention 1972, which to this day remains the cornerstone of the regime to combat the threat posed by 'public health in reverse' – a phrase used to characterise biological warfare. Little has been written

[5] TNA AIR 20/8727, Air Staff Target No. O.R./1065 Biological Warfare Agents and Weapons, and the Associated Problems of Their Storage and Handling under Service Conditions, Air Ministry (D.O.R. (A)), 16 July 1954.

[6] TNA CAB 21/4505, C.O.S. (58) 196, 19 August 1958.

[7] Peter M. Hammond and Gradon B. Carter, *From Biological Warfare to Healthcare: Porton Down, 1940–2000* (London, 2002).

on the immediate origins of the Convention based on The National Archive sources and what there is, does not fully reflect an in-depth review of all the relevant archival papers.[8]

Biological Disarmament: Phase One

The origins of the 1968 decision to pursue an initiative on BW disarmament can be traced to the Foreign Office's Arms Control and Disarmament Research Unit (ACDRU) paper on CBW written in 1966 and 1967, and especially work done in the Atomic Energy and Disarmament Department (AEDD) during late June 1968. ACDRU had been established following an initiative by the 1964 Labour government to develop new ideas on polices on arms control.[9] As the negotiations on the Nuclear Non-Proliferation Treaty (NPT) moved towards successful conclusion in January 1968, AEDD began to consider what next steps could be taken in UK disarmament policy. Fred Mulley, the Minister of State in the FCO, in particular was anxious that in addition to work in the nuclear field to follow up the NPT, the ENDC should also pay attention to the possibilities of progress in the non-nuclear field. A priority was an urgent need to take a fresh look at the CBW question.[10] CBW was in the news given US use of riot control agents and defoliants in Vietnam and the prominence that CBW matters had in the UN in 1967 and 1968. The UN Secretary-General, U Thant, had stated that, 'the international community was not sufficiently conscious of the danger inherent in this new type of weapon of mass murder' and 'that due attention had not been focussed on this very serious problem'.[11]

The first part of this chapter focuses on the original 1967 ACDRU paper and consequent discussion with the US during which the idea that BW should be separated from CW appears for the first time. Unfortunately AEDD files

8 Susan Wright, 'Geopolitical Origins', in Susan Wright (ed.), *Biological Warfare and Disarmament: New Problems/New Perspectives* (Lanham, MD, 2002) and Marie Isabelle Chevrier, 'The Politics of Biological Disarmament', in Mark Wheelis, Lajos Rozsa, and Malcolm Dando, edited, *Deadly Cultures: Biological Weapons Since 1945* (Cambridge, MA, 2006).

9 See TNA FO 366/3386, Creation of Research Unit for Disarmament, 1964 and TNA FO 371/181368, 181369 and 181370, FO Disarmament Research Unit and Advisory Panel on Disarmament, 1965.

10 TNA CAB 168/125, FCO telegram to Certain Missions, Guidance No. 156, 19 June 1968.

11 *Chemical and Bacteriological (Biological) Weapons and the Effects of Their Possible Use, Report of the Secretary-General* (New York, 1969), Foreword, p. viii. Thant was referring to comments made in 1968.

are unclear on the exact relationship between the 1967 discussions and the June 1968 decision to table a UK working paper proposing focus on BW in the ENDC, but the ACDRU paper appears to have shaped subsequent thinking in the Foreign Office's lead policy department. The second half of this chapter charts the internal Foreign Office, MOD and Cabinet Defence and Overseas Policy Committee deliberations between January and July 1968, which led to the landmark Working Paper on Microbiological Warfare.

The Arms Control and Disarmament Research Unit's Role

The first Wilson Labour Government established the Arms Control and Disarmament Research Unit (ACDRU) in 1965.[12] Lord Chalfont, Minister of State for Disarmament Affairs, decided to form a special advisory group on arms control policy. Hedley Bull, then a Reader in International Relations at the London School of Economics, was invited by Chalfont to be ACDRU's first Director. Under Bull the Unit embarked on a series of major policy research papers on arms control options, particularly on non-proliferation. CBW was also considered, and a substantial paper entitled 'Arms Control Implications of Chemical and Biological Warfare' was begun in early 1966 and drafted by Dr Deryck Viney.[13] A second draft was ready by 4 July 1966.[14] The final version was issued on 3 February 1967 and sent to the US Arms Control and Disarmament Agency on 15 February.[15] (A copy was also shared with Canada.[16]) The paper outlined the nature and status of chemical and biological weapons and the constraints on their use in the mid 1960s; and considered whether further

---

[12] Hansard 16 December 1965, Prime Minister, Foreign Affairs debate; TNA FO 371/181368, IAD 1027/1 Press Release, Arms Control and Disarmament Research Unit News Department, Foreign Office, 31 December 1965.

[13] TNA FO 371/187448, IAD 1039/2/G, Hedley Bull to Mr Street, Chemical and Biological Warfare, 17 March 1966.

[14] TNA FO 371/187448, IAD 1039/4/G, ACDRU (66) 2 (2nd Draft), Arms Control and Disarmament Research Unit, Arms Control Implications of Chemical and Biological Warfare Analysis and Proposals 4 July 1966.

[15] TNA FCO 10/179, folio 24, J.E.D. Street, Foreign Office to Mr R.I. Spiers, Embassy of USA, 29 June 1967; Ian Smart, Washington to Major-General R.E. Lloyd, Deputy Head of ACDRU, Chemical and Biological Warfare, 21 June 1967; TNA FCO 10/184, Biological and Chemical Weapons, 1967, folio 2, ACDRU (67) 1 (Final), 3 February 1967, Arms Control and Disarmament Research Unit, Arms Control Implications of Chemical and Biological Warfare Analysis and Proposals; H. Bull to Mr Street, 8 February 1967.

[16] TNA FCO 10/184, folio 17, R.E. Lloyd to Mr Street, 29 March 1967.

constraints were desirable and how they might be achieved. Three conclusions were of particular relevance to the BW problem.

First, it was thought desirable that CB warfare should be constrained. Second, there were a number of existing constraints on CB use: the non-suitability of some CBW agents; scepticism about their safety and effectiveness; their questionable utility in many circumstances; the effectiveness of counter-measures against them; fear of retaliation; the 1925 Geneva Protocol; and belief in a customary BW prohibition. However, since many of these constraints were perceived to be declining in effectiveness, they needed to be reinforced. Third, international agreements restricting possession of CBW were not thought feasible at that time because of verification difficulties – the dual-use problem and Soviet opposition to on-site inspection of any sort being the main challenges. Prospects for a CTBT had foundered earlier in the decade, in part, on recurring Soviet rejection of Western calls for some form of on-site inspection.

From this assessment the ACDRU paper concluded that the most promising way of reinforcing constraints on CBW was to promote further international agreements restricting their use. There were two options here: either canvass additional signatories for the Geneva Protocol and reform the Protocol either to include a right of reprisal or to permit the use of non-lethal agents; or to ban the use of biological agents unconditionally; and to provide for a 'reprisal code' or a more acceptably named equivalent. However, ACDRU did not think it opportune for any initiative in these directions given US policies at that time – use of CS and defoliant agents in Vietnam; instead an exchange of views between the UK and US and perhaps with other Western countries was the preferred option. This would help achieve some consensus before moving towards any international negotiations.

On 21 June 1967 Ian Smart, British Embassy Washington, wrote to Major-General Lloyd (Deputy Head of ACDRU) reporting US reactions to the ACDRU paper.[17] The US preliminary and informal view was positive; the Americans had appreciated the deference paid to US sensitivity about any initiatives on CBW arms control at that time. Although the US would have preferred no new initiatives at least until the situation in the NPT negotiations became clearer, they were favourably inclined towards 'the idea of engaging in informal discussions with us at research level'. The US asked whether sending a small team of CBW technical experts to London in October or November 1967 to engage in preliminary exchange of views would be acceptable. For reasons that are not entirely clear it took until 2 August 1967 for the UK to agree. This was

[17] TNA FCO 10/179, folio 28, Ian Smart, Washington to Major General Lloyd, ACDRU, 21 June 1967.

not because of any reluctance; on the contrary, the Foreign Office was 'highly gratified' by the US response as it was precisely this sort of close relationship, at the research level, that the UK hoped to develop with the US Arms Control and Disarmament Agency (ACDA).[18] This result was a good example of the way in which work done by ACDRU could stimulate US interest and which could eventually lead to recommendations for policy decisions. Time was needed to ensure that MOD experts would be able to take part.[19] Talks were subsequently arranged for 12–13 October 1967 in London.[20] It is possible too that the international communities' inability to follow-up effectively on Egyptian use of CW in Yemeni civil war between January and July 1967 may have been a factor in UK thinking.[21] (This conflict was fought in North Yemen between royalists of the Mutawakkilite Kingdom of Yemen and factions of the Yemen Arab Republic between 1962 and 1970. The royalists received support from Saudi Arabia, whilst the republicans were supported by Egypt and the Soviet Union.) Officials realised that the 1925 Geneva Protocol was not relevant for the situation faced in Yemen. Nor did it contain any enforcement provisions; and nor was any UN member state prepared to seek any condemnatory resolution on Egyptian use of CW in the General Assembly or Security Council. The Foreign Office's AEDD, as we have seen, was certainly looking for an effective way of dealing with CBW weapons in the arms control and disarmament context.[22] Yemen cannot have been far from their minds.

On 27 October 1967 Michael Cullis, now Director of ACDRU, informed the Minister of State, Fred Mulley, of the outcome of the ACDRU–ACDA talks.[23] Five principal conclusions emerged from the discussions. First, the CBW question in the arms control context had tended to be discussed too exclusively

[18] TNA FCO 10/179, folio 25, D.L. Benest to R.J. Abraham, D.S.12, MOD, Chemical and Biological Warfare, 12 July 1967; folio 28, D.L. Benest to Mr Street, 28 June 1967.

[19] TNA FCO 10/179, folio 29, J.E.D. Street to Sir J. Rennie and Mr R.J. O'Neill (Private Secretary Minister of State), 29 June 1967; folio 31, D.L. Benest to Sir J. Rennie, Chemical and Biological Warfare, 2 August 1967; folio 32, R.E. Lloyd, Foreign Office to I.M.H. Smart, Washington, 4 August 1967.

[20] TNA FCO 10/179, folio 39, I.M.H. Smart, Washington to Maj. Gen. R.E. Lloyd, ACDRU, 8 September 1967.

[21] See TNA FCO 8/712, Egyptian Use of Poison Gas in Yemen, 1967.

[22] TNA FCO 8/713, D.L. Benest to General Lloyd, ACDRU and Mr Blatherwick, Aden Department, 27 September 1967; see also D.J. McCarthy to Sir R. Beaumont, Poison Gas in the Yemen, 17 August 1967.

[23] TNA FCO 10/179, folio 43, Michael Cullis to Mr Mulley, 27 October 1967; Record of Informal Talks Between British and United States Officials on Arms Control Prospects for Biological and Chemical Weapons Held at the Foreign Office, London, 12–13 October 1967.

on a 'great power basis'. Second, the participants thought that there might be some utility in considering CW and BW separately from the arms control point of view. Third, there was little or no possibility that the US would adhere to the Geneva Protocol, even to a revised version.[24] Fourth, the prospects for some eventual international accord to replace the Protocol, however, would only be improved once the NPT had been completed – the Treaty would not be opened for signature until 1 July 1968. Finally, it was important to prepare for initiatives from other countries. Consequently, officials thought that it would be useful for both the UK and US to focus their research separately on drafting the framework of a 'new international instrument for controlling C and/or B weapons'.

What to do after the NPT?

Nothing further appears to have been done on CBW disarmament matters until January 1968. However, by the end of 1967 the UK was indeed keen to see action taken to deal with CBW in the arms control and disarmament context.[25] Foreign Office officials recognised that the chances of making practical progress at that particular time were slight. There were several reasons: the sensitivity surrounding CBW, which made it difficult to discuss the issues internationally and the virtual impossibility at that time of reaching agreement on satisfactory verification arrangements. US opposition to the 1925 Geneva Protocol was another major complicating factor.[26] Despite this background AEDD decided to prepare a paper on the next steps on arms control and disarmament that could be taken following completion of the NPT negotiations, which were then entering their end game. The paper itself was completed in conjunction with ACDRU in January 1968 and on 25 January a copy was sent to Ivor Porter (UK Ambassador to the ENDC) inviting comments.[27] AEDD wanted to submit this to the Minister of State recommending new objectives that might be pursued. FCO officials believed that the main target should be a CTBT; CBW, however, was consigned to the all too difficult category. AEDD's draft explained that this was because of several factors: US non-membership of the Geneva Protocol and opposition to attempts to improve it; US use of defoliant agents in Vietnam

[24] The US would not accede to the Geneva Protocol until 1975.

[25] TNA FCO 10/179, folio 53, Foreign Office telegram to UK Mission New York telno 7250, 8 December 1967.

[26] TNA FCO 10/179, folio 45, R.C. Hope-Jones, Foreign Office to Sir Leslie Glass, UK Mission, New York, 8 November 1967.

[27] TNA FCO 10/181, folio 58, R. Hope-Jones, AEDD to Ivor Porter, UKDis Geneva, 25 January 1968.

to destroy the forest cover for the Vietcong;[28] the secrecy surrounding CW and consequent lack of information; Soviet reluctance to accept the necessary intrusive inspection and similar attitudes in some Western and non-aligned countries. All these combined to suggest that the CBW issue was 'hardly a promising one for the ENDC'. Instead CBW arms control would be best pursued for the foreseeable future in private studies, preferably with ACDA. The draft paper was sent to the UK delegation to NATO and the Embassy in Washington. Neither questioned the CBW paragraphs. A copy of the paper was sent to the MOD in March; they offered no comments on the CBW section either.

The Next Steps on Disarmament: Paper Submitted to Mulley

On 10 June 1968 AEDD submitted a memorandum on 'The Next Steps in Disarmament' to Mulley in his capacity as Minister of State for Disarmament matters.[29] The covering note from Ronald Hope-Jones, head of AEDD, suggested an interdepartmental meeting to decide what the objectives should be and recommended that the paper should be put to the Cabinet Defence and Overseas Policy Committee (DOPC). Although, the paper itself did not propose any CBW measures, Hope-Jones believed that a UK initiative would be timely and would certainly have a good deal of internal political appeal given public and parliamentary unease over CB matters. This question had been discussed at a recent NATO Disarmament Experts' meeting during which the UK had half-promised to circulate a paper in advance of its next meeting in October. This envisaged a new Protocol, replacing the 1925 Geneva Protocol and would ban BW absolutely and the first use of CW. Some time between 11 and 20 June 1968 Mulley instructed AEDD to produce two possible revisions of the 1925 Geneva Protocol, making a distinction between CW and BW. However, Hope-Jones subsequently appears to have convinced Mulley that progress would only be possible if BW and CW were separated for arms control purposes and a Convention dealing entirely with BW pursued instead. Only

---

28 The US conducted a massive herbicidal programme during the Vietnam War over a period of about ten years aimed largely at South Vietnamese forests and to a much lesser extent against crops. The US eventually used a volume of some 72 x 103 m^3 containing almost 55 × 106 kg of active herbicidal ingredients – hormone-mimicking compounds and desiccants in what were known as Agents Orange, White and Blue. See SIPRI, *Ecological Consequences of the Second Indochina War* (Stockholm, 1976), pp. 24–45.

29 TNA FCO 10/181, folio 84, R. Hope-Jones, AEDD to Mr O'Neill, The Next Steps in Disarmament, 10 June 1968.

then was Hope-Jones able to prepare the first draft that ultimately became the UK's ENDC working paper on microbiological warfare tabled in August 1968. This was completed on the evening of Saturday 29 June. Mulley was able to take this with him for discussions with ACDA in Washington on 2 July and even then he only received Hope-Jones' paper on the plane.

Hope-Jones' paper argued that the Geneva Protocol was not a satisfactory instrument for dealing with CBW; for example, states had retained the right of retaliatory use; CBW production was still permitted; the term 'bacteriological' was not sufficiently comprehensive to include the whole range of microbiological agents that might be used in hostilities. CW, however, made the issue more complex. CW had been used in war and states would be reluctant to ban such weapons in the absence of effective verification, which was not then considered achievable given Soviet attitudes. In light of this it would be preferable to proceed with a new regime of obligations for BW rather than to do nothing. Despite the fact that 'fully effective verification is impossible' it would be better to constrain BW by placing more legal obstacles in the path of offensive programmes. To this end a new convention should be concluded that would ban BW production and offensive research; destroy any stockpiles; provide access to relevant military research to civil health authorities; and for the WHO to investigate allegations of use; and for the UN to investigate allegations of non-compliance.

UK–US Consultations: 2 July 1968

Following a conversation in Reykjavik between the US Secretary of State Dean Rusk and the Foreign Secretary Michael Stewart in the margins of the 24–25 June North Atlantic Council meeting, both agreed that Mulley should visit Washington to discuss BW disarmament with ACDA.[30] There was no single comprehensive brief for this visit, but Mulley planned to develop certain ideas put to him by Hope-Jones in his 10 June submission on BW disarmament.[31] This also included the Ronald Hope-Jones paper on microbiological warfare; however, as yet this latter paper had not been discussed with other Whitehall departments. The Washington meeting took place on 2 July. Mulley, Hope-Jones, the Ambassador Sir Patrick Dean and William Foster, Director of ACDA, and Dr Herbert Scoville were the principal participants. Mulley told the Americans that the UK was considering a separate Convention banning the

30 TNA FCO 10/181, folio 73, Note from Secretary of State – Dean Rusk meeting Reykjavik, 23 June 1968.

31 TNA FCO 10/179, folio 84, R. Hope-Jones to Mr O'Neill, The Next Steps in Disarmament, 10 June 1968.

use and possession of BW.[32] However, it would need to be accepted that it would not be possible to verify such a measure. That said, it might be stipulated that all relevant research should be transferred to public health authorities and that these would be obliged to publish all their results of their research. If agreement on these lines were worked out, then the WHO could be asked to nominate the 10–15 most advanced countries in microbiology, and that a Convention would not enter into force until it had been ratified by these, together with some 50–60 other states.[33]

The US team observed that any measures in the CW field would be extremely difficult for the US to accept. Foster pointed out that although the US could see advantages in a proposal to concentrate on BW as a diversionary tactic to distract attention from Soviet propaganda, any agreement without verification would be unacceptable to the US. Such an outcome would set an extremely bad precedent for future disarmament measures. Rather than tabling a draft convention, Foster wondered whether the UK could not confine itself to suggesting a number of proposals for further study. The US could live with this; if a treaty were to be tabled, the US would have no alternative but to attack it. Mulley replied that the UK felt politically obliged to table some proposals on CBW in light of domestic concerns over CBW. In any case, even if the UK did not present a draft treaty in July, it was inevitable that it would have to do so in January 1969 at the new ENDC session. Moreover, any working paper that was presented would in reality contain the basic provisions of a treaty anyway. Such a paper would in effect simply be a brief for lawyers to draft treaty language. Nevertheless, the UK was able to accept the US compromise provided the Americans could agree to the UN Secretary-General (UNSG) preparing a report on CW.[34] The Americans accepted this proposal. Hope-Jones revised the draft working paper in light of these discussions and then sent it to P. England (Head of Master General of the Ordnance (MGO) Secretariat) and Gerry Heath (Director Biological and Chemical Development Board in the MOD) on the morning of 5 July.[35]

Prior to a 12 July Cabinet Defence and Overseas Policy Committee (DOPC) meeting, AEDD provided defensive material for Mulley to deploy if

---

32 TNA FCO 10/179, folio 90, Record of Mulley's Discussion in Washington (ACDA), 2 July 1968.

33 TNA FCO 10/179, folio 90, Record of Mulley's Discussion in Washington (ACDA), 2 July 1968.

34 TNA FCO 10/181, folio 88, A Sir Edward Peck Comments on Draft Submission, 5 July 1968.

35 TNA FCO 10/181, folio 117, R. Hope-Jones to P. England, Head MGO (Sec) Arm. Dept, 5 July 1968.

other Ministers contested Foreign Office BW proposals.[36] AEDD argued that the proposals of the working paper – that a new convention would proscribe the use of microbiological agents for hostile purposes, their production, and offensive research, and require destruction of any stockpiles and the opening of research to civil uses – remained the best ones, especially since the UK would only be tabling them as a basis for discussion. In the event that the DOPC did not agree to these proposals, a fallback position was prepared. This brought them into line with MOD policy and was achieved by amending the core elements in the original proposals. States would now undertake never to engage in BW, even in retaliation; undertake to destroy within six months after the convention entered into force all BW agents over and above the type and quantities required for civil/passive/defence research purposes, and not to produce, or permit the production of such agents in future, or assist other states in their production; and, undertake not to conduct, finance or otherwise support research aimed at or involving the production of agents banned above, and to place under the control of the appropriate civil medical or health authorities all such research in this field as compatible with this obligation.

Hope-Jones proposed that states parties would undertake to destroy weapons within six months after a Convention entered into force, and thereafter not to produce or permit the production of any BW agent of types and in quantities that would be incompatible with the prohibition or to assist other states in their production. This was Hope-Jones' preferred option since it did not specifically refer to defence research, though the implication was felt to be sufficiently clear once its meaning was scrutinised. A revised paper was then redrafted for the DOPC that outlined UK disarmament proposals for the ENDC.

Defence and Overseas Policy Committee: 12 July 1968[37]

The Committee met on Friday 12 July. 'Further measures on disarmament' was the second item on the agenda. Mulley, speaking to his memorandum, noted that the ENDC would reconvene on 16 July and would probably have a wide-ranging discussion of various possibilities opened up by the NPT. It would aim to establish priorities for its January 1969 session. There had not been much time since the NPT's signature on 1 July to consider the UK's approach to the ENDC's next session, but he had discussed the main issues with the Director

36 TNA FCO 10/181, folio 107, R. Hope-Jones to Mr Moss, Further Measures of Disarmament, 10 July 1968.

37 TNA FCO 10/181, Cabinet Defence and Overseas Policy Committee OPD (68) 13th Meeting, Item 2: Further Measures on Disarmament, 12 July 1968.

of ACDA on 2 July. The UK, in his view, should be prepared to adopt a more independent role than previously and should take initiatives even if these were not wholly acceptable to NATO allies. Three initiatives might be pursued: one each on the CTBT, Peaceful Nuclear Explosions (PNEs) and BW. Mulley informed the Committee that he wanted to propose in the ENDC on 16 July that a Convention should be concluded, going beyond the provisions of the 1925 Geneva Protocol and ban the use, production and offensive research of BW.

Mulley informed his colleagues that the US did not support this proposal since Washington considered that it contained no adequate provision for verification; nevertheless, the Americans recognised that it was impossible to devise a verification system, and that by concentrating on BW, attention would be diverted from CW on which they were vulnerable to domestic and international criticism. Since the US would oppose a draft Convention, Mulley now proposed tabling a working paper suggesting that BW should be considered separately because of US unwillingness to ratify the Geneva Protocol, the difficulty of distinguishing between lethal and non-lethal chemical agents, and past and current use of riot control agents and defoliants in Vietnam.

Denis Healey, Defence Secretary, said that whilst he welcomed the proposal that the UK should play a more positive role at Geneva, the need to co-operate with the US in other fields (although he did not say so, he clearly had in mind Polaris Improvement[38]) set a limit to the extent to which the British could take an independent line. Some of the proposals on BW appeared, in Healey's mind, to be unsound. There was no basis for a distinction between biological and chemical agents on the grounds of lethality; a more relevant distinction was that biological weapons were more suitable for use against large civilian concentration than for tactical use. Healey went on to argue that there were verification problems – primarily the dual-use nature of materials and facilities. Many states produced biological agents for agricultural and veterinary use that would be suitable for military application. If such agents continued to be allowed to be produced, the states concerned would inevitably have a military capability. The proposal that research involving production of BW should be placed under the control of civil authorities would provide no safeguard in Soviet bloc countries where the problems of verification were most acute. BW production plants were inexpensive to construct and easily disguised. There would also be difficulties about allowing international organisations access to civil establishments on commercial confidentiality grounds.

---

[38] See John R. Walker, *British Nuclear Weapons and the Test Ban 1954–1973* (Farnham, 2010), chapter 11.

Following these principal contributions, the DOPC went on to discuss the BW proposal in detail. Several other points emerged. The World Health Organisation (WHO) would be unable to undertake any investigations or inspections under its present constitution, and the organisation would probably be unwilling to alter its mandate to do so. It was also important to ensure that legitimate production of vaccines should continue and defence research results should not be automatically made available to civil health and medical authorities. Any tabling of proposals on BW alone, it was argued by some, could be interpreted as an attempt to divert attention from the use of CW and would tend to stimulate criticism of the US and others who were using or developing such weapons. Nonetheless there were great advantages in a British initiative in view of the mounting public and parliamentary interest in these weapons; there had been considerable public and parliamentary concern in the early summer of 1968 over work done at MRE and CDEE.[39]

The Prime Minister, Harold Wilson, summarised the Committee's discussions. It was clear to him that the Committee was in favour of the UK taking a positive role and adopting a more independent attitude in the ENDC summer session. Furthermore, it was of great importance that the UK should be able to take initiatives despite the difficulties that had been identified. Mulley would be authorised to make a statement to the ENDC on 16 July declaring the UK's intention to submit proposals on CBW.[40] The Committee agreed that these should be prepared urgently by an interdepartmental committee of officials, which would seek the advice of appropriate scientific experts. This would be a reconstitution of the official interdepartmental committee on disarmament, which would report within two weeks so that if approved by DOPC, a working paper could be tabled early in August. This decision appears to have been occasioned by MOD concerns, or more precisely Denis Healey's complaints about lack of consultation on the original Foreign Office ideas.

Before the 12 July DOPC meeting, Healey had written to the Prime Minister on 10 July following circulation of Mulley's 5 July memorandum 'Further Measures on Disarmament'.[41] Healey protested that insufficient time had been allowed him or MOD for reflection, not to mention collaboration on a paper that raised questions of considerable importance to national security. It is not clear whether this objection was occasioned by the CBW aspects or nuclear items in the paper such as CTBT, PNEs and cut-off, but it seemed to him that

[39] TNA PREM 13/3464, FCO Guidance telegram to Certain Missions telno Guidance No. 169, 27 June 1968; MRE (Microbiological Research Establishment) and CDEE (Chemical Defence Experimental Establishment), both at Porton Down.

[40] ENDC PV/381, 16 July 1968.

[41] TNA FCO 10/181, folio 95, Denis Healey to Prime Minister, 10 July 1968.

Mulley was impatient and continually trying to push his ideas forward without proper consultation with relevant departments. However, given the start of UK attempts to obtain assistance from the US for what would become known as the Polaris Improvement Programme at roughly the same time, any suggestion that the UK might take steps that would annoy Washington was something of which Healey and MOD officials in the know would be exceptionally wary. In any case, the Defence Secretary now proposed that better interdepartmental machinery should be established, ideally a committee of officials. This would, in his view, ensure full consultation on disarmament matters between government officials, particularly the Foreign Office and MOD. Healey's recommendation appears to have been accepted; and on 12 July Sir Burke Trend (Cabinet Secretary) circulated details of the revised compositions and terms of reference for the DOP (O) Committee sub-committee on disarmament. Its terms of reference, however, were unchanged to consider questions of policy, arising from negotiations on disarmament. The original committee had been established in 1962, the same time as the ENDC.[42]

The Committee's composition was revised. Sir Edward Peck (Deputy Under-Secretary, Foreign Office) was Chairman with representatives from the Foreign Office, MOD and Commonwealth Office. Representatives of other departments and the United Kingdom Atomic Energy Authority (UKAEA) would be invited to attend meetings at which questions affecting their interests were to be discussed. One of the first things that the sub-committee did was to establish a sub-group of officials and scientific experts to consider and report on BW proposals for the ENDC. This was chaired by Sir Solly Zuckerman, the government's Chief Scientific Advisor and its membership included Dr C.E. Gordon-Smith (Director Microbiological Research Establishment, Porton Down), Ronald Hope-Jones (Head Atomic Energy and Disarmament Department, Foreign Office), Dr J.W. Howie (Public Health Laboratory Service, Ministry of Health), Sir Harold Himsworth (Medical Research Council), Gerry Heath (Director, Biological and Chemical Development Board, MOD), Dr Frank Panton (Defence Science 6, MOD), Colonel M. Bailey (Assistant Director (D), Defence Policy Staff) and Dr Alan Cottrell and Dr Bob Press (Cabinet Office).[43]

[42] TNA FCO 10/181, folio 96, Cabinet DOP (O) C Sub-Committee on Disarmament Composition and Terms of Reference, Note by Cabinet Secretary, 12 July 1968.

[43] TNA FCO 10/181, MISC 214 (68) 1 Cabinet Working Group on Chemical and Biological Warfare Work of the Group, Note by the Secretary, 16 July 1968.

Working Party on CBW: First Meeting

The first meeting in the Cabinet Office on 17 July considered the Foreign Office paper on Microbiological Warfare drafted by Hope-Jones on 29 June.[44] This meeting appears to have been successful in garnering support for the initiative.[45] Zuckerman's role was crucial; he was able to extract the relevant factual information from the technical experts that was needed to consider the Foreign Office draft paper. Zuckerman told the working party that it needed to identify and report on any technical difficulties inherent in the Foreign Office proposal for disarmament initiatives in the field of biological warfare and which might be taken in the ENDC. Participants agreed that separation of biological and chemical warfare was sensible, but as drafted the Foreign Office paper did not draw any clear distinction between BW research carried out for defensive and offensive purposes. They also noted that the facilities necessary to produce BW agent could be so widespread as to prevent any effective method of verification of a ban being devised. Zuckerman in summing up noted that they generally agreed that it was possible to separate a declaration of the objectives for an initiative from detailed proposals for actually verifying the achievement of those objectives. There would be no problem in opening MRE for inspection, however. Participants agreed that the Foreign Office would re-draft the paper in light of discussions and that a further working group meeting on Friday 19 July would consider this.[46]

Zuckerman's group duly reconvened.[47] There were six principal alterations to the draft paper.[48] First of all two additional points were added dealing with the disadvantages of using the terms 'bacteriological' – an archaic term that could imply exclusion of viruses and 'in war' – a phrase that might limit the prohibition to warfare and not armed conflict, which was seen as a broader concept. Second, the section dealing with potential bans on all aspects of chemical warfare was re-written to reflect the anticipated difficulty in securing such an agreement internationally. Third, the paper was changed to make it much clearer that what

44 TNA FCO 10/182, MISC 214 (68) 2 17 July Cabinet Working Group on Chemical and Biological Warfare Foreign Office Proposals, Note by the Secretary; CAB 130/389, MISC 214 (68) 1st Meeting Cabinet Working Group on Chemical and Biological Warfare, 17 July 1968.

45 TNA FCO 10/182, 143 R. Hope-Jones, AEDD to Sir Edward Peck, 18 July 1968.

46 TNA CAB 130/389, Cabinet Committee Working Group on CBW, Meeting Note, 19 July 1968.

47 TNA CAB 130/389, MISC 214 (68), 2nd Meeting Cabinet Working Group on Chemical and Biological Warfare, 19 July 1968.

48 TNA FCO 10/182, MISC 214 (68) 2 (revise), 17 July 1968.

was involved in a separate agreement on BW entailed obligations that could not be adequately verified. Fourth, the proposal's prohibitions were completely re-written on the basis of the phrase 'undertake not to produce biological agents for hostile purposes'. This was done to allow purely defensive work against the use of BW. Moreover, while the earlier version only referred to biological agents, the paper was now changed to include a reference to the equipment that would be needed to deliver such agents. Fifth, although the group believed that while the WHO might reasonably be expected to investigate complaints that BW had been used, it was not the proper body to investigate complaints of infringements of obligations on the central prohibitions of an envisaged Biological Weapons Convention.[49] Alternatively, they thought it better to propose a 'body of competent experts' to investigate allegations that the BW prohibition had been breached. Finally, a brief definition of BW was introduced: 'biological agents causing death or disease by infection in man, animals or crops.' MOD, however, suggested that there was a danger in the UK taking an initiative, or becoming a party to a Convention, banning a field of warfare for which no means of verification were foreseen. Such an action might have adverse repercussions in other areas. There was, moreover, a possibility of damaging the UK's close relationship with the US.

A meeting of the sub-committee was arranged for 23 July to finalise the draft ENDC working paper considered by Zuckerman's sub-group.[50] Foreign Office officials hoped that Mulley would sponsor the paper at the Cabinet DOPC meeting planned for 26 July, or in the week commencing 29 July. However, the MOD decided to circulate a 'unilateral paper' to the sub-committee outlining all the difficulties it believed would attend the sponsoring of a working paper on microbiological warfare in Geneva.[51] This document appears to have represented Healey's basic objections.[52] Zuckerman told the Prime Minister that in his view it was crystal clear that the points raised in this paper were directed

---

[49] Note at this stage UK thinking did not include toxins – these were to be included later in the draft Convention at US and Swedish suggestions once the negotiations began – see Cmnd 4725, *Further Documents on Disarmament: The Disarmament Negotiations 1970* (London, 197), pp. 7 and 12.

[50] TNA FCO 10/182, folio 153, R. Hope-Jones to A.M. Warburton, United Nations (Political) Department, 24 July 1968.

[51] TNA CAB 168/125, K.T. Nash, AUS (Pol) MOD to Sir Edward Peck, Foreign Office, Further Measures of Disarmament: Microbiological Warfare 23 July 1968 and Attachment Disarmament – Proposed Initiative on Biological Warfare Ministry of Defence Points.

[52] TNA FCO 10/182, folio 152, Edward Peck to Lord Hood, Biological Warfare: Disarmament Sub-Committee, 22 July 1968.

against the initiative of tabling the working paper.[53] Although the Foreign Office and other Whitehall departments did not disagree that substantial difficulties were involved, all expect the MOD considered that it was tactically wrong to put these counter-arguments to Ministers at the same time as the proposed working paper. MOD believed that the draft working paper needed further consideration by other departments; it would need to be recast together with suggestions as to how the objections (on the verification problems) should be answered when the issues were inevitably discussed in Geneva. Zuckerman, however, strongly opposed such a paper going to the DOPC. He was a strong advocate of arms control and disarmament measures and was clearly impatient with MOD technical objections – much the same had applied in discussions on the CTBT.[54] However, the sub-committee did agree the text of a working paper as produced by Zuckerman's group.[55] This version contained six substantive alterations from the 11 July version.[56]

The first of these concerned the first paragraph that was expanded to include two additional reasons why the Geneva Protocol needed to be revised. Second, paragraph 2 was re-written to highlight more clearly the point that the disagreement on whether the Protocol covered all agents or only lethal ones would make it difficult to secure agreement to revise the existing ban on CW. A proposal to ban BW was acceptable, but on grounds of their relative lethality. Third, paragraph 3 was changed to illustrate more clearly that the UK recognised the impossibility of devising effective verification in either the chemical or microbiological fields. However, DOPC did not think that the UK should be deterred from trying to secure agreement on new obligations that went beyond those in the Geneva Protocol.

The fourth change concerned paragraph 9, which contained the principal recommendations. The basic prohibition was now on the production of microbiological agents for hostile purposes rather than on agents of biological warfare that had no independent peaceful justification. This, it was hoped, would make clear that production of agents for defence research purposes was not prohibited. Although the 17 July draft made no reference to delivery systems, the new text now extended the proposed prohibition to the production of BW munitions, weapon systems and stockpiles for hostile purposes. This

---

53 TNA CAB 168/125, Zuckerman to Prime Minister, Further Measures of Disarmament (OPD 968) 54, 25 July 1968.

54 See Walker, *British Nuclear Weapons and the Test Ban*, chapter 10.

55 TNA CAB 148/71, OPDO (D) 68 1st Meeting Cabinet Defence and Overseas Policy (Official) Committee Sub-Committee on Disarmament, 23 July 1968.

56 TNA FCO 10/182, R. Hope-Jones to Mr Moss, Microbiological Warfare, 24 July 1968.

draft also contained an obligation to place all research in this area under the control of the appropriate civil or health authorities. However, the sub-group had believed that the term 'control' was unsatisfactory partly because civil health authorities did not, and could not, control the work conducted by the private sector. Accordingly, the paper was amended so that civil or health authorities could have access to research results of potential relevance to them.

The fifth change dealt with the suggestion in the draft that all investigations of complaints would be undertaken by the WHO. Following some discussion, the WHO's role was altered so that it would only have the task of investigating allegations that microbiological methods of warfare had been used. Furthermore, a 'competent body of experts' under UN auspices would investigate allegations that the ban on production and research had been infringed. This change was made because officials recognised that the WHO was not qualified to investigate BW production.

Finally, a new proposal was also added. Following an idea from Ronald Hope-Jones, an additional paragraph was included suggesting that a Biological Weapons Convention should contain an undertaking to support, in accordance with the UN Charter, efforts to counter the use, or threatened use of microbiological methods of warfare. Hope-Jones had originally had in mind a United Nations Security Council Resolution (UNSCR) coupled with declarations from individual states roughly analogous to the UNSCRs declarations that embodied the Security Assurances following the NPT.[57] However, both Foreign Office Legal Advisers and UN (Political) Department argued that such an arrangement would be inappropriate where there were no distinctions between the haves and the have-nots. The UN Political Department was content with the revised paragraph. MOD also took the opportunity to advance amendments to the paper that would have lessened the impact of UK proposals by suggesting that an ENDC Committee of Experts might examine the possibility of devising an adequate verification system. Zuckerman observed that the UK initiative was posited on the assumption that it would not be possible to devise such a system, and that in making this proposal, MOD were acting in bad faith. MOD eventually withdrew their suggestions, but officials agreed to annex a note to the main paper to be circulated to the DOPC outlining objections to the UK initiative.[58]

---

[57] Nuclear weapons states undertook to refrain from attacking non-nuclear weapons states unless they were acting in alliance with a nuclear weapons state – these were known as negative security assurances.

[58] TNA FCO 10/182, OPD (68) 54 24 July 1968, Cabinet Defence and Overseas Policy Committee, Further Measures of Disarmament, Note by the Minister of State for

The MOD's objections came down to two key issues and these were major defects that could not be remedied in its view. There were no satisfactory arrangements for verification, which would set a poor precedent for other disarmament measures. Nor were there satisfactory proposals to deal with the fact that many countries maintained stocks of biological agents for medical research large enough for a small offensive capability. Moreover, a country with an adequate capacity to produce agents for medical purposes could produce within a short period of time sufficient material virtually to wipe out the population of a country such as the UK.[59] There was also the matter of depriving the UK of a means to retaliate and the effect on the UK's military relations with the US if it were to table this unwelcome initiative with no real advance consultation. MOD worried that the US might retaliate, or at least be much less forthcoming on exchanges of technical information.[60] Despite MOD objections the Disarmament Sub-Committee accepted the paper from Zuckerman's working group for transmission to Mulley for approval before being circulated to Ministers for consideration at the DOPC meeting on 26 July. Mulley duly circulated the draft ENDC working paper on 24 July 1968.[61]

Defence and Overseas Policy Committee, 26 July: Mulley's Paper

'Further Measures of Disarmament' was item five on the Committee's agenda on that day.[62] Mulley informed his colleagues that in accordance with the decision reached at its last meeting on 13 July, he had made a statement to the ENDC in general terms that the UK intended to submit proposals on CBW. There were now three main changes to the proposed UK draft working paper; first, the definition of the microbiological agents to be banned had been changed from those with no independent peaceful justification to those intended for hostile

Foreign Affairs, 24 July 1969; Annex A set out the MOD note on the implications of the proposed initiative.

59 TNA DEFE 11/672, P.T.E. England, Head of MGO Secretariat to DUS (P) Working Party on Chemical and Biological Warfare – Disarmament Proposals, 17 July 1968; and K.T. Nash, AUS (Pol) to PS to Secretary of State, Further Measures of Disarmament: Progress Report, 19 July 1968; D.W. Fraser, Acting Assistant Chief of the Defence Staff (Policy) to Chief of the Defence Staff, 24 July 1968.

60 TNA DEFE 11/672, F. Cooper, DUS (P) to Secretary of State, OPD – Friday 26 July, 25 July 1968.

61 TNA FCO 10/182, folio 167, Sir Edward Peck to Mr Mulley, 24 July 1968; OPD (68) 54 Note by the Minister of State for Foreign Affairs, 24 July 1968.

62 TNA FCO 10/182, Extract from OPD (68) 14th Meeting, Item 5: Further Measures of Disarmament, 26 July 1968.

purposes; second, the appropriate civil medical or health authorities would have access to and not control of microbiological research; and third, the WHO role would be limited to investigating allegations that a party to the proposed Convention had used microbiological agents against another party, leaving the investigation of allegations about possession or production or research to a body of experts to be established under the UN. Although these changes would weaken the paper, he was still prepared to accept them.

The main problem arose on the note prepared by the MOD on the initiative's implications. These concerned verification, non-retaliation and consultation with other countries. Mulley could not accept that the admitted difficulties of BW verification constituted an adequate reason for the UK not to take an initiative; or that such an initiative would prejudice the UK's position on arms control verification more generally. Britain, Mulley argued, should go as far as possible in establishing procedures under which complaints against breaches of a convention would be publicised. If states parties to a convention refused to allow such complaints to be investigated, then there would be a right for other states parties to withdraw from the Convention. Moreover, there would be an advantage in adding a provision on the lines of security assurances similar to those given in connection with the NPT. As for non-retaliation, British interests lay in minimising the likelihood of further development of microbiological weapons to the point at which they became a military risk, and the best hope of this was to include a non-retaliation clause in a convention. Without such a clause states could argue that they had the right to retaliate and to stockpile BW. Finally, the US was consulted on 2 July. Although William Foster (Director ACDA) had said that the Americans did not favour the UK taking an initiative on CBW, they would not oppose the tabling of a working paper. It might be that the UK would not be supported by the US or Soviet governments, but this was not necessarily against UK interests.

Denis Healey said that he fully supported the terms of the statement that Mulley had made to the ENDC on 16 July. Despite this it was nevertheless important that any paper tabled in Geneva should not commit the government as firmly as the draft envisaged on a number of points. Effective verification against BW, in Healey's view, was not possible because the necessary agents could be produced easily and cheaply. Techniques for their distribution were not difficult or expensive. If the UK sought inclusion of a provision against retaliatory use, it would be difficult to argue against one for CW; nor was it clear what the appropriate response would be if any state used BW against the UK. It was important too that the UK should not forfeit exchanges of information on CBW with the US. The draft paper had the appearance of an outline convention rather than a working paper; it would therefore need to be altered to indicate clearly that the UK recognised the verification difficulties. The paper would also have to exclude the possibility that knowledge available to MRE of the particular forms of BW to which the UK was vulnerable would become public. This could be best achieved by avoiding firm proposals that civil, medical or health authorities would have full access to MRE's research work. Some of this material was important to UK defence. Finally, the entry into force provisions should be made contingent on ratification by a larger number of states than proposed.

The Minister of Health, Kenneth Robinson, said that he was advised that it would be contrary to the WHO's constitution for it to undertake investigations of allegations that one party to the Convention had carried out a microbiological attack on another. Moreover, it would open the way to discussion of the Vietnam War at WHO Conferences, which to date had been successfully resisted. It would be better if the proposed UN Experts' panel were to undertake the whole task of investigating complaints. Ministers agreed that some amendments would be necessary before the paper was tabled in the ENDC. However, these should be such as not to destroy the proposed initiative's impact. Ministers believed that the best prospect for *détente* between East and West lay in the disarmament field. They recognised that BW verification presented great difficulties, and that it would not be possible in the sense normally understood at that time. Nevertheless, a BW Convention would have moral authority; and if BW were actually used, it should be possible in practice to establish such use.

In the Prime Minister's absence, Michael Stewart (Foreign Secretary) summarised the meeting's outcome. He observed that the draft working paper should be revised in light of the discussion and then cleared by Healey, Robinson, Mulley and himself without further reference to the Committee. In particular, the first sentence of paragraph 3 should be amended to refer to the exceptionally difficult problem of CBW verification. It would be preferable to retain the reference to access by civil, medical or health authorities to relevant research in microbiological agents as part of the UK's firm proposals, but redrafted to safeguard UK defensive interests. The text on verification would need to be altered to indicate that the methods outlined were put forward as possibilities rather than as firm proposals, whilst the text on entry into force would also have to be changed to accommodate Healey's view that entry into force would require ratification by most, if not all, states rather than the smaller number suggested by the draft paper. The Committee noted with approval Stewart's summary and invited him to arrange for the draft working paper to be revised as proposed and cleared, if possible, without further reference to the Committee.

Cabinet Secretary's Redraft and the Final Version: 31 July 1968

On 29 July Sir Burke Trend (Cabinet Secretary) wrote to Stewart with a redrafted paper in light of the 26 July DOPC meeting.[63] The paragraph on entry into force was omitted entirely since it was thought to be politically unrealistic to increase the number to 40 or 50 states mentioned in the draft, it was therefore better to delete the reference to ratification. This letter was circulated to Healey, Mulley and Robinson. However, the Foreign Office did not regard the draft as it now stood as 'even a satisfactory basis for discussion'.[64] Despite this Mulley circulated an alternative version that was acceptable with some modification by other Ministers. The most important change here was that civil authorities should only have access to relevant research work, not responsibility for it. Stewart then wrote to Burke Trend on 31 July informing him that agreement had now been reached on a text for Mulley to table in Geneva on 6 August.[65] The Prime Minister was informed between 31 July and 2 August when Trend acknowledged Stewart's note and advised him that Wilson had been notified.[66] Wilson's only interest was in securing suitable publicity for the UK paper – 'care should be taken to get a good press'.[67] The Washington embassy gave a copy to ACDA on 5 August.[68] Alva Myrdal – Swedish disarmament ambassador in Stockholm – received her copy on 6 August. The paper was also distributed to NATO delegations in Brussels on 9 August.[69]

[63] TNA CAB 168/125, Sir Burke Trend, Cabinet Office to Foreign Secretary, 29 July 1968.

[64] TNA FCO 10/182, folio 172, R. Hope-Jones, AEDD to Mr Williams and Mr Day, Further Measures of Disarmament, 31 July 1968.

[65] TNA FCO 10/182, folio 174, Michael Stewart to Sir Burke Trend, Cabinet Secretary, Further Measures of Disarmament, 31 July 1968.

[66] TNA FCO 10/182, Burke Trend to Foreign Secretary, Further Measures of Disarmament, 2 August 1968.

[67] TNA FCO 10/182, D. Gruffydd Jones, Cabinet Office to D.I. Morphet, Further Measures of Disarmament, 6 August 1968.

[68] TNA FCO 10/182, folio 157, D.K. Timms, AEDD to Ian Smart, Washington, 1 August 1968 and A.M. Wood, Washington to D.K. Timms, Chemical and Biological Warfare, 5 August 1968.

[69] TNA FCO 10/182, R. Thomas, UKDel NATO to D.K. Timms, AEDD, UK Working Paper on Chemical and Biological Warfare, 13 August 1968.

ENDC Geneva, 6 August: UK Working Paper on Microbiological Warfare

Mulley, as foreshadowed in his 16 July statement, tabled the working paper in the ENDC on 6 August as ENDC/231. He informed the Conference that he was setting out in more detail his proposals for a Convention to ban or proscribe the use for hostile purposes of microbiological agents causing death or disease by infection in man, other animals or crops. Mulley stressed that the UK was not submitting a draft Convention but a working paper to stimulate discussion with a view to clarifying certain key points. However, come November 1968 Hope-Jones made clear to the Defence Overseas Policy (Official) Committee's Sub-Committee on Disarmament that the UK should pursue British proposals for a ban on microbiological methods of warfare when the ENDC reconvened in March 1969 and before the United Nations Secretary-General was asked to prepare a technical study on CBW and the effects of their use.[70] And perhaps as Denis Healey had feared, momentum was thus building for the UK to present a draft treaty.

Conclusion

The proposal to separate BW from CW and concentrate on BW first was now firmly on the ENDC agenda directly as a result of UK drive and determination – particularly in the Foreign Office and by Sir Solly Zuckerman, and after a good deal of internal debate and argument too. The UK was not blind to the limitations inherent in its proposal, but Ministers and officials ultimately recognised that there comes a time when risks must be taken if progress is to be made in multilateral disarmament; they also were keen too to see a prominent role for the UK. In 1969 the UK duly went on to table a draft treaty based on the working paper.[71] And it was this initiative that led directly to the Biological and Toxin Weapons Convention concluded in Geneva in September 1971.[72] There was, however, much opposition from the Soviet Union and Warsaw Pact countries to the UK proposal in the Conference of the Committee on

---

[70] TNA CAB 148/71, OPDO (D) 68 2nd Meeting, Cabinet Defence and Overseas Policy (Official) Committee Sub-Committee on Disarmament, 4 November 1968; OPDO (D) (69) 1, Biological Warfare, Note by the Secretaries, 29 January 1969.

[71] ENDC/225, 10 July 1969 and ENDC/255/Rev. 1, 26 August 1969.

[72] See TNA CAB 148/92, OPD (69) 29 12 June 1969, Biological Warfare, Memorandum by the Minister of State for Foreign Affairs and Commonwealth Affairs; and CAB 148/93, Chemical Warfare and the Disarmament Negotiation Memorandum by the Minister of State for Foreign and Commonwealth Affairs, 22 September 1969.

Disarmament (CCD) as the ENDC had now become; the neutral and non-aligned states were not that enamoured either and little progress was made even with the US announcement in November 1969 that it would be abandoning its offensive BW programme. The key year would be 1971. How did the UK shape the debates and arguments, and more importantly influence the final outcome? What were the limits of UK influence? What issues were important to the UK and why? The next chapter seeks the answers to these questions in a detailed review of the twists and turns of the decisive six months from March to September 1971 when the Convention was finally concluded.

Chapter 6

The UK and the Biological and Toxin Weapons Convention Negotiating History: March–September 1971: The Key Months

Introduction

The year began with the Conference of the Committee on Disarmament (CCD) biological weapons convention negotiations in stalemate. The Soviet Union and its allies, and indeed the non-aligned states (the G12), still opposed separation of BW from CW as envisaged in the draft UK Convention first tabled in 1969 and subsequently amended.[1] For the UK there were two key issues that had to be covered in the Convention: prohibition of use and investigation of complaints of alleged use, which the UK believed were the only sort of meaningful compliance provisions that could be devised for biological weapons. British pursuit of effective verification provisions in disarmament treaties had been longstanding and remained a bone of contention between East and West. These were critical provisions in what transpired to be the decisive year of negotiations for the UK and were relentlessly pursued. The negotiations can be divided into two periods: March through to August 1971, and August to September when the final and crucial negotiations were completed.[2] This chapter will also highlight just how tortuous any text based multilateral negotiations are – words and phrases matter a great deal, hence the concentration and anxieties surrounding them.

---

[1] The non-aligned or G12 were: Argentina, Brazil, Burma, Ethiopia, India, Mexico, Morocco, Nigeria, Pakistan, Sweden, the United Arab Republic (UAR) and Yugoslavia.

[2] For further details on the content of the Biological and Toxin Weapons Convention see Nicholas Sims, *The Diplomacy of Biological Disarmament: Vicissitudes of a Treaty in Force, 1975–85* (London, 1988) and by the same author, *The Evolution of Biological Disarmament: SIPRI Chemical and Biological Warfare Studies 19* (Oxford, 2001).

Soviet Turnaround: March 1971

There was no progress at all in the first three months of the year. The positions of the three political groups – Western, Eastern and non-aligned – in Geneva were clear. Soviet and non-aligned opposition to a separate biological weapons convention was the principal stumbling block. London, as a consequence, was starting to focus on possible ways of advancing the negotiations.[3] However, before this could make any impact Alexey Roshchin, the Soviet Ambassador to the CCD, called on Henry Hainworth, UK Ambassador, on 29 March to inform him that on the following day he would be tabling in the name of the seven Socialist Group countries a draft Convention on 'CBW'.[4] In effect it was only a biological weapons convention with a commitment to negotiate on a Chemical Weapons Convention (CWC) set out in its Article IX. It did not cover 'use'. Roshchin said that this text was being tabled in order to prevent the CCD from running into an impasse. The new Soviet draft was sent to the FCO. As its title implied – 'Draft Convention for the Prohibition of the Development, Production and Stockpiling of Bacteriological (Biological) Weapons and Toxins and on their Destruction' – it represented a significant step towards key elements in the longstanding UK position.[5] Initial UK comments were:

- The wording of operative Article I of the new Soviet draft was similar to that of operative Article II (A) of the UK draft, and went further than the earlier Soviet position in that it covered agents and not simply weapons. There was no equivalent of the ban on research in Article II (B) of the UK draft.
- Unlike Article I, Article II referred to weapons and not to agents. The wording of Article I appeared to close any loophole, but this aspect required further clarification.
- The complaints procedure in Article VII was similar to that in Article III (2) of the UK text.[6]

---

[3] TNA FCO 66/298, D.K. Timms, Disarmament Department to Mr Benest, Mr Summerhayes, Meeting with the MOD: 12 March, 12 March 1971.

[4] The Socialist Group were: Bulgaria, Czechoslovakia, Hungary, Mongolia, Poland, Romania and the USSR.

[5] CCD/325/Rev.1, 15 April 1971. (Revised text issues for technical reasons.) The text of the Soviet draft and subsequent versions are in the Annex.

[6] TNA FCO 66/298, Soviet Draft BW Convention and D.L. Benest, Disarmament Department to Sir T. Brimelow, Mr Godden, Chemical and Biological Warfare (CBW), 1 April 1971.

- There was a veiled reference in Article VIII, and also in the tenth preambular paragraph, to United Nations General Assembly Resolution 2603A (XXIV) on the scope of the 1925 Geneva Protocol on which the UK had abstained.[7]
- The commitment to further CWC negotiations went beyond that in the UK draft, although the opening part of Article IX with its reference to 'effective measures' was taken from the UK draft Convention. In the new Soviet document, the commitment was to pursue negotiations on measures to prohibit CW altogether whereas in the UK version the commitment was to pursue negotiations on measures to strengthen the existing constraints on CW. The commitment to prohibit CW altogether was also emphasised in the Preamble.

Western Group states were unclear as to why the Russians had suddenly switched positions, but speculation centred on a Soviet wish to have a treaty for that year's United Nations General Assembly, to keep the CCD in business and add substance to recent Soviet peace initiatives.[8] This was the period of growing East–West rapprochement and the first signs of *détente* that would come to shape international relations through the early to mid 1970s.[9] Despite this, the new draft treaty was still a dramatic *volte-face*. The Western Group generally felt that the new Soviet text provided a serious basis for negotiation, but there were still several awkward elements, and the problem would be how to secure their removal. The US delegation drew particular attention to the reference to UN General Assembly (UNGA) Resolution 2603 A (XXIV) on CBW and the 1925 Geneva Protocol's status under customary international law at the very moment when the Senate Foreign Relations Committee had this problem so clearly before it.[10] The Americans also told the UK CCD delegation privately that some thought would need to be given to future draft Convention texts; they

---

7 UNGA Res. 2603 A 1969 had declared that the Geneva Protocol embodied the generally recognised rules of international law prohibiting the use in armed conflicts of all biological and chemical methods of warfare, regardless of any technical developments. This was adopted by a vote of 80 to three with 36 abstentions. The US voted against (as did Portugal and Australia) because of its use of CS in Vietnam; the UK did not accept the statement either, in part because of its own recent policy change to exclude CS from the scope of the Geneva Protocol.

8 The Western Group consisted of the UK, US, Canada, Italy, Japan and the Netherlands.

9 See Raymond L. Garthoff, *Détente and Confrontation: American–Soviet Relations from Nixon to Reagan* (Washington, DC, 1985), chapter 3.

10 TNA FCO 66/298, J.T. Masefield, UKDis Geneva to D.K. Timms, Disarmament Department, CBW: Western Six Meeting 31 March, 31 March 1971.

hoped that the next one to be tabled in the CCD would be a Chairman's draft, in other words a near final version. Procedurally it was easier for the CCD's co-chairmen (the US and USSR) to draft and send a treaty to the UN General Assembly for its consideration and adoption; this had occurred in the NPT and Seabed Treaty negotiations in 1968 and 1971 respectively.

The UK delegation felt that the UK should respond fairly soon to the Soviet move, specifically in a plenary statement on 6 April. British diplomats thought that it would be important to give a welcoming response since a separate Convention had of course been a UK idea.[11] Keeping the Union Flag associated with any emerging text would be an important consideration. Moreover, the UK should not be seen following US reactions as one in a line of states echoing welcoming US statements. Disarmament Department, the new name for the Atomic Energy and Disarmament Department, back in London argued instead that Geneva should only give the Soviet draft a more general welcome, and make clear that the UK would need to study the Soviet draft carefully.[12] Furthermore, any statement would have to draw attention to some of the outstanding points of difficulty, but the Department did not want to make too much of these lest the UK appear ungracious – finding the right balance was always a difficult objective in negotiations. Two key problematic points, however, were identified: the omission of a provision prohibiting 'use' – a key UK requirement, and the manner in which the draft dealt with the commitment to further negotiation of a Chemical Weapons Convention. Lord Lothian, the FCO Minister of State, agreed that Henry Hainworth should make a brief statement along these lines, which he did on 6 April.[13] For the UK it was sufficient at this early stage simply to register that there were certain points of difficulty. This statement also lamented the absence of an assistance Article that would enable humanitarian assistance to be provided to a state party attacked with BW.[14] London decided that a more considered response would be given on 22 April when Lord Lothian himself would address the CCD. However, even on this latter occasion Lothian only made a few general remarks, reinforcing those made by Hainworth on use and the Geneva Protocol, and the need for a virtually automatic investigation of complaint procedure. This was one of the key UK requirements and was seen as the best way of addressing the verification problem posed by BW and dual-

---

11 TNA FCO 66/298, J.T. Masefield, UKDis Geneva to D.K. Timms, Disarmament Department, CBW: Western Six Meeting 31 March, 31 March 1971.

12 The Disarmament Department was now the lead FCO department on multilateral arms control, having taken over the role of the Atomic Energy and Disarmament Department.

13 TNA FCO 66/298, D.L. Benest, Disarmament Department to Sir T. Brimelow and Mr Goden, Chemical and Biological Warfare (CBW), 1 April 1971.

14 CCD/PV.507, pp. 12–15, 6 April 1971.

use as described in the previous chapter. Lothian also commented on assistance provisions.[15]

The Disarmament Department now prepared a comprehensive assessment of the Soviet draft. This included detailed analysis of each of its preambular paragraphs, the substantive articles and suggestions for amendments, which would improve the draft from a UK perspective.[16] Tactics were considered separately. One of the more immediate issues concerned the question of alleged BW use. The Soviet draft had no equivalent of the UK procedure for submitting complaints of BW use direct to the UN Secretary-General. However, at this early stage the FCO felt that the UK should not make this a sticking point or hold up any agreement by trying to insist that use should be covered by a Convention. As for negotiating tactics, the Disarmament Department believed that there might be three possible ways in which the UK could respond to the Soviet draft. First of all, it could stick with the UK draft, but express readiness to consider amendments to it; second, accept the new Soviet draft as a basis for negotiation, but seek to have it amended in ways which suited the UK; and third, work for production of a new draft. Of these options the last was not thought to be the best since neither the US nor the Soviet Union would feel inclined to move away from their drafts. Both might be prepared to accept a 'new draft' as this would draw on the existing two versions, which in many respects covered the same ground. As we can see, care needs to be taken on elaborating negotiating positions and this requires time for careful consideration on substance and strategy – things cannot be rushed. It is a bit like chess – several moves down the line need to be weighed carefully.

US Delegation Views: Early April 1971

On 2 April the US Ambassador and members of his delegation discussed the Soviet draft with Henry Hainworth and his delegation.[17] We can see from this meeting that the Americans were already thinking in terms of a joint US–Soviet draft Convention. The wider context of evolving superpower relations was crucial here. Views were exchanged on various points and issues arising from

---

15 CCD/PV.510, pp. 5–8, 22 April 1971; TNA FCO 66/298, D.K. Timms, Disarmament Department to J.T. Masefield, UKDis Geneva, 15 April 1971. Lord Lothian agreed that he would only say a few general words; the Ambassador would give a more detailed response at a later stage.

16 TNA FCO 66/298, Soviet Draft Convention.

17 TNA FCO 66/298, J.T. Masefield, UKDis Geneva to D.K. Timms, Disarmament Department, Soviet Draft BW Convention, 7 April 1971.

the Soviet draft. Four points are worth noting here. First, the Soviet draft did not refer to 'research' in Article I. The Americans, some of whom believed that even legitimate research might be classified as falling within the prohibitions of Article II, thought that the absence of a ban on research might not be a problem and perhaps even an advantage. Both sides agreed that the Russians should be asked why there was no definitions Article; it was also thought desirable that the definitions in Article I of the UK draft should be retained in any future text – 'microbial or other biological agents or toxins causing death, damage or disease to man, other animals or crops'.[18]

Second, the UK and US agreed that the question of use should not be ignored; however, the Americans did not seem unduly keen on retaining the reference from the UK draft Convention. They agreed, nevertheless, that if the complaints procedure for investigating cases of alleged BW use were made too full, the Russians might well object. However, the UK and US delegations thought that if such a procedure was linked firmly to the UN Security Council (UNSC) and of a limited nature with no reference to any initiating role for the UN Secretary-General, then the Russians might eventually be persuaded to accept it. A perceptive insight – this is, of course, was happened in the final text, but we are getting ahead of events. Third, the Americans believed that the time-scale for destruction of BW stocks was too short. By the time a Convention entered into force, however, there would be no practical problems for the US. Although the US delegation still favoured a procedure for reporting destruction, UK officials pointed out that this might raise difficulties, particularly in the form of requiring verification or permitting an unsatisfactory unverified action on trust.

Fourth, the Soviet text carried no comparable provisions to those outlined in Article III (I) of the UK's own draft Convention, which would allow any party that believed that biological methods of warfare had been used against it to lodge a complaint with the UN Secretary-General, submitting all evidence available in support of its complaint and a request that an investigation should be carried out and a report submitted to the UN Security Council (UNSC). British officials were unclear at what stage in the Soviet approach in its draft Convention that a veto could be used in the UNSC, or whether it would apply when no action but only 'consideration' was involved. The difference hinged on the word 'consideration' in the second sentence of Article VII (I) and 'the investigation' in the following sentence of the Soviet draft – 'Such a complaint should include all possible evidence conforming its validity, as well as a request for its consideration by the Security Council. The Council shall inform the

---

18 CCD/255.Rev.2, 18 August 1970.

States Parties of the result of the investigation.' The parallel UK provision used the phrase 'and request that the complaint be investigated'. This ambiguity would need to be cleared up with the Russians – as noted above, words and phrases matter.

Following the discussion with the Americans, the UK delegation met the Soviet Ambassador on 7 April.[19] Roshchin was quite firm in defending those elements where the Soviet draft touched on points of principle and emphasised that the Soviet Union had moved a long way towards the British position, which was true. He was adamant too that 'use' should not be covered in the draft. If it were, it would weaken the Geneva Protocol by overlapping its prohibitions on use of BW, but omitting the ban on the use of CW. There was also some consideration of the same issues that had been aired in the UK–US bilateral on 2 April.[20]

On the question of including a definitions Article, such as the one used in the UK draft's Article I, Roshchin said that he thought the British definitions were not quite wide enough. On Article VII, the Russians said that their preliminary reaction was that they had no strong feelings about the difference between 'considered' and 'investigation'. However, they took the view that a state might wish to have a complaint considered, but not necessarily investigated; similarly, the UN Security Council might not consider an investigation justified. In light of this meeting the delegation concluded that the UK would need to accept that the Soviet draft was to the 'right of centre' of the UK positions. Accordingly, it would prove difficult to secure widespread CCD support for anything very different from it, especially on the question of reference to United Nations General Assembly Resolution 2603 A and on further commitments to negotiations for a Chemical Weapons Convention.

Dealing with the Americans

In mid April the Embassy in Washington reported that since US policy on BW had been the subject of a Presidential decision and announcements – Nixon's decision in 1969 to abandon the US offensive BW programme – further US action would have to be cleared at the highest levels in the interested departments.[21]

19 TNA FCO 66/298, Speaking Notes for HE's Call on Ambassador Roshchin, 4.00 P.M., Wednesday 7 April 1971.

20 TNA FCO 66/298, J.T. Masefield, UKDis Geneva to D.K. Timms, Disarmament Department, Soviet Draft BW Convention, 8 April 1971.

21 TNA FCO 66/298, L.J. Middleton, Washington to D.M. Summerhayes, Disarmament Department, FCO, CBW, 15 April 1971.

Alan Neidle (ACDA) pointed out in confidence that there was a consensus emerging within the Administration on what its general response to the Soviet draft should be. Specifically, this was that the US should react positively and that they should not spoil this reaction by trying to be perfectionist. The Nixon Administration would not wish to press for every possible improvement, but instead seek those that were of greatest importance. Even on some of the more important points, Neidle said that the US would possibly adopt a fairly flexible attitude for the sake of a joint agreement, provided national security interests were not affected. Interestingly enough, Neidle believed that the UK was not worried about not being associated with any revised draft treaty produced by the US and USSR. London had always been sensitive to the need to maintain UK parentage of the emerging Convention and still was at the point in time. Neidle noted that when more progress had been made in developing the US response, he hoped to consult the UK and give London a chance to comment on the emerging US position. The Americans would take this into account in adopting the final US position, but that was as much as he could promise. UK views were important it would seem, but not decisive.

On 19 April the UK CCD delegation passed the UK's detailed views on the Soviet draft to the US delegation.[22] In response, the Americans said that their basic attitude was still emerging, but they knew that Washington's inclination was to recognise that the Russians had made 'a giant step towards the Western position'. Accordingly, they would try to accept Soviet language and articles whenever possible if they did no real harm. Final decisions would be taken at the highest levels, probably including submission to the President and anything that emerged from this once agreed would be very difficult to change. Henry Hainworth observed that the UK's main concern was the absence of any satisfactory provision on investigations of alleged use. Although the Americans undertook to report the strength of UK views, they made it clear to Hainworth that they did not attach as much importance to this issue as the UK. When Hainworth pointed out that the UK planned to outline its detailed comments in a speech to the CCD, the Americans asked him to reconsider. By the end of April the US would be putting Washington's views privately to the Russians, and would not wish the UK to say in public what was being put in private to the Russians. Washington naturally hoped that the UK would not put forward in public views differing from those put privately. In response, the delegation pointed out that the UK position was rather different from that of the Americans since the UK already had a draft Convention before the CCD and a well-known

22 TNA FCO 66/298, W.N. Hillier-Fry, UKDis Geneva to D.M. Summerhayes, Disarmament Department, UK Views on the Soviet Draft BW Convention of 30 March, 20 April 1971.

public position that it might be expected to defend. Disarmament Department noted that the US seemed to feel that the UK was being over-fastidious about the differences between the Soviet draft and the UK text.[23] Clearly there were worries in London that US willingness to reach agreement quickly would be at the expense of substance and UK interests and the two were clearly related in British views. Here we can see a core problem in arms control and disarmament negotiations: the tension between the desire for progress and the need to address the real technical realities, particularly over compliance provisions. For once it seemed that the US was being a great deal less fussy on an area that it had hitherto been a strong advocate – not least because of Congressional obsessions.

Evolving Attitudes on Tactics: The Way Ahead

Disarmament Department considered the UK's position in light of US thinking at the end of April. Officials now agreed that the UK would need to await details of the promised considered US comments that were to be put to the Russians. British views had already been made clear to the US – both in Geneva and in Washington. The US position would also be considered in the North Atlantic Council (NAC), NATO and Henry Hainworth would, in light of this, recommend which points the UK should put to the Russians. On the central issue of the prohibition on use, FCO officials recognised that the UK might eventually have to fall back on some compromise; however, the department hoped that at least some of the original UK investigation of alleged use complaints procedures could be retained. This point was important to the MOD whose officials believed that there could be no retreat from the initial UK position on investigations unless Ministers were consulted and agreed to any change in the delegation's instructions. MOD officials attached particular importance to the UK exerting every effort to include procedures relating to investigation of use in any compromise draft.[24] This was the only meaningful compliance measure in the BW context as had been noted in the Whitehall discussions the previous summer as described in the previous chapter.

On 23 April the delegation set out its views to the Disarmament Department, including its ideas as to what might be provided to the Russians to consider

---

23 TNA FCO 66/298, D.K. Timms, Disarmament Department to Mr Benest, Mr Summerhayes, Soviet Draft BW Convention, 21 April 1971.

24 TNA FCO 66/298, D.K. Timms, Disarmament Department to J.T. Masefield, UKDis Geneva, 15 April 1971.

including in a new compromise text.[25] These were discussed briefly with Lord Lothian during his visit to the CCD the previous day and he agreed that the UK should follow the US tactic of not overburdening the Russians with comments. This might take the form of two separate sets of observations: one consisting of the key substantive points with the second focusing largely on drafting details. Following a US suggestion, the UK saw advantage in the two delegations not adopting identical positions. The US could make the running on the references to UNGA Resolution 2603 A – the one on the Geneva Protocol and its status under international law – and the UK could concentrate on the use issue.

London believed that the Soviet draft could be interpreted in such a way as to veto the UN Security Council from even considering a complaint concerning alleged BW use. Washington did not share this assessment. Although the blocking of even consideration of a complaint was deemed highly unlikely in practice, the wording of the Soviet Article VII on this issue remained unsatisfactory for the British. There was no apparent link between a request that a complaint be considered, and any subsequent investigation. Such ambiguities matter and would create a loophole and detract from the proposition that any investigation into non-compliance should be automatic. For this reason the UK preferred its original version of these procedures; there simply had to be some sort of automaticity to any request for an investigation otherwise the procedure's utility and deterrent effect would be seriously undermined.

On 28 April Phil Farley, Acting Director of ACDA, asked the Minister at the Embassy in Washington to call to receive the detailed US comments on the Soviet draft Convention.[26] Farley said that the US wished as far as possible to facilitate negotiations in Geneva and had thus gone as far as it could in accepting the Soviet draft, even if it was not ideal. Washington had tried to take into account matters of concern to the UK. For the US the idea of a ban on use might be adequately dealt with in a preambular paragraph since an operative Article was apparently unacceptable to the Russians. As for strengthening the investigation of alleged use complaints procedure, the US left open the possibility of supporting an amendment if others in the CCD proposed one at a later stage of the negotiations. Since a joint UK/US/USSR draft was unlikely to be acceptable to the Russians, an alternative was to have separate but identical drafts tabled by the Soviet Union and its allies and the US and the UK. This idea had originally been floated by the Hungarians at the UN General Assembly and had even figured in internal UK thinking on future options. Provided the UK could accept the general US approach, Washington hoped to discuss

25 TNA FCO 66/298, W.N. Hillier-Fry, UKDis Geneva to D.L. Benest, Disarmament Department, UK Comments on Soviet Draft BW Convention, April 1971.

26 TNA FCO 66/298, Washington telegram No. 1463 to FCO, 28 April 1971.

its proposals in NATO as soon as possible; and, if feasible, pass US comments to the Russians before the CCD went into recess on 6 May. However, the Americans accepted that this timetable was too tight. Farley gave the Minister a copy of the revised draft Convention that the US intended to show the Russians. On 29 April Jim Leonard, US Ambassador to the CCD, handed over copies of the US paper containing comments on the Soviet draft, including the amended draft to the UK CCD delegation. He urged the desirability of a joint UK–US approach to the Soviet delegation; he emphasised too that US instructions were approved by the White House and that it would be very difficult to obtain any modification.[27]

Hainworth observed that the difficulty with the US paper and draft Convention was the omission of points of importance to the UK; especially a procedure for automatic investigation of allegation of use and assistance to victims of BW attack. However, Hainworth shared the US assessment that it was most unlikely that the Russians would agree to the inclusion of an operative Article in the Convention on either issue. On negotiating procedures generally, Hainworth strongly recommended that London should accept the US suggestion of a joint approach to the Russians leading to the tabling of identical but separate texts of which the Western one would be co-sponsored by the UK and US alone.[28] If the UK decided to press the Americans to make further changes even after British views had been considered in Washington, Hainworth argued that such action would only risk encouraging the US to revert to the practice of the CCD's bilateral co-chairmen negotiating draft agreements themselves. Given the UK provenance of a separate Convention, Hainworth thought that it would be important to have a British imprint on any final draft.[29]

Meanwhile in Washington Alan Neidle (ACDA) was explaining to the Head of Chancery – Lawrence Middleton – why the UK's views on complaints procedures and an express prohibition on use could not be accepted.[30] There were five reasons. First, initial US acceptance of an investigation of complaints procedure on use in the UK draft had been somewhat reluctant, but Washington had accepted the logic of its inclusion given the presence of an Article on non-

27 TNA FCO 66/298, UKDis Geneva telegram No. 40 to FCO, Draft Biological Warfare Convention, 29 April 1971.

28 TNA FCO 66/298, UKDis Geneva telegram No. 40 to FCO Draft BW Convention, 29 April 1971.

29 It is worth noting here that Henry Hainworth had been closely involved in the original test ban treaty talks in the late 1950s, early 1960s and was thus very familiar with the problems in negotiating with the US. See also John R. Walker, *British Nuclear Weapons and the Test Ban 1954–73* (Farnham, 2010).

30 TNA FCO 66/298, Washington telegram No. 1510 to FCO, 1 May 1971.

use. Without such an Article, a complaints procedure lost some of its point. Second, it was not essential to preserve US or UK security; it was of most interest to small states and that being so, it was for them to suggest it. Third, there was a distinction between the investigation of complaints about production and investigation of complaints about use. The former added something to the rights of States Parties to the Convention whilst the latter was related to an act of aggression and as such added nothing to the rights states already enjoyed under the UN Charter. Fourth, a complaints procedure written into the Convention could prove embarrassing if allegations of use were made in a region of great political sensitivity such as the Middle East. Circumstances might be such that a reference of a complaint to the UN Security Council might be undesirable. There could be occasion where it might be more difficult to prevent such a referral if the right of recourse to the UNSC was set out in the Convention. Finally, to some extent the Soviet Article VII already covered the question: an allegation of use must assume a breach of the obligation on production.

Tactically, the Americans also had reservations about the UK's approach. US officials believed that serious objections to the Soviet draft should be limited to matters affecting the security of the US and its allies. The Russians would not accept that a complaints procedure was a matter of vital importance to the US. Furthermore, there could be even less chance of acceptance of such a procedure being accepted if the UK were to insist upon it without US support. If the Russians rejected it in the forthcoming discussion, it would be much more difficult to persuade them to reverse their position later. However, the Russians might well listen more sympathetically to the non-aligned at a later stage in the negotiations. As a compromise Neidle suggested that the UK should declare that such a provision on a complaints procedure would be useful, but not an essential interest and would reserve its right to support it should another delegation request its inclusion later in the negotiations. MOD reacted strongly against the suggestion that the UK should accept the omission from the Convention of a complaints procedure, which given the internal debates on verification in 1968 discussed in the previous chapter is hardly surprising. For MOD officials such a move would represent a departure from established British policy that could only be made with Ministerial sanction. Instead the MOD believed that the right course was to urge the US to reconsider. The Disarmament Department and Lord Lothian agreed. The US had not done enough to push the point.[31] So the UK tried again with a suggestion for the US to consider on a complaints procedure for allegations of use in the text the Administration planned to hand

31 TNA FCO 66/298, D.M. Summerhayes, Disarmament Department to Sir T. Brimelow, Draft BW Convention, 30 April 1971.

to the Russians. London had no luck. Farley told the Embassy in Washington that Washington had come to a different conclusion.[32] A complaints procedure based on the UK Article III was rejected because the reference to the UN Secretary-General would make it unacceptable to the Russians and any attempt to use a more indefinite formula such as 'appropriate international procedures' would be viewed with the gravest suspicion. US minds were made up and no amount of logic from the UK could shift the Americans, their own foreign policy imperatives demanded an opening to the Russians and a perfect BW Convention was not going to get in the way.

Jim Leonard called on Henry Hainworth on 3 May.[33] He confirmed that an Article on complaints of use had been set on one side in US thinking, but definitely not rejected. In the US view, the whole history of its negotiations with the Russians suggested that it would be counter-productive to insist on a procedure for investigation of complaints of use at that particular stage. Such a provision would be submitted in accordance with all other US views to the Soviet government's Council of Ministers. If the Russians were looking for a point to reject in any US re-draft, they would almost certainly choose this one if it were to be included. It would then be extremely difficult to reverse a Council of Ministers' decision. A more useful tactic would simply put the issue down as a matter that would make eventual acceptance of the point much easier if pressure for it developed in the next CCD session. Leonard argued that the UK could rely with quite a high degree of confidence on some of the non-aligned and some Western delegations pressing the point. This turned out to be a misplaced assessment.

The US was anxious to inform NATO of its position on the Soviet draft before 6 May, and also hoped that the UK would support the proposed US response.[34] Hainworth recommended to London that the UK should align itself with the US position.[35] Parting company with the US at that juncture would, in Hainworth's opinion, 'in no way increase our chances of getting what we want in the final treaty'. However, London was far from convinced, nor indeed of the force of US arguments. Officials saw no real grounds on which they could recommend Ministers to authorise a change in the UK position. Both Washington and Geneva were tasked to press the Americans yet again. But there

---

32 TNA FCO 66/298, Washington telegram No. 1538 to FCO, 4 May 1971.

33 TNA FCO 66/298, UKDis Geneva telegram No. 43 to FCO, 3 May 1971.

34 In the days when the Western Group in Geneva was actually very small, it was customary for US and UK proposals to be discussed in NATO in order to ensure that NATO security interests were not overlooked and for political reasons to reassure key players such as West Germany who were not in the CCD.

35 TNA FCO 66/298, UKDis Geneva telegram No. 44 to FCO, 4 May 1971.

was no change in the Administration's thinking; the best that emerged was an amended paragraph in the US paper that left the issue slightly more open than before.[36] The main obstacle to a co-ordinated US–UK approach was the US wish to present a new draft Convention in the CCD. London believed that this was premature: improvements to the existing texts made more sense. Offering specific comments could do this. By tabling their own new draft, the Americans would only highlight the differences between the US and UK positions. However, the UK CCD delegation subsequently pointed out that the so-called 'new draft' was actually the Soviet draft with the minimum changes required at that stage. The US believed that this saved time and avoided mistakes and that the UK really ought to join the US in its approach by associating British views with the US comments paper, at least orally, and hand over a UK paper on the use issue – promising language at a later date.

There is some indication that continuing UK pressure on the use and complaints issue exasperated the US. On 5 May Jim Leonard underlined to Henry Hainworth that he was concerned that the UK should understand the basic policy rationale for Washington's reaction to the 30 March Soviet draft. The Russians had made a 95 per cent step towards the UK–US position for which they were being exposed to humiliating attacks in the CCD from the non-aligned.[37] A major political response by the UK and US was called for in order to attain their common objective – a ban on BW. If the Americans were now to respond by giving the Soviet delegation a series of observations instead of a consolidated text, which all previous experience in Geneva would lead them to expect, Roshchin would be totally perplexed as to what significance to attach to them. The CCD recess would, therefore, be wasted. Indeed there was a danger that the Russians would conclude in such circumstances that the Americans were not serious and would drop the whole idea of a separate BW Convention. Despite these protestations, London decided that it could not agree to a joint UK–US approach as the US draft fell well short of the sort of Convention that the UK wanted to see tabled as the basis for the next phase of the CCD's negotiations. Convincing the US was the key to preserving the chance of securing UK objectives, and that was proving an uphill task.

The main weakness in UK eyes was the US draft's failure to deal adequately with a prohibition on BW use and in particular the omission of a procedure

---

36 See also Foreign Relations, 1969–76, Volume E-2, Documents on Arms Control, 1969–72 at http://www.state.gov/r/pa/ho/frus/nixon/e2/c22095.htm (accessed 15 August 2011) for the US account of a UK–US meeting in Washington. 228. Telegram 78171 From the Department of State to the Mission in Geneva, May 6, 1971, 0149Z.

37 TNA FCO 66/298, UKDis Geneva telegram No. 49 to FCO, 5 May 1971.

for investigation of complaints of use.[38] Both of these went to the heart of the BW problem and international control. There were five compelling reasons, in the UK view, for including provisions on use in any draft. First, since there would in practice be virtually no possibility of checking on complaints of production and possession, the only aspect of the complaints procedure that would have any practical deterrent effect against non-compliance would be the one on investigations of alleged use. A procedure for investigation of use would constitute an effective piece of verification machinery. Without it the draft Convention risked being little more than a piece of paper. Second, the absence of a complaints procedure relating to use would be a dangerous precedent for future negotiations on a ban on chemical weapons, where verification was a key requirement for any sort of meaningful ban. Third, since this procedure was such a prominent feature in the UK draft Convention and had recently been emphasised by Lord Lothian in his 22 April speech to the CCD, officials would need to seek Ministerial authority before they could alter the delegation's instructions. Finally, the UK believed that there was considerable CCD support for including a complaints procedure; and, moreover, there were still problems given the repetition in the proposed US text of many of the Soviet draft's weaknesses, especially the absence of any definition of the substances to be prohibited.

Disarmament Department now believed that inclusion of an operative Article ruling out use would not be strictly necessary. A suitable preambular paragraph plus an appropriate complaints Article would, it was thought, have an equally binding effect. Foreign Office Legal Advisers however took a different view. Without a substantive Article on use, it would not be possible to claim that a complaints procedure could be so extended. There was thus no doubt in British minds that inclusion of a complaints procedure on use would significantly strengthen the Convention.

Since the British attitude precluded a joint US–UK approach to the Russians, the preferred tactic was now for separate but mutually reinforcing approaches.[39] Even though the question of use was the main and perhaps only substantive point, officials thought that any discussion with the Russians would need to emphasise the large measure of common ground between the UK and US. The UK would give the Russians a non-paper that had earlier been agreed between Disarmament Department and the UK CCD delegation. Jim Leonard expressed regret that the UK was not prepared to take joint action with the

---

38 TNA FCO 66/299, FCO telegram No. 5 to UKDis Geneva and Washington, 30 April 1971.

39 TNA FCO 66/299, FCO telegram No. 1270 to Washington and UKDis Geneva, 5 May 1971.

Americans, but he hoped that the US could now circulate a paper to NATO so that the Permanent Representatives in Brussels would have time to consider the US position and concur in the proposed action.[40] This would enable him to speak to the Russians before Roshchin returned to Moscow at the end of the CCD's session on 13 May.

The UK paper set out the basis for the British response to the Soviet draft Convention and naturally gave prominence to UK attitudes to a ban on use. This in turn drew on points made in Lord Lothian's 22 April speech to the CCD. The paper's first annex incorporated most of the US paper and the revised Alan Neidle formula and contained detailed drafting and substantive comments on specific aspects of the Soviet draft – Articles IV, V (both on aspects of national implementation) and VIII (relationship to the Geneva Protocol). It also proposed new language on, for example, Article IV and V where the Soviet text did not quite meet all UK concerns.[41]

British officials also wished to preclude any possibility that a Convention could be circumvented in the manner that the Soviet Articles IV and V only appeared designed to prevent. However, experience indicated – especially in the NPT – that it was extremely difficult to formulate provisions on national implementation that adequately took into account the great diversity of potential parties' legal systems. Accordingly and to simplify greatly the negotiations, the UK suggested that the legal formulations from previous multilateral arms control treaties should be used in the Convention since these had already proved acceptable. Therefore a new Article IV should make use of key phrases contained in paragraph 1 of the NPT's Article III and paragraph 1 of the Partial Test Ban Treaty. The revised composite Article IV would thus read:

> Each State Party to this Convention shall, in accordance with its constitutional process, take any necessary measures to prohibit and prevent development, production and stockpiling, acquisition and retention of the agents, toxins, weapons, equipment and means of delivery specified in Article I of the Convention, within the territory of such state, under its jurisdiction or under its control anywhere.

This was the formulation that appeared in the final version of the BTWC – Article IV.

Alternative language was proposed to cover what were later to become known as the Convention's non-proliferation aspects, namely the prohibition

40 TNA FCO 66/299, UKDis Geneva telegram No. 52 to FCO, 6 May 1971.

41 TNA FCO 66/299, D.M. Summerhayes, Disarmament Department to Sir T. Brimelow, 7 May 1971; UKDis Geneva telegram No. 55 to FCO, 6 May 1971.

on the transfer of BW to other states. The UK believed that the comparable provisions here should be based on appropriate NPT text. Officials thought that such language should pose no difficulty in a BW Convention between parties that possessed BW weapons and those that did not – it would not be discriminatory, everyone would be treated the same. All parties would agree not to acquire or retain biological weapons. It would thus be in each party's interest that all other parties were precluded by unambiguously clear provisions from helping any recipient whatsoever, whether party or non-party, to manufacture or acquire biological weapons. An appropriate text for such a provision read thus:

> Each State Party to this Convention undertakes not to transfer to any recipient whatsoever, directly or indirectly and not in any way to assist, encourage or induce any state, group of states or international organisation to manufacture or otherwise acquire any agent, toxin, weapon, equipment or means of delivery specified in Article I of the Convention.

This was now virtually the final version of what was to be Article III in the completed Convention. The second annex of UK comments, which was not to be handed over, contained the Speaking Notes for use with the Deputy Head of the Soviet delegation. This outlined proposals on language, phrasing and positioning of material in the preambular paragraphs and operative articles in the Soviet draft. The most notable of these suggestions were replacement of 'microbiological' with 'microbial', 'ancillary' for 'auxiliary' in Article I. 'Stockpiling' in English need not apply to weapons already deployed and 'possess' should appear after 'produce' and before 'stockpile'. It was also be made clear to the Russians that 'development' covered 'research' as the two concepts were separate in English. On 10 May the UK paper was circulated to the North Atlantic Council (NAC).[42] The Council discussed this and the US comments and agreed that the UK and US should proceed and present their views to the Russians before the CCD went into recess on 13 May, though without commitment by the NATO's member states to support everything said.

[42] TNA FCO 66/299, UKDel NATO Brussels telegram No. 162 to FCO, 11 May 1971.

UK Efforts to Strengthen the Use and Complaints Provisions

On 12 May the Disarmament Department drafted a possible alternative for incorporating a complaints procedure on use in any new draft Convention.[43] This consisted of two operative articles. States Parties would undertake not to use BW or toxins for hostile purposes in accordance with the prohibitions on the use of such weapons in the Geneva Protocol. This textual proposal would hopefully enable both the Convention and Geneva Protocol States Parties with or without reservations to accede to the new Convention. The complaints procedure directed against use was similar to one proposed earlier by the UK delegation, but with some alterations; for instance, the word 'inquiry' was substituted for 'investigation' on the grounds that the latter was a technical term used in Article 34 of the UN Charter, which would point directly to action by the UNSC, and this was not what was required in a BW Convention context.

Hainworth had discretion to put these alternative formulations to the Russians, but he was instructed to emphasise that they should be left in no doubt that the UK much preferred the formulations in the existing UK draft Convention. Hainworth met Roshchin on 12 May to hand over copies of the UK's comments on the Soviet draft.[44] Roshchin emphasised that the Soviet draft of 30 March represented a substantial step and the Soviet Union and its allies now expected a substantial step from the West. However, the UK points on use and the Geneva Protocol meant that the UK was not going far enough towards the Soviet position. Use was the crucial point since mention of it would weaken the rules of international law established by the Geneva Protocol because CW would not be dealt with at the same time. Omission of use in the US draft showed that the Americans appreciated the Soviet argument. Given the Soviet response Hainworth saw little point in handing over the revised UK texts on use and complaints procedures.[45] That particular door was formally and perhaps finally bolted. There comes a time to stop banging one's head against a brick wall.

Following these fruitless exchanges Hainworth wrote a long letter to Disarmament Department assessing the way ahead for the negotiations in general and how to further UK objectives in particular.[46] He concluded that the UK would not succeed in obtaining its objectives; indeed he observed, 'we shall

[43] TNA FCO 66/299, FCO telegram No. 10 to UKDis Geneva, 12 May 1971, Draft BW Convention; FCO telegram No. 11 to UKDis Geneva, 12 May 1971.

[44] TNA FCO 66/299, UKDis Geneva telegram No. 59 to FCO, 12 May 1971.

[45] TNA FCO 66/299, H.C. Hainworth, UKDis Geneva to D.M. Summerhayes, Disarmament Department, Soviet Draft BW Convention, 14 May 1971.

[46] TNA FCO 66/299, H.C. Hainworth, UKDis Geneva to D.M. Summerhayes, Disarmament Department, 12 May 1971.

be lucky if we even retain a reference to a prohibition of use such as now figures in the preamble to the American redraft'. Worse still, persistence would only lead to the Russians ignoring UK views since the US position would be the one that mattered. The delegation would then find itself obliged, 'to haul down the Union Jack, which our brief instructed us to keep flying from any eventual draft'. Although it was not inconceivable that the Russians could make yet another *volte-face*, this was unlikely. Since they were under considerable attack by the non-aligned for supporting BW's separation from CW, they would not be in any hurry to repeat the exercise. One possible option might be the addition of an Article expressly withdrawing any reservations to the Geneva Protocol, which signatories might have in respect of BW. Nothing more became of this idea, however.

In spite of Hainworth's protestations, the Department believed that it had little option but to part company from the Americans since the UK was effectively being asked to drop its own draft and support another one which officials felt unable to recommend to Ministers.[47] The US latest text failed to cover use satisfactorily because it had no procedures aimed at deterring BW use – deterrence of non-compliance was still a key UK objective. Nevertheless, officials recognised that the time would inevitably come when a choice between the drafts would need to be made, but this was not it. It was still important to secure an effective deterrent to use, and hence BW production and consequently some form of language in the draft to ensure this. Since the alternative language for use had not been put to the Russians, it would be worthwhile now to show it to the non-aligned where it might receive a more sympathetic response. The Disarmament Department had concluded nevertheless that it was very doubtful that the Russians would respond to UK representations on complaints procedures, especially since the US was lukewarm. If in the end it was impossible to secure general agreement on inclusion of a complaints procedure, then the Department had already decided in mid May that it would not recommended to Ministers that the UK should refuse to join such a Convention, merely because it did not address this issue to UK satisfaction. However, this point had not yet been reached and if the FCO was to carry the MOD with it when the time came, it was essential to show that the FCO had in the meanwhile put up a good fight in favour of dealing with use of BW in the way that was originally proposed in the UK draft Convention of July 1969.[48]

---

47 TNA FCO 66/299, D.M. Summerhayes, Disarmament Department to H.C. Hainworth, UKDis Geneva Draft BW Convention, 17 May 1971.

48 TNA FCO 66/299, D.M. Summerhayes, Disarmament Department to Sir T. Brimelow, Mr Godden, Private Secretary, Biological Warfare (BW), 14 May 1971.

Continuing Consultations

Following discussions with the US and several non-aligned delegations in early June 1971, the UK CCD delegation proposed some alternative formulations to London on the use question on 2 June.[49] These were variations on the department's earlier proposals that had not been shown to the Russians following Hainworth's meeting with Roshchin on 12 May. An amended possible Article on use was prepared if there were no takers, an alternative option would be to add a phrase to make clear that the use must have been a consequence of a violation of the prohibitions in the present Uk draft's Articles I and II. Similar formulations were elaborated for investigations and complaints procedures. All of these were sent back to London for departmental consideration. However, by the end of June it had become evident that the US was not really interested in covering use, and the gestures they had made in their approach to the Russians were just that. Adverse US reactions to the revised UK texts brought this position home. The only course now for the UK was to interest one or more non-aligned delegations. Disarmament Department considered the various formulations provided by the delegation on use and complaints procedures. Several of these were still open to the same likely objection; for instance, they would reinterpret the Geneva Protocol by default, and remove the chances of the Soviet Union and its allies accepting radical changes to their Article VIII.

CCD Summer Session 29 June – 30 September

The penultimate negotiating stage came with the opening of the CCD's summer session. Both the UK and US were eagerly awaiting the considered Soviet response to the comments and texts provided on 12 and 13 May. By mid July the indications were that the Russians had not given their allies a new draft Convention until after the beginning of the session.[50] Delegations in Geneva were well aware that the Romanians were delaying presentation of the new Soviet draft, claiming that they had had insufficient time to study it. Romania was also reported to be against making further concessions to the West, arguing that the Socialist Group had made quite enough concessions already. However, the UK delegation understood that the new version of the text did not contain any operative Article on prohibition of use.

49 TNA FCO 66/299, H.C. Hainworth, UKDis Geneva to D.M. Summerhayes, Disarmament Department, Draft BW Convention, 2 June 1971.

50 TNA FCO 66/300, W.N. Hillier-Fry, UKDis Geneva to D.L. Benest, Disarmament Department, Draft BW Convention, 13 July 1971.

On 22 July, before the regular CCD plenary, Roshchin gave Hainworth the Soviet reply to US representations made at the end of the spring session.[51] This contained a redraft of the Convention that had also been provided to the US delegation the previous day. This particular text was not to be tabled in the CCD itself; instead it was for discussion as a preliminary to agreement on parallel identical drafts. In handing this over, Roshchin said that the Soviet side had in effect accepted all the suggestions that the West had put forward except for the prohibition on use. The Russians had two reasons for this: first, the Geneva Protocol was sufficient as it stood; second, since the Protocol prohibited both CW and BW, to include a prohibition on only one in the new Convention would mean that there were different approaches for the prohibition of use of the two types of weapon.

The Western Group met on 22 July to be debriefed by Hainworth and Leonard on the outcome of their discussions with Roshchin.[52] Leonard reported that it would probably not be too difficult to table an agreed US–Soviet draft Convention. He had underlined to Roshchin that any text tabled would be a basis for further discussion in the CCD and he had urged Washington to react quickly to the positive Soviet response. Hainworth advised London that the UK should be ready to co-sponsor any joint US–Soviet draft since it was unlikely that there would be any intermediate draft between the impending joint draft and one that could be submitted to the UNGA. If the UK did not co-sponsor, then its role in the Convention's elaboration would be lost from view. There were no other disarmament measures in sight with which the UK would be likely to be associated by name.

ACDA considered accepting the Soviet draft as presented without change.[53] The main concern was to place a negotiating document on the table as soon as possible, hoping that improvements would follow from subsequent discussions. There was only one section, however, that ACDA really wanted changed and this was preambular paragraph 10, which contained a reference to UN General Assembly Resolutions that the US had opposed. On 28 July the Joint Chiefs of Staff agreed to accept the Soviet draft and the Administration then hoped to table a joint draft in the CCD with the Russians on 3 August.[54] Hainworth thought that since the Americans had decided unilaterally to destroy their BW capability, they were anxious to 'get some quid for their pro'. They were less interested in the content of an agreement than in the fact of reaching quick

51 TNA FCO 66/300, UKDis Geneva telegram No. 77 to FCO, 22 July 1971.

52 TNA FCO 66/300, UKDis Geneva telegram No. 81 to FCO, 22 July 1971.

53 TNA FCO 66/300, UKDis Geneva telegram No. 83 to FCO, 26 July 1971.

54 TNA FCO 66/300, UKDis Geneva telegram No. 86 to FCO, 29 July 1971.

agreement with the Russians. This explained their reluctance to see substantive amendments introduced.

Biting the Bullet: Accepting Reality

FCO and MOD officials met on 22 July to consider the revised Soviet text and to see what changes had been made following UK and US comments on CCD/325/Rev.1 – the Soviet March draft.[55] Preliminary thought was also given to future UK tactics. The various revisions appeared to be generally acceptable. Officials also discussed the drafting of Articles III and IV, which were designed to prevent proliferation without affecting the free transfer of those materials used for peaceful purposes. Officials were concerned about the replacement of the phrase 'as a means of warfare' with 'for military purposes' at the end of Article VIII of the new draft, corresponding to Article VIII of the original, which also incorporated a commitment to further negotiations on a Chemical Weapons Convention.

Overall the revisions made in the Soviet draft were rather as expected; indeed officials felt that the outcome could have been a great deal worse. They regretted that the Russians had been unwilling to accept UK views on use and on a complaints procedure apart from a minor concession in this direction in a new preambular paragraph 2 – 'desiring thereby, for the sake of all mankind, to exclude completely the possibility of bacteriological (biological) agents and toxins being used as weapons'. However, since UK views had been put with some force and rejected with equal force, it was unlikely that there would be any further Soviet move towards acceptance of an Article on use. Furthermore, since the Russians had met the US proposals almost completely, the Americans would see little difficulty in accepting the draft with no more than perhaps minor revisions. Consequently, agreed chairmen's parallel drafts would no doubt be presented to the CCD shortly. As for UK co-sponsorship, however, FCO and MOD officials believed that this was undesirable; it was important not to be seen to be too willing to abandon UK views. Future negotiations on a chemical weapons ban were relevant here. Any realistic CWC would require a prohibition on use. Soviet views on this score caused some unease. Since, the Russians argued, CW and BW were dealt with together in terms of use in the Geneva Protocol, they should not be treated separately by making special provision on use in a Convention dealing solely with BW. Should the Russians persist with this line, it would have unfortunate consequences in a CWC context. Its logical extension

[55] TNA FCO 66/300, J.M. Hay to Mr Benest and Mr Timms, 23 July 1971.

would mean that the Russians could not accept a CWC incorporating express provisions prohibiting use. Any suggestion that such a Convention should cover use would be no doubt countered by arguments that the UK accepted that the Protocol was adequate to deal with BW use. If the UK were to co-sponsor the present draft Convention, it would be difficult to rebut such an argument and officials agreed that the UK should not tie its hands in this way, but should continue wherever possible to put the case for expressly banning use in the BW Convention.

Officials also realised that, given the likely timetable of events following US acceptance of the Soviet draft, they would need to submit to Ministers recommending that HMG agree to accept the terms of the draft even though it did not include provisions on use or a complaints procedure. Disarmament Department put a submission along these lines to Lord Lothian on 29 July.[56] This argued that on balance the UK should not co-sponsor. Four days later on 2 August Lord Lothian discussed the UK position and tactics with officials as well as Henry Hainworth's personal message to him urging that the UK co-sponsor.[57] Although the balance between the options was a fine one, the arguments against co-sponsoring were felt to be stronger. To sponsor a text that failed to meet UK requirements on use would not appear consistent with British policy since it would imply a greater degree of approval for the revised Soviet draft than Britain was prepared to give. Such a move would also complicate the task of trying to win support in the CCD for the changes the UK wanted.[58]

Officials accepted that there were practical difficulties in securing amendments on use. Consequently, UK efforts would now have to concentrate on the complaints procedure for cases of non-compliance where there was some support for British views. Nevertheless, officials recognised that the consideration that would apply at the end of the summer CCD session would probably be very different. At that point, when negotiations were over, there would be strong grounds for arguing that the UK should be prepared to join consensus in commending the Convention to the UNGA as the best text available, despite the absence of any amendment on use. For the moment – 3 August – the UK regarded the tabling of a revised Convention as a basis for

---

[56] TNA FCO 66/300, D.L. Benest, Disarmament Department to Mr Wiggin, Mr Goodenough, BW Convention, 29 July 1971; D.L. Benest, Disarmament Department to Mr Wiggin, BW Convention, 29 July 1971.

[57] TNA FCO 66/300, D.L. Benest, Disarmament Department to Mr Wiggin, Mr Goodenough, Draft BW Convention, 2 August 1971; H.C. Hainworth, UKDis Geneva to The Marquess of Lothian, FCO, 30 July 1971.

[58] TNA FCO 66/300, FCO telegram to UKDis Geneva No. 15, 3 August 1971.

negotiation as a positive step, even though it needed strengthening in some important respects.

Inclusion of Toxins in Article I

Alva Myrdal, Swedish Ambassador, proposed that the Soviet draft Convention should cover toxins.[59] These were described in the Swedish working paper CCD/333, 6 July 1971 as 'man-made chemically synthesised substances' or 'by chemical means modified substances of biological origin or occurrence', in other words those toxins and chemically closely related compounds which were not of biological occurrence or origin.[60] Sweden concluded that the concept of toxins had to be clearly and unambiguously defined in a treaty obligation and a definition could be selected from the different criteria listed in CCD/333. The Americans had proposed similar language in February 1970 as an amendment to the UK draft. At that point the UK could not accept such a wide amendment, although it did accept a simple amendment on toxins. Officials were aware that it was not thought to be practical or economic in 1971 to produce toxins by chemical synthesis. Even if this could be done, difficulties might arise over the definition of the word 'toxins'. Inclusion of agents of chemical origin could well increase pressure to extend the Convention still further in the direction of CW. However, London did not feel strongly and was thus able to accept Myrdal's suggested amendment to both the American and Soviet draft texts.[61] Toxins were included in Article I, but there was no specific definition other than the general statement 'whatever their origin or method of production'.

US and USSR Table Identical but Separate Draft Conventions: 5 August 1971

Ambassadors Leonard and Roshchin introduced identical, but revised draft Conventions for CCD consideration on 5 August.[62] Both outlined the reasoning

[59] TNA FCO 66/300, W.N. Hillier-Fry, UKDis Geneva to D.L. Benest, Disarmament Department, 23 July 1971.

[60] CCD/333, Sweden Working Paper on Aspects of the Definition of 'Toxins', 6 July 1971.

[61] TNA FCO 66/300, D.L. Benest, Disarmament Department to W.N. Hillier-Fry, UKDis Geneva, 3 August 1971.

[62] CCD/337 and CCD/338. See also TNA FCO 66/300, UKDis Geneva telegram No. 94 to FCO, 5 August 1971; CCD 527th Meeting 5th August.

behind the draft conventions. Roshchin, speaking on behalf of the Socialist Group, noted that the Geneva Protocol had established a universally recognised rule of international law. It was impossible to include in the draft a prohibition of use without detracting from the effectiveness of the Protocol; to do so would admit different approaches to the prohibition of BW and CW. Nevertheless, a preambular provision that was not a repeated declaration of the prohibition on use in the Geneva Protocol would need to reflect the natural result of the conclusion of the Convention: namely, ceasing production and destroying stockpiles would exclude completely the possibility of the use of BW.[63] The new drafts obliged states to conduct negotiations in good faith on effective measures on CW. Provision for a Review Conference after five years would also probably play an important part in achieving progress towards the prohibition of CW. The procedures for lodging a complaint with the UN Security Council in the case of an infringement ensured the best way for an objective investigation, and the taking of prompt and effective measures for suppression of any infringement.

Leonard played tribute to the UK role in the elaboration of the Convention that would support and strengthen the Geneva Protocol, which was specifically and unambiguously protected by Article VII of the draft. This precluded any possibility that the new Convention might in any way be interpreted as limiting or detracting from obligations assumed under the Protocol. The addition of toxins added significantly to the breadth of this first multilateral disarmament agreement in CBW. The US believed that the Biological and Toxin Weapons Convention should enter into force at an early date and the required number of ratifications should therefore be relatively small.[64]

Intention behind UK Article IV: Assistance in the Event of CW Use

Both the Nigerian and Burmese delegations were unclear about the meaning and significance of the phrase 'appropriate assistance' in the UK Article IV. Both were concerned that under this Article military assistance might be imposed upon the victim of a BW attack against its wishes. However, no public statement explaining the exact significance of this term appears to have been made. UK officials believed that medical or relief assistance, which would be provided on request, would be made available to the victim.[65] This understanding was shared with the Nigerians and Burmese. In private, officials in London and Geneva did

63 CCD/PV.527, 5 August 1971, pp. 5–13.

64 CCD/PV.527, 5 August 1971, pp. 13–18.

65 TNA FCO 66/300, D.L. Benest, Disarmament Department to W.N. Hillier-Fry, UKDis Geneva, 6 August 1971.

not think that action against any BW aggression should be entirely excluded since this might fatally damage the idea of deterrence.[66]

Non-aligned Attitudes to the Joint Drafts: July–August 1971

During the G12 meeting on 28 July to consider amendments to the draft Convention on the table, no agreement was reached.[67] Some new ideas in addition to Yugoslav proposals were considered, but a member of the Brazilian delegation felt that a common view would not emerge for another two weeks. It was still far from clear that Mexico and Sweden had decided to endorse the principle of a separate BTWC. The United Arab Republic (UAR) and Moroccan delegations were considering a new Article that would involve acceptance of the principle of the prohibition of CW production and destruction of stockpiles. They also worked out language to strengthen the Soviet Article on the commitment to further CWC negotiations. The principal proposer of amendments was Yugoslavia, and these emerged in a co-sponsored paper by most of the G12 in CCD/341.[68] Overall the Western delegations were able to accept most of the changes except those numbered eight and ten.[69] Number eight called for a new preambular paragraph that would have read as follows: 'Affirming the principle that a substantial portion of the savings from measures in the field of disarmament should be devoted to promoting economic and social development, particularly in the developing countries'. Number ten sought a re-draft of Article VII thus:

> Each State Party to the Convention accepts the principle of complete prohibition of chemical weapons and undertakes to continue negotiations in good faith with a view to reaching early agreement on prohibition of their development, production and stockpiling and on their destruction, and on appropriate measures concerning equipment and means of delivery specifically designed for the production and use of chemical agents for weapons purposes.

---

66 TNA FCO 66/300, W.N. Hillier-Fry to D.L. Benest, Disarmament Department, 21 July 1971.

67 TNA FCO 66/301, R. Houliston to Mr Masefield, Mr Hillier-Fry, 28 July 1971.

68 CCD/341, 17 August 1971, Brazil, Burma, Ethiopia, India, Mexico, Morocco, Nigeria, Pakistan, Sweden, the UAR and Yugoslavia; TNA FCO 66/301, R. Houliston to Mr Masefield, Mr Hillier-Fry, Draft BW Convention, 29 July 1971.

69 TNA FCO 66/301, UKDis Geneva telegram No. 99 to FCO, BW Convention, 18 August 1971; and J.T. Masefield, UKDis Geneva to D.K. Timms, Disarmament Department; BW Non-Aligned Working Paper CCD/341, 18 August 1971.

The UK Reaction: Henry Hainworth's Speech to the CCD

Five days after the Americans and Russians had tabled their identical draft conventions, Hainworth outlined UK views on the new draft.[70] Overall its appearance was welcomed. Hainworth told the CCD that this new draft contained most of the key elements that the UK believed to be important and it represented an improvement on the earlier draft – CCD/325/Rev.1. In particular, the inconsistencies in the references to 'weapons' and 'agents' were ironed out in Article I, which as it transpired was in its near final form, and there was now clear provision for the destruction of agents used in weapons – now Article II. There were, however, some weaknesses; for example, the UK remained convinced that a comprehensive convention dealing with biological warfare should ensure that the parties undertook never in any circumstances to conduct, assist or permit research for purposes of the sort that were being prohibited under the draft Convention. Furthermore, an ideal Convention should include an Article providing for the express renunciation by all parties of BW use: such a provision was missing from the 5 August drafts. Although there were several counter-arguments, the UK believed that these were invalid. First, a prohibition on use would not undermine the Geneva Protocol; on the contrary, it was entirely relevant to repeat that earlier promise in such an appropriate Convention. Second, a ban on use would not lessen the prohibition on CW use in the Geneva Protocol since Article VIII of the new drafts stated that nothing in the Convention was to be interpreted as in any way limiting or detracting from the obligations assumed by any state under the Geneva Protocol. Third, the present drafts required parties not to develop, produce, possess or acquire biological means of warfare and the appropriate equipment, and thus parties would be in no position to use them. Although this position had an undoubted appeal as an argument in logic, it was important to recall that none of the drafts – including the UK's – would prevent the production and stockpiling of BW agents for peaceful purposes. Indeed Article X offered positive encouragement for such activities. It was theoretically possible that such agents might be used for hostile purposes. The Convention's validity would, therefore, not obtain in all circumstances if such a loophole were left unplugged. The CCD's negotiations, however, offered a useful opportunity for mending the hole in the Geneva Protocol left by reservations regarding the right to use BW in certain circumstances.

Hainworth reiterated the significance of verification of use and the great importance of including an appropriate mechanism in the Convention to secure

70 CCD/PV. 528, 10 August, 1971 pp. 19–28.

this objective. This should include a prompt, impartial and effective system, which in all probability would include direct examination on an attacked state's territory. Such an investigation would provide very strong evidence that another state had violated the ban on the production and possession of BW. In the CCD's margins the UK continued to propose improvements to the draft language on investigation of use.

Following Hainworth's speech, the delegation proposed some options to the Russians on 13 August; a revised version of the UK Article VI on investigations was handed over.[71] This version differed slightly from earlier text with the references to the Geneva Protocol deleted and with the addition of the second sub-paragraph from Article VI of the 5 August draft in its place. This in turn had been derived from Article III (3) of the original 1969 UK draft; the delegation stressed that this draft was without status. It was not an amendment, but rather an illustration of the sort of language the UK thought desirable. After six days consideration the Soviet delegation gave its views on the UK papers; they made clear that they were not speaking on instructions.[72] In short, the Russians were opposed to any attempt to re-draft Article VI. The proposed non-aligned addition to Article V was, however, acceptable as a satisfactory broad and general formulation that would allow for preliminary steps that could include an inquiry or any other appropriate procedure. This read, 'consultation and co-operation pursuant to this Article may also be undertaken through appropriate international procedures within the framework of the United Nations and in accordance with its charter'. This phrase was included in what became the final version of the Convention.

Disarmament Department's detailed comments on the Geneva redraft of Article VI were despatched by telegram on 23 August.[73] This noted that if the draft Convention were to contain no substantive prohibition on use, it could be argued that since the prohibition in Article I did not include use, the revised Article VI was legally inconsistent with Article I. Moreover, inclusion of the reference to use would strictly speaking not extend the obligations under the Convention so as to cover use. That said, it might be argued that use could be aired internationally as evidence of a breach of Article I. One possible way around this difficulty, which would have the effect of including use amongst the activities that could be dealt with by the complaints machinery but without explicit mention of the Geneva Protocol, would be to include in Article VI a

---

71 TNA FCO 66/301, UKDis Geneva telegram No. 99 to FCO, 18 August BW Convention; W.N. Hillier-Fry, UKDis Geneva to D.M. Summerhayes, Disarmament Department, 16 August 1971.

72 TNA FCO 66/301, UKDis Geneva telegram No. 102 to FCO, 20 August 1971.

73 TNA FCO 66/302, FCO telegram No. 19 to UKDis Geneva, 23 August 1971.

suitably worded reference to the second preambular paragraph. This paragraph then read as follows: 'Desiring thereby, for the sake of all mankind, to exclude completely the possibility of bacteriological (biological) agents and toxins being used as weapons.'

However, the UK CCD delegation replied on 24 August that the time left for persuading the co-chairmen to accept any amendment on verification that might meet UK concerns even in a part was limited.[74] Indeed the delegation felt that the UK would have to recognise that it would be impossible to secure an explicit reference, or even an imprecise formula that would allow for automatic impartial investigation prior to consideration of any complaint by the UN Security Council. However, for the first time the US appeared to lend its weight to an active effort to persuade the Russians to accept language designed to meet British concerns here. Despite this they were not prepared to provide positive support for a procedure aired outside the Security Council.

The UK Rethinks Its Position: End of August 1971

By now it was clear that London recognised that the UK would have to settle for a lot less than its preferred position on verification.[75] Indeed it was now time for 'second string solutions' and in this case the non-aligned amendment to Article V provided a suitable basis for the two critical elements in the UK position: namely, to ensure that the complaints machinery could cover complaints of use and that the setting in hand of enquiries could not in practice be prevented by exercise of the veto in the UNSC. Article VI was the key here, especially the details of Security Council involvement in the consideration of any complaint of non-compliance. In the UK view it would be better to obviate the Council's permanent members' veto by means of an accompanying resolution that would include a provision by which it would be agreed that the concurring votes of the permanent members would not be required. This could either be made to apply to all of the actions contemplated in operative paragraph 1 of Article VI or only to the second paragraph that referred to 'necessary measures for the investigation of a complaint'. Accordingly the UK CCD delegation was instructed to pursue these amendments with other delegations as they thought best, but it was not to make any commitment to co-sponsorship. Since other delegations attached importance to the UK co-sponsoring the text, officials felt that this leverage should be exploited to secure the amendments being sought.

---

74 TNA FCO 66/302, UKDis Geneva telegram No. 104 to FCO, 24 August 1971.

75 TNA FCO 66/302, FCO telegram No. 21 to UKDis Geneva, 25 August 1971.

Likely Soviet Position: End of August 1971

The Soviet attitude was central to the UK's chances of success in securing its objectives. By 26 August the position was reasonably clear.[76] The Soviet delegation would not accept the amendment to Article VI and nothing further on the prohibition of use, but they were prepared to consider a revised Article V. Nor would they accept inclusion of 'research', but some provision on assistance in the event of a BW attack was acceptable.[77] The Russians had made it abundantly clear throughout that they would not envisage any verification procedures that might alter the UNSC's rights and procedures. Henry Hainworth concluded that since the UK could not formally lay the requirement upon the UN Secretary-General, the proposed re-wording of Article VI was the nearest the UK was likely to get to ensuring that the Security Council would normally arrange a fact-finding mission into complaints. It would be difficult to provide for an investigation of complaint of use in a Convention that did not specifically prohibit use. If it were possible to convince the Russians to accept the non-aligned amendment to Article V and a revised Article VI, then this would probably be as far as they would move. Such an outcome would in practice give the UK what it wanted.

US–Soviet Discussions: Late August 1971

On 27 August Leonard informed the Western Group that he hoped to work out with Roshchin early the following week the package of amendments to recommend to their capitals.[78] This timing was likely to slip because of the delay in reaching agreement with the G12 on a revised Article VIII on future work towards a CWC. This was the hardest problem confronting the Americans. They had tried several formulae on the delegations responsible for the non-aligned draft – United Arab Emirates and Morocco – but had sent none of these to Washington and ended up being embarrassed when the Moroccans convened a meeting of the G12 on 26 August to seek comments on these formulations. Re-introduction of the word 'complete' before 'prohibition' was the problem. Hainworth was told privately that members of neither the Western Group nor of G12 would be asked to co-sponsor the next draft Convention.

The position of UK co-sponsorship hinged on the outcome of the co-chairmen's recommendations. Ministers could not be asked to take a decision

---

76 TNA FCO 66/302, UKDis Geneva telegram No. 108 to FCO, 26 August 1971.

77 See discussion in next chapter on the question of 'research'.

78 TNA FCO 66/302, UKDis Geneva telegram No. 109 to FCO, 27 August 1971.

until these were seen.[79] Prospects for UK co-sponsorship depended on whether officials could point to some significant advance towards UK positions. Were the co-chairmen to recommend favourable amendments, Hainworth was authorised to indicate that he would make a very strong recommendation to HMG that co-sponsorship was now possible, and that he had every reason to expect that it would be sympathetically received. The US delegation commented that the effect of using the words 'having as their objective' in preambular paragraph 2 and of referring to this 'objective' of the Convention in Article V and VI would be to bring out again much too provocatively the whole problem of use. Not only the Soviet bloc, but many non-aligned delegations would be strongly opposed too. Leonard thought that the UK was more likely to gain what it wanted by concentrating on less explicit formulae.[80]

The UK and Soviet delegations held what was to prove to be a crucial meeting on 31 August.[81] This considered the various formulations for the preambular paragraphs, Article V and VI. UK diplomats asked the Russians if they had yet produced language on assistance. I.I. Tcheprov handed over what he stressed was only a delegation draft and could not be shown at this stage to any other delegation except the Americans:

> Each State Party to the Convention undertakes to provide support or assistance, in accordance with United Nations Charter, to any party to the Convention, if the Security Council decided that bacteriological methods of warfare or toxins have been used against that party.

Tcheprov emphasised that it was based on the Article in the UK draft Convention. However, 'affirms its intention' was replaced with 'undertakes' and 'appropriate' removed to emphasise that parties would be taking on a new obligation. The assistance envisaged was humanitarian. The Russians agreed with the UK view that it would probably be necessary to include a phrase to make it clear that assistance would only be given at the request of the injured party.

As for Article V, Tcheprov opened by saying that the Soviet delegation was divided as to whether to recommend acceptance of the phrase 'in relation to the objectives of' to Moscow and would suggest something shortly. UK diplomats noted that in order to meet earlier Soviet comments they proposed to make 'objectives' singular and wondered whether it would help to say 'having as their objective' in the second preambular paragraph. The Russians said that the first

79 TNA FCO 66/302, FCO telegram No. 24 to UKDis Geneva, 27 August 1971.

80 TNA FCO 66/302, UKDis Geneva telegram No. 111 to FCO, 30 August 1971.

81 TNA FCO 66/302, UKDis Geneva telegram No. 113 to FCO, 31 August 1971.

suggestion might be an improvement, but the second was one that they were not even prepared to report to the rest of the Soviet delegation. In the Soviet view it was not possible to limit the objective to one preambular point; and, furthermore, it would be completely unacceptable to the non-aligned since one of their objectives was progress towards a CWC. So far as the Soviet delegation was concerned it would make it very difficult to accept the proposed change in Article V. And on this issue, they pressed the UK to accept the US view and the relative priority of consultation in relation to the objective of the Convention, or in application of its provisions. In Russian apparently words appearing before 'or' always have a slight connotation of priority over the words that follow. This was not so in English and the UK delegation offered to reverse the order if this would help. The US attitude on these questions was one of benevolent neutrality, and chiefly depended on the Soviet reaction to UK ideas.

The UK delegation underlined that it was desirable to parallel the reference to the 'objective of the Convention' in Article V with a similar reference in Article VI. Tcheprov said that he would have to consider the point, but it seemed to raise similar problems and would widen the range of matters that could be taken before the Security Council. He volunteered, however, that as it was, any state could appeal direct to the Security Council on such a point. The Russians would think over the addition of phrases referring to the objective of the Convention in Articles V and VI, but they would not entertain the thought of such a reference in the preambular paragraph 2. Nevertheless, they agreed that the effect of the preambular paragraph 2 might be strengthened if, as the Americans had proposed, it was put as the penultimate preambular paragraph and the run-in changed to read, 'desiring by this Convention'.[82] Hainworth believed that this meeting was helpful since he had also impressed on Roshchin the political significance of the UK package. If the assistance Article emerged in its then form, the UK could make considerable play of the explicit assumption that the Security Council may consider complaints that biological methods of warfare had been used. Summarising the position, the delegation pointed out to London that if the Russians could accept the UK points just raised on Article V and VI, then it would be in a good position to recommend that the UK should co-sponsor the next draft presented to the CCD. The delegation informed the US of the Soviet views and interest in the US reaction to the UK proposal for Article V and VI; they also mentioned that American assistance in obtaining a favourable Soviet reaction would be timely and valued. The Americans, however, had some reservations both on 'in relation to the objectives', particularly in

---

[82] TNA FCO 66/302, J.T. Masefield, UKDis Geneva to J.M. Hay, Disarmament Department, 25 August 1971.

Article VI and on the assistance Article quoted above, but undertook to consider them urgently.

The US Response: 1 September 1971

The US delegation's considered view on these topics was given to the UK on 1 September.[83] The draft assistance Article was also discussed with the Russians on the same day. American diplomats could now support addition of 'in relation to the objective' in Article V, although they considered that the plural would have been better logically and more widely acceptable. In Article VI the Americans objected to the expression 'breach of the objective'. An improvement might consist of 'constitute a breach of the obligations assumed under the provisions of this Convention or are contrary to its objectives'. However, the US saw problems in referring to an undefined objective and considered that such a phrase was not necessary to provide for the submission of complaints of use to the Security Council since this could already be done under the UN Charter. On assistance, the US delegation had alternative approaches, which they sent to Washington after discussing them first with the Russians. The first version was a revival of their earlier suggestion on the UK draft to add 'medical, protective or other similar' before 'assistance' and to supplement this with a statement in the UNGA on security assistance similar to the one used by the UK and US as nuclear weapons states in the NPT. A second change would see addition to Article VI of 'Each State Party to the Convention undertakes to support immediate Security Council action to provide assistance, in accordance with the Charter, to any State Party against whom the UNSC decided biological methods of warfare and toxins have been used'. The US delegation, also accepted a UK suggestion to incorporate a phrase like 'at the party's request'. On reflection London thought that the exchanges with the Russians were helpful in furthering UK views.[84] Officials attached particular importance to the successful amendment of Articles V and VI. If the desired changes were made, then there would be a convincing case for recommending co-sponsorship to Ministers; if not, co-sponsorship would be unacceptable to everyone in Whitehall. Much depended on US attitudes, but the UK delegation advised against any further approaches in Washington lest this complicate its relations with its US counterparts in Geneva.[85]

[83] TNA FCO 66/302, UKDis Geneva telegram No. 114 to FCO, 1 September 1971.

[84] TNA FCO 66/302, FCO telegram No. 25 to UKDis Geneva, 1 September 1971.

[85] TNA FCO 66/302, UKDis Geneva telegram No. 115 to FCO, 2 September 1971.

The Soviet Reply: 2 September 1971

By 2 September the Soviet delegation had accepted – ad referendum to Moscow – the re-ordered and slightly reworded preamble as proposed by the UK on 27 August.[86] However, they were still considering the addition of 'in relation to the objective' in Article V, which the US delegation had urged the Russians to consider sympathetically. The Russians could accept the re-ordered version of Article VI with, in paragraph 2, addition of the Polish wording 'investigation' and the substitution of 'initiate' for the second 'undertake'. These changes appeared in the final version of Article VI, but the Russians were still not disposed to consider the addition of 'the objective' to Article VI. On 6 September the Americans called a meeting of the Western Group to report Soviet final views on most of the outstanding issues in the operative Articles:

- Article I: the Soviet delegation agreed to add a comprehensive definition of toxins as proposed by Sweden, but employing language prepared by the Americans. (This would be 'whatever their origin or method of production'.) They were also ready to recommend to Moscow the addition of the word 'protective'. Article I was now in its final form.
- Notification of destruction: the Soviet delegation would not accept an addition on timing to Article II, but did agree to recommend to capitals that the co-chairmen would announce that the USSR and the US would make public statements when all the requirements of Article II had been fulfilled.
- Article V: both the co-chairmen could accept the non-aligned and UK additions to this Article.
- Article VI: thus far the co-chairmen were unable to agree on the Article. The Russians were not prepared to accept the UK text on the grounds that this language implied a waiver on the right of veto in the UNSC in respect of complaints. In paragraph 2 'investigation', 'initiate' and the final sentence on UNSC notification were unchallenged. It was agreed to make 'result' in the last sentence plural, which the US felt would imply that there were several stages to an investigation.[87]

---

86 TNA FCO 66/302, UKDis Geneva telegram No. 117 to FCO, 2 September 1971.

87 TNA FCO 66/302, UKDis Geneva telegram No. 119 to FCO, 6 September 1971; J.T. Masefield, UKDIS Geneva to J.M. Hay, Disarmament Department Western Group Meeting of 2 September: Draft BW Convention; UKDis Geneva telegram No. 122 to FCO, 8 September 1971.

On assistance, the Russians were still prepared to see an Article on this issue, but wished to restore the word 'appropriate.' The draft Article now read:

> Each State Party to the Convention undertakes to provide or support appropriate assistance, in accordance with the UN Charter, to any party to the Convention which so requests, if the Security Council decides that bacteriological (biological) methods of warfare and toxins have been used against that party.

The Russians, however, doubted whether the new Article would increase support among the non-aligned for the Convention. Both co-chairmen agreed to seek contingent authority for the Article on the basis that if they discovered that support for the BTWC would be increased by this addition, it would be included. The Americans had little doubt that such a provision would be so included, but a lobbying effort in capitals would be required.

As for Article VIII, the US and Soviet Union agreed to recommend to their respective capitals the following language:

> Each State Party to this Convention affirms the recognised objective of effective prohibition of chemical weapons and, to this end, undertakes to continue negotiations in good faith with a view to reaching early agreement [as soon as possible] on effective measures for the prohibition of their development, production and stockpiling and for their destruction, and on appropriate measures concerning equipment and means of delivery specifically designed for the production or use of chemical agents for weapons purposes.

They recognised that Washington might prefer to reject non-aligned language 'as soon as possible'. This phrase was not accepted into the text, which then became the final version of Article VIII.

The Co-chairmen Prepare the Final Draft for Capitals: 7 to 8 September

A further long round of negotiations between the Americans and Russians was held on 7–8 September to resolve the remaining problems. Agreement was reached on all the outstanding points in the text to be referred to Washington and Moscow. The crucial agreements concerned the second preambular paragraph of the 5 August draft; Article VI; assistance now appearing as Article VII. The second preambular paragraph now read, 'Determined for the sake of all mankind, to exclude ....'. Despite late Soviet pressure, it retained its new place as the penultimate preambular paragraph, a fact very important for the UK.

Article VI was now in its almost final form: the only changes introduced subsequently concerned minor editorial points such as replacement of 'this' with 'the' in the first sentence of paragraph 1 and replacement of the 'United Nations Charter' in paragraph 2 with the 'Charter of the United Nations'. However, the Russians were not prepared to include language from a Polish draft on the UNSC unless it was balanced by a phrase stating that it was only 'after the Security Council had decided' that States Parties 'would assist the Council to take all necessary measures'. In the end, after consulting the UK CCD delegation, the Americans felt that they had no alternative but to drop the Polish text. They were nevertheless pleased with the new text for the opening sentence of paragraph 1 since the US attached importance to the implicit reference to 'use' in the word 'deriving', namely, an 'obligation deriving from the provisions of the Convention'.

Roshchin had been very difficult on the assistance Article and had for some time threatened not to include any such Article in the final draft. In the end the US agreed to accept the following:

> Each State Party to the Convention undertakes to provide or support appropriate assistance, in accordance with the United Nations Charter, to any party to the Convention which so requests, if the Security Council decides that such Party has been exposed to danger as a result of violation of this Convention.

The US believed that this formulation provided at least a 90 per cent connotation that assistance would be dependent upon biological weapons or toxins having been used.[88] This Article, now numbered VII, was in its final substantive form; the only subsequent alteration saw the replacement of the penultimate word 'this' with 'the'. The Russians were, however, adamantly opposed to any reference to savings from disarmament and the Americans felt that they had to choose between pressing for a pre-empting reference to savings to avoid adverse non-aligned reactions and urging an assistance Article. For the sake of their allies they chose the latter.

Washington and Moscow Consider the New Draft Mid September 1971: UK Reactions

As for timing, Leonard told Henry Hainworth that he thought that it would take at least ten days for Washington's comments to arrive and he expected

---

88 TNA FCO 66/302, UKDis Geneva telegram No. 122 to FCO, 8 September 1971.

Moscow to be no quicker.[89] The Russians confirmed this assessment and the earliest that Washington's comments would appear would be the weekend of 17–20 September. At this point the final text might be sent to NATO for its views. Tabling in the CCD would be about 23 September, but the 28th was more likely. In any case, there would now be no intermediate draft between that of 5 August and that put forward to the UNGA in the name of the CCD as a whole.[90] Leonard reported that the Russians had raised the question of co-sponsorship of the new draft, as they wanted to avoid totally the problem of co-sponsors and present a 'committee draft', namely, one which all members of the CCD endorsed.[91] When Leonard had mentioned the possibility of a draft co-sponsored by the co-chairmen alone, Roshchin insisted that all his allies would have to appear as co-sponsors. Leonard told the Western Group that he saw virtue in the idea of a committee draft. It might for example appear as an annex to the CCD's draft annual report to the UNGA. However, Hainworth noted that informal discussion of a document without parentage might lead to delegations seeking to insist on their points of view, thus prolonging and confusing the proceedings.

The latest co-chairmen's draft Convention did not represent the best possible BTWC that the UK could have devised. However, Henry Hainworth was convinced that it was the best attainable in the circumstances and certainly, subject to only very minor changes, the only Convention likely to be secured. Its provisions made it impossible practically and legally for a party to use BW. This objective was made clear in the revised second preambular paragraph and more particularly by its new position as the penultimate paragraph. The Soviet statement on 5 August, which was to be further strengthened when the new draft was tabled, would provide confirmatory evidence that the USSR regarded the BTWC as preventing the practical possibility of BW use in any circumstances, including in warfare. The addition of 'never in any circumstances' to Article I reinforced such a view.[92] The final formulation of Articles V, VI and VII made it clear that parties could complain of use under Article V, which would allow for independent enquiry prior to Security Council consideration where the parties agreed. Alternatively, a party could complain under Article VI directly to the UNSC with the presumption that the Council would initiate an enquiry before taking a political decision on a complaint. This interpretation was shared by the

[89] TNA FCO 66/303, UKDis Geneva telegram No. 124 to FCO, 9 September 1971.

[90] TNA FCO 66/303, UKDis Geneva telegram No. 125 to FCO, 9 September 1971.

[91] TNA FCO 66/303, UKDis Geneva telegram No. 124 to FCO, 9 September 1971.

[92] TNA FCO 66/303, FCO telegram No. 30 to UKDis Geneva, 10 September 1971; TNA FCO 66/303, George S. Vest, Acting Permanent Representative, US Delegation to NATO to Manlio Brosio, Secretary-General, NATO, 24 September 1971.

US and Soviet Union. Hainworth concluded that the UK had thus succeeded to a reasonably explicit extent in attaining its two principal negotiating objectives. Both the US and Soviet Union attached some importance to UK co-sponsorship of the new draft. Further comments from the UK, such as those in his speech to the CCD on 10 August, might undermine its chance of success at the UN. Now was the time for compromise. Furthermore, given the UK's pioneering role it was appropriate for the UK to be one of the depository powers. All things considered, Hainworth recommended that the UK should now co-sponsor the final draft when it appeared; and moreover, in the absence of any unwelcome amendments, co-sponsor the UNGA resolution of commendation in New York.[93]

Disarmament Department outlined UK views on the 8 September text in a telegram to Geneva on 10 September:

- Preambular paragraph on use: since this no longer contained a reference to the Convention, either directly or by means of the word 'thereby' the view was that addition in Article V of the phrase 'objective of this Convention' indirectly covered by referring back to this preambular paragraph, was somewhat weakened. However, the rewording and now promised position of this paragraph as the penultimate one reinforced the UK view that it was impossible practically and legally to use BW. The main alteration in the 8 September draft preamble was that it now progressed from expression of general principles to those dealing with specific issues in regard to BW. The revised agreement thus emphasised the importance of paragraphs relating to use that now appeared immediately before the operative provision of the Convention.
- Article I: the addition of 'never in any circumstances' strengthened the obligations assumed under this Article. The way in which 'protective' was included instead of 'peacefully' ensured that passive defensive measures were permitted. There was no objection to the definition of toxins.
- Article V: this now included two important changes over the 5 August draft. At UK insistence, the first sentence now included a reference to the 'objective of this Convention', which in UK eyes was set out in preambular paragraph 9 namely, the complete exclusion of the possibility of bacteriological (biological) agent and toxins being used as weapons. The use of prohibited agent and weapons was now covered by the undertakings of this Article. These undertakings were also extended by the addition of a second sentence that referred to 'appropriate international

93 TNA FCO 66/303, UKDis Geneva telegram No. 126 to FCO, 9 September 1971.

procedures within the framework of the UN and in accordance with its charter'. There was no precise definition of what these were, but it could be argued that the phrase embraces UN machinery other than the UNSC. Action under this Article could not therefore be prevented by use of the veto.

- Article IV: the US interpretation of paragraph 1 was that it now implicitly referred to 'use'; London agreed, indeed this was an unexpected gain.
- Article VII: although the UK was naturally satisfied that the co-chairmen had agreed on wording of an assistance Article, it was disappointing that the explicit reference to use had been dropped. It was none too clear what the term 'exposed to danger' meant. If the Article was interpreted as before, only being re-drafted to omit the reference to use, which was unwelcome to the Russians, then there were no difficulties. However, if it were to be strictly construed as referring to potential danger as a result of violation of Article I, in other words as the result of another state's development, production, but not of BW use, then it might be felt that the 'appropriate assistance' could only be military in nature.
- Article IX: although 'as soon as possible' would have been better than 'early', the draft was acceptable as it stood.[94]

Although the Department believed that the UK had 'sustained slight losses during this last round', particularly in preambular paragraph 2 and in the assistance Article, there was a welcome gain in Article VI. The present text represented a considerable advance towards the UK's position over that tabled on 5 August. Officials in London shared Hainworth's assessment that the present text was the best obtainable and that it was now incumbent upon the UK to support it. Should the opportunity arise, the Department also felt able to recommend to Ministers that the UK should co-sponsor this draft. MOD accepted the arguments too.[95]

Ministerial Endorsement: 16–19 September

Disarmament Department submitted to Lord Lothian on 16 September recommending that as soon as agreement of MOD Ministers was secured, Lord Lothian should authorise Hainworth to make a speech welcoming the final

94 TNA FCO 66/303, UKDis Geneva telegram No. 126 to FCO, 9 September 1971.

95 TNA FCO 66/303, R. Preston, DS22, MOD to PS/Secretary of State, Geneva Disarmament Conference: Biological Weapons (BW) Convention Draft under Cover of Letter from R. Preston to J.M. Hay, Disarmament Department, 13 September 1971.

draft and to co-sponsor should the opportunity arise.[96] David Summerhayes, Head of Disarmament Department, noted in this submission that Article V, with the addition of the phrase 'to the objective of the Convention', taken with preambular paragraph 9, extended the obligations of this Article by implication to cover the question of use. This submission's recommendations were agreed by the Minister on 19 September and instructions were sent to Geneva to this effect on 20 September.[97] London accepted that the draft that had emerged was the best attainable in the circumstances and that more criticism from the UK now could only be counterproductive. Hainworth's welcoming speech was cleared by both MOD and the FCO and endorsed by Lord Lothian. London's only slight reservation on the UK CCD delegation's draft concerned its tone.[98] It seemed to give rather too warm a welcome to the draft since the UK would have preferred to see a more effective coverage of the question of use. London inserted an important change in the final paragraph where the following was inserted in the opening sentence; '... the view that the present draft represents a sound and realistic basis for an agreement. I believe that we have succeeded in this hard and purposeful negotiation in working out the best draft attainable in present circumstances.'[99]

The Final Steps

Meanwhile back in Geneva, Roshchin told Leonard on 18 September that Moscow accepted the revised draft except for two points.[100] First, the Russians required removal of the second 'biological' in preambular paragraph 8 on the grounds that it was illogical given the subject matter of the Convention; and, second, the deletion of 'appropriate' in Article VII. Leonard took the view that the first of these points was something the Russians would have to argue with the non-aligned. The second question was to be covered by an interpretative statement saying that the Soviet Union interpreted assistance as being medical and humanitarian. The Russian points were accepted. On 23 September the US delegation told the UK that the US Administration had now approved

96 TNA FCO 66/303, D.M. Summerhayes, Disarmament Department to Sir T. Brimelow, Mr Goodenough, Biological Warfare Convention, 16 September 1971.

97 TNA FCO 66/303, FCO telegram No. 331 to UKDis Geneva, 20 September 1971.

98 TNA FCO 66/303, J.M. Hay, Disarmament Department to Mr Goodenough, Draft BW Convention, 21 September 1971.

99 CCD/PV.542, 28 September 1971.

100 TNA FCO 66/303, UKDis Geneva telegram No. 132 to FCO, 18 September 1971.

the 8 September draft, although they expressed a preference for the 5 August version of the first preambular paragraph. A telegram was on its way to the US delegation in NATO to seek allied concurrence for the 8 September draft.[101] There were no objections and the US delegation gave clearance to their colleagues in Geneva to table the new draft in the CCD.[102] Meantime the US Geneva delegation would try to secure agreement for the above point, although it would not be made a sticking point. There was no agreement, so it reverted to the 8 September version.[103] An agreed joint draft Convention co-sponsored by the Socialist Group, the US and UK was tabled on 28 September in CCD/352. The principal negotiating task was complete. Geneva had produced a text for submission to New York on 28 September. This draft was for all intents and purposes the Convention's final version; only a few minor editorial alterations were made before the text was opened for signature on 10 April 1972. It had taken three years from the UK's first proposal for a separate ban on biological weapons to a completed Convention, with most of the key negotiations coming in the final six months.

Conclusion

The essential difference between the US position and the UK was that the British regarded the BTWC as an end in itself and therefore something to be made as practical and watertight as possible. The US, on the other hand, regarded it as something subordinate, although contributing to, the more important and broader objective of reaching agreement with the Soviet Union on nuclear arms control. SALT/ABM negotiations in Helsinki and Vienna were running in parallel with the CCD debate on biological weapons. This was more important for both the Russians and Americans and of course had a much higher political and public profile, with a BW Convention playing second fiddle. Consequently, the superpowers considered that the negotiations should be kept essentially in their hands – British persistence was not always welcomed. This would ensure the Convention's speedy conclusion and maintain the negotiating confidence and goodwill that would contribute to the SALT negotiations' success. However, this attitude gave rise to some resentment in the CCD, particularly the non-aligned

---

101 TNA FCO 66/303, UKDis Geneva telegram No. 135 to FCO, 23 September 1971.

102 TNA FCO 66/303, P. Lever, UKDel NATO, Brussels to D.L. Benest, Disarmament Department, Draft BW Convention, 28 September 1971.

103 TNA FCO 66/303, UKDis Geneva telegram No. 135 to FCO, 23 September 1971.

participants – chiefly Sweden and India. In the early days the co-chairmen were quite welcoming of the UK's BW initiative as something that could be turned to their advantage. That said, by 1971 both had rather lost patience with UK insistence on perfectionism and apparent failure to recognise the greater importance of the broader bilateral superpower negotiations. The UK's stalwart defence of its interests and the attention to detail on the meaning of agreed text demonstrates clearly that in international negotiations words and phrases really do matter; and it helps that the drafting is done in English. Following the at times tortuous debates and arguments over ostensibly recondite and legalistic phrases is, however, essential to understanding the scope and meaning of the agreement eventually reached. The failure to secure the sort of investigation mechanism into cases of alleged use favoured by the UK would come back to haunt the Convention in the 1980s when the shortcomings of its compliance mechanisms were exposed by the anthrax outbreak in the Soviet city of Sverdlovsk caused by an accident at a BW facility.[104] A decade was subsequently spent looking into possible verification measures and negotiating a 'legally binding instrument' otherwise known as the verification protocol; this ended in failure with the collapse of negotiations in 2001 following US rejection of the draft protocol presented by the Ad Hoc Group's Chairman Tibor Toth.[105] Some 40 years after the conclusion of the negotiations lack of effective compliance provisions still remain the BTWC's fundamental limitation.

---

[104] See TNA FCO 66/1520, Sverdlovsk (inc. bilateral with Americans) 1980 and TNA FCO 66/1521, Sverdlovsk – follow up 1980.

[105] See Jez Littlewood, *The Biological Weapons Convention: A Failed Revolution* (Aldershot, 2005) for detailed discussion of the Verification Experts (VEREX) process and the BTWC Protocol negotiations that ran from 1995 to 2001; see also *Strengthening the Biological and Toxin Weapons Convention: Countering the Threat from Biological Weapons*, Cm 5484. Presented to Parliament by the Secretary of State for Foreign and Commonwealth Affairs by Command of Her Majesty April 2002, pp. 9–13.

Chapter 7

The UK and Why 'Research' Was Dropped from the Draft Biological and Toxin Weapons Convention

Introduction

Article I of the Convention prohibits development, production, stockpiling, acquisition or retention of microbial or other biological agents or toxins whatever their origin or method of production, of types and in quantities that have no justification for prophylactic, protective or other peaceful purposes. Weapons, equipment or means of delivery designed to use such agents or toxins for hostile purposes or in armed conflict are also banned. It is sometimes forgotten that the first draft Convention tabled by the UK in the summer of 1969 also included a ban on offensive research. This proposal remained on the table until the end game of the negotiations in August and September 1971. So why did it disappear from the final text despite having wide support? Unfortunately the negotiating record – UK state papers and ENDC/CCD PV records and Working Papers – is not absolutely clear on this point, but we can piece together the evolution of this particular provision and identify the time it dropped off the agenda. Seemingly, for the UK at least, there were other more pressing battles to fight and win in the negotiations' last months, notwithstanding the importance of a ban on offensive research during 1968 and 1969. As we have seen, ensuring that a prohibition on use was covered and that there were effective provisions for the investigation of alleged BW use were British priorities. What this episode shows is how key issues can be overtaken by events and that negotiation outcomes do not always lead to clarity in some areas, either by design or oversight.

If we look back to the summer of 1968 and the UK working paper we can clearly see the central place that a prohibition on offensive research had in British thinking.[1] This suggested that a BW Convention, 'would also need to deal with research work. It should impose a ban on research work aimed at production

[1] ENDC/231, United Kingdom Working Paper Dated 6 August 1968 on Microbiological Warfare.

of (BW agents that had no independent peaceful purpose), as regards both microbiological agents and ancillary equipment.'[2] Article II (i) of the 1969 UK draft Convention required States Parties not, 'to produce or otherwise acquire or assist in or permit the production or acquisition of (a) microbiological agents of types and in quantities that have no independent peaceful justification for prophylactic or other purposes, (b) ancillary equipment or their vectors the purpose of which is to facilitate the use of such agents for hostile purposes'. Sub-paragraph (ii) obliged States Parties not to conduct or permit research aimed at production of the kinds prohibited in Article II (i). MOD, although not objecting to this idea, observed that the phrase 'aimed at' seemed to be potentially a little narrow since it implied a firm prior intention to produce. Instead, 'designed to create or facilitate the possibility of' was suggested.[3] However, the FCO preferred the original text as it thought that the MOD alternative would make the scope of the prohibition too wide.[4] A revised draft Convention was submitted to Mulley on 30 May 1969.[5] After some further discussion in the Cabinet's Defence and Overseas Policy Sub-Committee on Disarmament, none of which impacted on inclusion of research, the Prime Minister agreed on 23 June that a draft Convention could be tabled in Geneva.[6]

Draft Convention Tabled: July 1969

The relevant parts of Article II in the UK draft tabled on 10 July 1969 were:

> Each of the Parties to the Convention undertakes
>
> (a) not to produce or otherwise acquire, or assist in or permit the production or acquisition of
>
> (i) microbial or other biological agents of types and in quantities that have no independent peaceful justification for prophylactic or other purposes;

---

[2] ENDC/231, paragraph 7.

[3] TNA FCO 66/136, Peter Nailor, Head of DS 22 MOD to R. Hope-Jones, AEDD FCO, 22 May 1969.

[4] TNA FCO 66/136, R. Hope-Jones, AEDD, FCO to P. Nailor, Head DS 22 MOD, 29 May 1969.

[5] TNA FCO 66/136, R. Hope-Jones to Mr Williams, Convention for the Prohibition of Biological Methods of Warfare, 30 May 1969.

[6] TNA FCO 66/137, E. Youde, Downing Street to N. Barrington, 23 June 1969.

(ii) ancillary equipment or vectors the purposes of which is to facilitate the use of such agents for hostile purposes;

(b) not to conduct, assist or permit research aimed at production of the kind prohibited in sub-paragraph (a) of this Article.[7]

Initial ENDC reactions do not seem to have focused on the provision on research – other issues were more important to delegations, especially the Soviet Union and its allies and the non-aligned states. For them the main problems concerned the separation of CW from BW, the relationship of this new draft to the 1925 Geneva Protocol and the draft's provision on investigations of non-compliance. A revised draft Convention was tabled by the UK on 26 August 1969 in light of comments received.[8] There was only one change to Article II and this came in (a) (i) where the wording was amended slightly to refer to 'independent justification for prophylactic or other peaceful purposes' instead of 'independent peaceful justification for prophylactic or other purposes'.[9] Article II (b) remained untouched.

US Reactions

First detailed US comments were not given to the UK until February 1970. Washington suggested that Article II (b) should be amended to 'not to conduct or permit, or assist any recipient whatsoever to conduct research aimed at production of the kind prohibited in sub-paragraph (a) of this Article'.[10] HM Embassy in Washington had further discussions with the Arms Control and Disarmament Agency about the draft Convention in May, but there was still no sign of any opposition to a ban offensive research in Article II (b), or in any further debates over the drafting.[11] FCO files at this time contain only one

---

[7] ENDC/225, Draft Convention by the United Kingdom Dated 10 July for the Prohibition of Biological Methods of Warfare and Accompanying Draft Security Council Resolution.

[8] ENDC/225/Rev.1, 26 August 1969.

[9] TNA FCO 66/138, D.L. Benest, Disarmament Dept to Sir T. Brimelow and Mr N. Williams, 13 August 1969.

[10] TNA FCO 66/223, Convention for the Prohibition of Methods of Biological Warfare 1970, Attachments to W.N. Hillier-Fry to D.L. Benest, Disarmament Dept, 24 February 1970.

[11] TNA FCO 66/224, Convention for the Prohibition of Methods of Biological Warfare 1970, L.J. Middleton, Washington to D.K. Timms, Disarmament Dept, 21 May 1970.

reference to some doubts about any inclusion of research in the Convention, and this came from a non-governmental source. The distinguished US biologist Professor Joshua Lederberg argued in discussions with the UK delegation in Geneva that Article II (b) either banned all research or none. He stated that his students challenged his own work on the grounds that it could be picked up by the military and applied for offensive purposes. He would prefer a different wording that would ban the conduct of development work on BW weapons systems.[12] This does not seem to have had any impact on UK thinking. That said, given Lederberg's influence with the US Administration on BW matters, this might explain the later US willingness to accept Soviet language limited to development. But this is conjecture.

The Soviet Position: First Reactions

The Soviet Union and its allies tabled their own draft Convention on 30 March 1971 as we saw in the last chapter, thus changing their position, which until then had been emphatically opposed to the separation of BW from CW. This text made no mention of research even though the Russians had not raised it as a problem in their initial reactions to the UK draft on 22 July 1969.[13] On 2 April the US Ambassador and members of his delegation discussed the Soviet draft with the UK CCD Ambassador Henry Hainworth and his team.[14] We can see from this meeting that the Americans were by then thinking in terms of a joint US–Soviet draft Convention. Views were exchanged on various issues arising from the Soviet draft, but these did not refer to 'research' in its Article I. The Americans, some of whom believed that even legitimate research might be classified as falling within the prohibitions of Article II of the UK version, thought that the absence of a ban on research might not be a problem and perhaps even an advantage. The UK stuck to its guns. Separately the US met the Soviet delegation on 5 April 1971 to elicit some clarifications on the 30 March draft Convention. Jim Leonard the US Ambassador specifically asked his Soviet counterpart whether the word 'development' in the Soviet text included the concept of research, to which Roshchin replied 'yes, absolutely'.[15]

---

[12] TNA FCO 66/225, UK Draft BW Convention Discussion with Professor Lederberg, 5 August 1970.

[13] ENDC/PV.421, 22 July, pp. 23–7.

[14] TNA FCO 66/298, J.T. Masefield, UKDis Geneva to D.K. Timms, Disarmament Department, 7 April 1971.

[15] http://www.state.gov/r/pa/ho/frus/nixon/e2/83666.htm (accessed 15 August 2011), telegram from US Mission to SecState Washington 8901, 5 April 1971, CCD –

Following the discussion with the Americans, the UK delegation met the Soviet Ambassador on 7 April.[16] Roshchin pointed out that in Russian 'development' was a broad notion that covered 'research'. This explained why there was no reference to research in the Soviet draft. In preparing further written materials for use in subsequent negotiations with the Russians the FCO's Disarmament Department argued that, 'the two concepts of development and research are separate in English. While the UK delegation appreciated that this distinction might not exist in other languages, in the interest of precision, they suggest that any revised draft ... should include wording on the lines of Article II (b) of CCD/255/Rev.2.'[17] Neither the Soviet delegation members Masterkov nor Abarenkov or Ambassador Roshchin responded to this point when they met W.N. Hillier-Fry and J.T. Masefield and Henry Hainworth respectively on 13 and 14 May.[18]

Working Towards the Joint Drafts: Summer 1971

The US saw the Soviet March draft convention as a major concession.[19] From then on the task was to prepare a joint text that the Americans and Russians could table in the CCD. However, the UK alternative – a further revised version that had been tabled on 18 August 1970 with Article II (b) unaltered – remained on the table.[20] Throughout the summer the US and Soviet Union worked on a joint draft Convention and eventually tabled two separate but identical drafts on 5 August 1971.[21] Although the US had consulted the UK extensively, the new text made no mention of research. The activities to be prohibited in a new Article I included development, production, stockpiling, acquisition and retention, but not research. The language on the basic prohibitions had been taken from the Soviet draft Article I with the addition of the word 'retention'.

SOCIALIST BW DRAFT CONVENTION – MEETING WITH SOVDEL, National Archives, RG 59, Central Files 1970–73.

16 TNA FCO 66/298, Speaking Notes for HEs Call on Ambassador Roshchin, 4.00 P.M., Wednesday 7 April 1971.

17 TNA FCO 66/302, FCO telegram No. 132 to UKDis Geneva, 7 May 1971.

18 TNA FCO 66/299, W.N. Hillier-Fry, UKDis Geneva to D. Benest, Disarmament Department, FCO, 14 May and UKDis Geneva telegram No. 59 to FCO, 12 May 1971.

19 TNA FCO 66/298, L.J. Middleton, Washington to D.M. Summerhayes, Disarmament Department, FCO, CBW, 15 April 1971.

20 CCD/255/Rev.2.

21 CCD/337 and CCD/338.

Neither the US nor Soviet ambassadors explained the omission of 'research' when they introduced their texts to the CCD.

Henry Hainworth, however, noticed the omission. That the text did not address all the UK's concerns had been one of the main reasons why the UK had not co-sponsored the US draft. Hainworth told the CCD on 10 August that the UK still believed that a comprehensive Convention dealing with BW should ensure that the parties undertook never, in any circumstances, to conduct assist or permit research for purposes of the sort that are being prohibited under the draft Convention.[22] Such a provision was particularly important in a Convention which as drafted encouraged under its then Article IX the fullest possible exchange of information for the use of biological agents and toxins for peaceful purposes. He noted too that the UK was not alone as the United Arab Republic, Pakistan and the Netherlands had all expressed their clear support for such an inclusion.[23] On the evening of 19 August 1971 Soviet delegation members Tcheprov and Toulinov called on the UK mission for a two-hour discussion on a range of outstanding issues. On research Tcheprov explained that the Russian concept of development covered research, but UKDis Geneva gained the impression that he might be prepared to consider the possibility of an amendment.[24]

By 26 August, however, the Soviet position had hardened.[25] It would not now accept inclusion of 'research'. Tcherpov told the UK mission that he could offer the delegation no encouragement that it would prove possible to cover

---

22 CCD/PV 528.

23 UAR (CCD PV/516 paragraph 16) '... we wonder why the useful paragraph (b) of Article II of the UK draft has not been reflected in the socialist draft convention as well ...'; Pakistan (CCD PV/519 paragraph 26) '... we are of the view, however, that sub-paragraph (b) of article II of the UK draft, in accordance with which parties would undertake not to conduct, assist or permit research aimed at production of biological agents should be included in the convention'; Netherlands (CCD PV/525 paragraph 130) '... it would be of great value if the draft introduced by the Soviet Union and others were to reflect the content of sub-paragraph (b) of Article II of the UK draft, which seeks to rule out research directed to weapons development in this field. In our view, a ban on such research would constitute a logical complement to the ban on production.'

24 TNA FCO 66/299, UKDis Geneva telegram No. 102 to FCO, 20 August 1971. The Russian Без перевода means both to research and to develop, but this is not the word used in the Russian language version of the final text of the BTWC. This uses разрабатывать, which means work up, work out, design, exploit, develop, operate, work, devise, elaborate, schedule. An English definition of research as a verb is to carry out investigations into a subject, problem etc., whereas develop means to come or bring to a later or more advanced or expanded stage; grow or cause to grow gradually.

25 TNA FCO 66/302, UKDis Geneva telegram No. 108 to FCO, 26 August 1971.

the point in the next draft Convention. UK diplomats commented that clearly that Russians were not prepared to meet the UK on a point that the US also now appeared to oppose.[26] This implies that the US was not now in favour of an inclusion of research. On the following day (27 August 1971) the UK delegation informed the Western Group of the outcome of its discussions with the Russians. They reported the negative Soviet response to the UK's request that research should be covered. No other delegation commented.[27] And that was that: there appears to have been no further effort to secure inclusion of research. For the UK the remaining weeks were spent, as we saw in the last chapter, trying to ensure that BW use was covered by the Convention and that there would be some meaningful investigative provisions in cases of alleged use; coverage of research was a secondary issue lost in the final bargaining.

Conclusion

Although the UK sought a prohibition on conducting, assisting or permitting what we would now characterise as offensive BW research, there was no express provision in the final text of the Convention agreed in September 1971. It is not entirely clear from the archival record whether the UK believed that the Russian comments to the effect that development and research were interchangeable terms in Russian meant that the original UK objective was secured. There was certainly some support in the CCD for a clear statement on the inclusion of research right up until August 1971, but it is also apparent that following the Soviet draft Convention of 30 March the Russians opposed any mention of the word research. Was this simply a genuine linguistic problem, or did it betoken something more sinister, such as a wish to exclude offensive research from the reach of the prohibitions? In terms of timing, the answer is perhaps no – the secret Soviet decree establishing a new BW development programme – Enzyme – was not issued until 1973.[28] Under the same decree a civilian cover organisation – Biopreparat was created to mask the offensive

---

26 TNA FCO 66/302, UKDis Geneva telegram No. 107 to FCO, 26 August 1971, BW Draft Convention 5 August; see also UKDis Geneva telegram No. 108 to FCO 26 August 1971: 'they do not want the addition of research'.

27 TNA FCO 66/302, J.T. Masefield, UKDis Geneva to J.M. Hay, Disarmament Department, Western Group Meeting of 27 August: Part II, 27 August 1971.

28 Ken Alibek with Stephen Handelman, *BioHazard* (London, 1999), p. 41.

development programme.[29] This implies that concealing illegal activities was not a consideration, but we cannot be sure.

In English, as the UK pointed out at the time, there is a very clear distinction to be made between 'research' and 'development' – the two terms are not synonymous. There were also no unilateral or common interpretative statements on the meaning to be given 'to develop' as it appeared in the final Convention text for Article I. In the absence of any such statements, one might conclude that offensive research is not expressly prohibited by the Convention. However, the UK could argue, on the basis of statements made by members of the Soviet delegation during the negotiations to both the US and UK that research aimed at an offensive capability could fall under the heading of development. That said, this point has never been tested and certainly not reflected in any of the Convention's Review Conference Final Declarations held in 1980, 1986, 1991, 1996, 2002 and 2006. Words matter. Negotiating records of a specific text helps us understand the objectives as well as the rationale for various compromises and formulations. This is not a trivial point in this context as an exclusion of offensive research on BW from the Convention's prohibitions could mask significant activities, given the dual-use problem, and lay the basis for a break out from the Convention. It could also allow a state intent on such a programme to argue that its activities were permitted and thus not subject to any censure or investigation. The core problem is that the 'underlying science is unalterably dual-use',[30] a fact recognised by UK officials back in the summer of 1968 when they first contemplated a separate ban on biological weapons. Dual-use also presented itself as an issue in verification of a cut-off in the production of fissile material for weapons nuclear weapons. However, although much less of a problem compared to the biological sphere, as we will now discover in the next chapter, it still posed difficult negotiating challenges for the UK.

---

[29] Igor V. Domaradskij and Wendy Orent, *BioWarrior: Inside the Soviet/Russian Biological War Machine* (New York, 2003), p. 151; see also David E. Hoffman, *The Dead Hand: Reagan, Gorbachev and the Untold Story of the Cold War Arms Race* (London, 2011), p. 132.

[30] Hoffman, *The Dead Hand*, p. 132, citing Joshua Lederberg – US Nobel Prize-winning biologist.

Chapter 8
The UK Nuclear Weapons Programme, Fissile Material Cut-off and Safeguards 1956–1973

Introduction

Arms control could not hamper UK offensive BW programmes for the simple reason that there were no such programmes to hamper. Nuclear weapons were a different matter, so it is time to turn to the UK's nuclear weapons programme and disarmament. Apart from the comprehensive nuclear test ban treaty no other arms control measure threatened the development of the UK weapons programme more than the proposed cut-off in the production of fissile material (plutonium 239 and uranium 235) for weapons purposes. Both in the late 1950s and again in the 1960s, US initiatives in this area presented an unwelcome dilemma for the UK. Ministers and officials did not find it at all comfortable to stand out against disarmament measures, but at the same time they wanted to ensure that the UK weapons programme as planned could meet its targets without interference or delay. When it came to a choice, the demands of the weapons programme took precedence over arms control and disarmament. We can see that in early 1956 the UK position that the military programmes of the UK, US and USSR should be excluded from any measures taken to prevent the manufacture of nuclear weapons by other countries.[1] The chronic shortages and expense of fissile material in the UK were recurring problems throughout the period covered in this book, and in the early period this shortage meant that scarce fissile material was devoted primarily to the build up of the strategic deterrent.[2] It was a factor in shaping the contents of the nuclear weapons testing programme and was instrumental in the creation of the 1959 barter agreement with the US on exchanges of fissile material whereby UK plutonium was

---

1 TNA EG 1/115, Measures to Prevent the Manufacture of Nuclear Weapons by Countries not Already Possessing Them, UK position undated paper, but probably Jan/Feb 1956.

2 TNA DEFE 24/304, 1969–70 Healey to Prime Minister Tactical Nuclear Weapons, 16 June 1969.

exchanged for US highly enriched uranium (HEU). The prospects of a fissile material cut-off accelerated the growing dependence of the UK programme on US assistance. Officials spent a good deal of time worrying about a cut-off and how to accommodate such an initiative whilst meeting UK production requirements. As early as 1955 it was felt that the stock of fissile material in the UK was never likely to be sufficient to meet all requirements.[3] [4]

An essential part of the deliberations on a possible cut-off was the pioneering work done by the United Kingdom Atomic Energy Authority (UKAEA) on what inspection procedures the UK could accept which would give reasonable assurance that a cut-off on production of nuclear material for weapons purposes could be verified with confidence. This was a particularly difficult problem for the UK, as distinct from the US and USSR, in that the UKAEA, for economic reasons, used the same facilities for both peaceful and military purposes with the resultant intermingling of the two production campaigns. At the time, the UK was not prepared to build separate facilities for exclusively peaceful and exclusively military uses or to run separate campaigns; it would have been expensive and time consuming, and have made it even more difficult for the UK to meet the military production time scales it had set itself. This chapter provides a chronological overview of the cut-off debate and its potential impact on the UK weapons programme; it also looks in passing at the UK's pioneering work on safeguards procedures and approaches. As with the CTBT the UK embarked on original research programmes into the problems posed by arms control and disarmament verification; it was not enough to say that effective verification or control was required, finding out how to make it so was a recurring theme in UK arms control policy.

Cut-off Inspection: Early Thinking

The UKAEA had started thinking about the problem of controls on fissile material production as far back as 1956. In fact, it went beyond mere thinking and a practical exercise, known as 'Operation CIRCUS', was set up at Windscale

---

[3] TNA AIR 2/13745, Record of a Meeting to Discuss Megaton Warheads and Their Vehicles, 7 September 1955.

[4] TNA CAB 134/756, A.E. (O) (55) 127, Fissile Material for Civil Purposes. Note by the Atomic Energy Authority Table IV and I, 9 November 1955. This refers to the decision taken by A.E. (O) (55) 66 in TNA AB 134/755, which remains a closed file at the time of writing. In May 1955 the Atomic Energy (Official) Committee had agreed that the military programme for 1955/56 to 1960/61 would be for the production of 3,065 kg of HEU and 723.5 kg of plutonium.

in Cumbria to test the efficacy of control regimes to see how feasible it was to divert plutonium according to two different sets of rules.[5] Plan 1 involved an audit only, whereas Plan 2 involved resident inspectors from the Atomic Energy Research Establishment (AERE) at Harwell. The key objective was to establish practically what accuracy could be expected from two possible methods of control.[6] This exercise was scheduled to begin on 11 April 1956 and to last no more than six months, depending on the experience obtained as the exercise proceeded. It ended at the end of July.

CIRCUS was in essence conducted because the UKAEA had for some time been asked by Whitehall to give advice on the best way of controlling the production of nuclear material to prevent its diversion to military uses. This was needed both to assist in international discussions on disarmament, but more particularly in connection with discussions on the control mechanisms that might be instituted by the proposed International Atomic Energy Agency (IAEA) following Eisenhower's Atoms for Peace programme launched in December 1953.[7] CIRCUS confirmed that control of plutonium production at Windscale was feasible, but it would require an inspection staff numbering about 20. Resident inspection would not interfere with the general flow of work and the production of the chemical reprocessing plant could probably be assessed to an accuracy of better than five per cent. Chemical plant inspection would have to be allowed to take samples. Sampling points and possibly the whole plant would need to be inspected during construction, to verify that 'faking' of samples was impossible. If these latter two points were satisfied, then an inspection during plant operation would not involve any further great loss of commercial secrecy. Inspectors would need to be chosen for their tact, patience and linguistic ability.[8] As for auditing, CIRCUS concluded that remote auditing alone had no value, but coupled with occasional site inspection and sampling at more than 24 hours notice, it could probably limit diversion to 25 per cent of the amount of material present and or moving through the plant. Short notice would cut this to ten per cent.

---

[5] TNA AB 38/515, J.C.C. Stewart to W. Strath, UKAEA, 7 February 1956. Windscale was the site of the UK's first two plutonium production reactors and spent fuel reprocessing plant.

[6] TNA AB 38/515, J.C.C. Stewart to Mr W.L. Owen and Mr K.B. Rose, 12 March 1956.

[7] TNA AB 38/515, Minutes of a Meeting Held at St Giles Court to Discuss Operation Circus, 3 August 1957.

[8] TNA AB 38/515, International Control of Atomic Energy Installations: Report of an Exercise Note by Dr B.F.J. Schonland, 8 November 1956.

After CIRCUS the UKAEA began work in February 1957 on preparing UK views on the detailed structure of an inspection and control system for the control of fissile material.[9] Sir William Penney (Director of the Atomic Weapons Research Establishment [AWRE] at Aldermaston) was the chair of what became known as the Working Party on Inspection and Control. This body agreed that it would begin with the UK programme and deal with the then present plant and technology and assume that there would be no secret or hidden plant. Members also agreed that any atomic energy information required by the inspection control team would be made available, although in fact there might be unwillingness to disclose such information to international observers.[10] A key assumption for the Working Party was that the inspection system would have to account for all fissile material from the moment the system became fully operational. In particular, it would have to be able to detect deliberate small 'losses' of high-grade material, the purpose of which would be to supply a clandestine nuclear weapons programme. The maximum possible scale of such losses would need to be estimated and inspection systems would have to be devised to ensure that material losses were kept to a minimum without making an inspection system unreasonably large and frustrating to the facilities under inspection. As noted above, the CIRCUS cut-off inspection exercise had taken place at Windscale and it was clear to the UKAEA that this work was much more solid than any that the US had undertaken up until that time as reported under the Inventory Control Study – the Vitro Report.[11] MOD views on a cut-off were clear enough: there could be no question of supporting any agreement before a satisfactory control system was in place.[12]

Late in the summer of 1959 the UKAEA was still completing its practical study of the problems associated with a cut-off.[13] Although much of the work was done, the report was never completed because of other pressing Authority commitments. The UKAEA Board concluded in early July 1959 that a control

[9] TNA AB 16/2172, Sir William Penney's Working Party on Fissile Production Control Scheme, W.G. Penney to Sir John Cockcroft, Sir Christopher Hinton and Mr W. Strath, Disarmament – Fissile Production Control Scheme, 13 February 1957.

[10] TNA AB 16/2172, Sir William Penney's Working Party on Fissile Production Control Scheme Working Party on Inspection and Control, W.G. Penney Director's Office, AWRE Aldermaston, 1 March 1957.

[11] TNA AB 16/2172, Sir William Penney's Working Party on Fissile Production Control Scheme, Secretary to Mr Walker, 14 March 1957.

[12] TNA FO 371/129836, UN1192/828 H.C. Budden, MOD to Miss G.G. Brown, Foreign Office, 2 August 1957.

[13] TNA FO 371/140520, IAD 29/1G/G Record of a Conversation between S/State and Mr Herter on 28 August 1957.

system along the lines proposed would not be tight enough to be effective and it was unlikely that any further study by the Authority would produce any better system, although Penney's report would need to be written up in final form.[14] Work was put in hand to complete the report by the end of September. This work concluded that it was not thought possible to inspect and control plutonium production in any country within better than five per cent of material present and U-235 within better than two per cent, and a country which set out to evade the control would know that cheating within these limits was very unlikely to be detected.[15] This would mean enough material could be squirreled away for roughly six kiloton bombs a year or more depending on weapons design expertise. But as further work was done on this problem, taking into account technical advice from the Authority's Production Group at Risley – Ned Franklin and Tom Tuohy – the margin of plutonium accounting was reduced from five per cent to two per cent and systematic observation of production plants over a reasonable period of time would reduce the margin of undetectable error.[16] In fact, the UKAEA now thought that in the longer term, if an historical series of measurements could be obtained for particular plants, the limits for both plutonium and U-235 might be reduced to something like one per cent.[17]After some further editing and checking, the final version of the report was submitted to Penney on 7 October 1959.[18]

Whitehall wanted Penney to take advantage of his trip to Washington in October to discuss nuclear weapons related issues to include exchanges on cut-off issues too. This was essential because both the Foreign Office and MOD had agreed that in preparing the UK position for the forthcoming Ten Power Disarmament working group meeting in Geneva in 1960, technical discussions on a cut-off should form part of the West's proposals.[19] The Foreign Secretary Selwyn Lloyd and his US counterpart Christian Herter agreed on 30 July 1959 that there should be joint technical discussions in Washington on the

---

14 TNA AB 16/3301, Extracts A.E.X. (59) 23rd Meeting on 9 July 1959.

15 TNA AB 16/3301, The Technical Feasibility of Inspection and Control of the Future Production of Fissile Material throughout the World Summary of Conclusions, 22 July 1959.

16 TNA AB 16/3301, A.C.W.V. Clarke, Economics and Programming Branch, UKAEA to Sir Alan Hitchman, 10 September 1959.

17 TNA AB 16/3301, E.A. Hitchman to Sir William Penney, Cut-Off, 18 September 1959; UKAEA International Control of Fissile Material Production, Report of a Working party of the UKAEA, October 1959, p. 11, paragraph 15.

18 TNA AB 16/3301, L.B. Kidd, Economics and Programming Branch to Mrs M. Hardaker, Sir William Penney's Office, AWRE, 7 October 1959.

19 TNA FO 371/140520, IAD 29/1G/G H. Hainworth to Mr O'Neill, Anglo-American Technical Discussions on the Cut-off, 25 August 1959.

subject of control of a cut-off agreement. Lloyd had mentioned that the UK had been considering disarmament proposals, including a cut-off as Harold Macmillan in Moscow had raised this with Khrushchev in February and the Soviet premier had shown some interest.[20] The technical meeting was eventually set for 29 to 30 October. This appears to have agreed that further technical work would be undertaken and that the UK's contribution would focus on clandestine centrifuge plants and the detection of clandestine plants by sensitive radiochemical analysis. The talks would resume in mid March 1960.[21] As for Penney, he thought that it might suit the UK to have a cut-off in four to five years time (1963–64), provided it was universal and reliable. It was quite another matter to contemplate a cut-off position comparable to the one applying in the test ban context, in other words no more production from the day that the talks started.[22]

Verification Working Paper for the Eighteen Nation Disarmament Committee (ENDC): 1962

The work undertaken in the late 1950s provided a basis for a paper submitted by the UK to the ENDC in 1962. This was designed to demonstrate that, on the basis of UK experience, there could be a margin of error between ten and 20 per cent in the assessment of past production even under a stringent international control system, the comparable figure for current production being from one to two per cent. This, however, was a purely technical assessment: it was also the view that in the early stages of disarmament cheating by a nuclear power would have to be very substantial to have any military importance.[23] This paper had originally started as an internal UKAEA exercise that had focussed on the Authority's industrial plants and looked at diversion scenarios, particularly at Capenhurst.[24] There was substantial input from staff from the Authority's Industrial Group – this was the part of the UKAEA responsible for the production facilities at Windscale, Capenhurst and Springfields. It was seen by both the UKAEA and MOD as an important paper and was not be used for a

20 TNA AB 16/3301, P. Dean, Foreign Office to Sir Richard Powell, MOD, 12 August 1959; D.E.H. Peirson, UKAEA to Sir Leonard Owen, UKAEA Production Group.

21 TNA AB 16/3301, Extract Form Minutes of A.E.X (60) 10th Meeting US–UK Discussions on Cut-off, 11 March 1960.

22 TNA AB 16/3301, W.G. Penney to The Lord Plowden, 4 July 1959.

23 TNA CAB 148/50, OPD (O) (D) (64) 5, A Guide to the Disarmament Negotiations Note by the Secretaries. Document Prepared by the Foreign Office,17 December 1964.

24 TNA AB 16/4778, W.G. Penney to Sir Solly Zuckerman, MOD, 13 July 1962.

mere 'pot boiling exercise'.[25] What these departments meant by this is unclear. MOD requested that an additional point should be made in the paper to bring out even more clearly the difficulties of verifying the quantity of fissile material held in weapons stocks. Three particular problems were identified. One was that fissile material was in an awkward shape and was not available for assay – unless the weapons were to be destroyed. Second, there was the inevitable loss of military security, which might well be a stumbling block, but this would depend on how a cut-off treaty was framed. Thirdly, there was the problem of non-dissemination of nuclear information – though this could be avoided by restricting inspection to the 'other side' namely those outwith any location that would provide a direct view of classified shapes of weapon components. These difficulties only added to the uncertainties in the assessment of past production and therefore strengthened the report's main conclusion, which was clearly that some kind of retained deterrent would be advisable.[26] Penney agreed and made some changes to reflect these points.[27] The paper was tabled in Geneva on 31 August as a contribution to the ENDC's discussion on 'Measures in the Field of Nuclear Disarmament, together with Appropriate Measures of Control'.[28] Foreign Office officials thought that this paper was intended to engender a realistic appreciation of the feasibility of controlling a cut-off and establishing the size of existing stockpiles; it was important to persuade the ENDC to adopt a more business-like procedure by devoting some of its time to technical studies.[29] The UK as ever saw the need to concentrate on the practical realities of disarmament rather than the generalities.

Cut-off in Production of Fissile Material for Military Purposes: Early Steps 1956–62

In July 1956 the MOD noted that since the UK had begun the development of nuclear weapons, service requirements were always far bigger than the numbers

25 TNA AB 16/4778, B.T. Price, MOD to A.D. Pemberton-Pigott, Foreign Office, 27 July 1962.

26 TNA AB 16/4778, B.T. Price, MOD to Sir William Penney, July 1962.

27 TNA AB 16/4778, The Technical Possibility of International Control of Fissile Material Production: Paper by Sir W. Penney for Conference on Disarmament 1962, W.G. Penney to A.D.F. Pemberton-Pigott, Foreign Office, 21 July 1962.

28 ENDC/60, The Technical Possibility of International Control of Fissile Material Production, 31August 1962.

29 TNA FO 371/171133, IAD 1036/22 A.D.F. Pemberton-Pigott to P. Wilkinson, Washington, 18 December 1963.

that could be produced quickly. The size of the military programme was restricted in fact by economics and the technical and scientific possibilities. The relatively early date by which the use of fissile material for the production of weapons would cease under the Anglo-French disarmament plan presented in the United Nations Disarmament Commission (UNDC) sub-committee meeting in London on 19 March 1956 was acceptable only because that plan envisaged political settlements alongside carefully devised stages of disarmament.[30] MOD was therefore prepared to rest on a rather low stock of nuclear weapons if the world situation justified moving to Stage Three of that plan. The US 3 April 1956 disarmament plan was not acceptable because it provided for the cessation of nuclear weapons production at a stage that would not be very much later than the other steps of the plan.[31] When preparing a UK one-stage disarmament plan for the Disarmament sub-committee, the Foreign Office only discovered late in the day from AWRE and the MOD that the UK needed more time than had previously been realised to build up its own stockpiles of weapons.[32]

In 1956 US Secretary of State John Foster Dulles proposed to the UNDC that all future production of fissionable materials should be for non-weapon purposes. Any commitment to a cut-off in the spring of 1958 would be subject to two UK conditions; it could only enter into force if a proper control system had been installed and that the Americans would provide the UK with the material needed to complete its weapons programme.[33] Anxieties were particularly acute at this time given the shortage of fissile material. In fact one of the key objectives in the nuclear testing programme was to develop warhead designs that made more economical use of fissile material. The Windscale piles had not produced as much fissile material as intended and were now shut down following the fire in one of the piles in October 1957.[34] The first of the four new plutonium production reactors at Chapelcross did not come on stream until February

---

[30] Anglo-French Working Paper Submitted to the Disarmament Subcommittee: Proposed Synthesis, March 19, 1956 in *Documents on Disarmament 1945–1959 Volume I 1945–1956*, Department of State, Historical Office Bureau of Public Affairs, August 1960, pp. 595–8.

[31] TNA FO 371/123702, UN 1192/512, Note by Ministry of Defence, 3 July 1956.

[32] TNA FO 371/123702, UN 1192/512, New York telegram 514 to Foreign Office, 3 July 1956.

[33] TNA FO 371/132686, IAD 122/26, Geneva Negotiations on the Suspension of Nuclear Tests Disarmament Policy Memorandum, Foreign Office, 15 September 1958.

[34] Margaret Gowing, *Independence and Deterrence: Britain and Atomic Energy 1945–1952*, Vol. 2 (London, 1974), p. 389. After 1952 the piles were producing roughly 75 kg of plutonium per year. TNA AB 48/335, Broadhurst to J.L. Gillams, 20 October 1964. By 1962 the eight production reactors at Calder Hall and Chapelcross were producing 500 kg of plutonium a year. TNA AB 48/195, Draft Paper for the NRDC Fissile Material Supplies and

1959, with all reactors up and running by May of that year. However, since it took apparently about 12 to 18 months to produce a billet of military grade plutonium, material from Chapelcross would not have become available until 1960.[35] Fortunately for the UK, the Russians were adamantly opposed to a cut-off agreement separate from other disarmament measures at that time. Like the British they too probably had insufficient material for their planned weapons programme.[36]

In January 1957 the Foreign Office Minister of State (Commander Noble) met Harold Stassen, then Special Assistant to President Eisenhower for Disarmament Policy to discuss a range of disarmament issues; one of these was the cut-off. It struck Ministers and officials that this US idea entailed risks for the British weapons programme.[37] At this time the UK's first thermonuclear tests were still five to six months away and it was unclear whether the new designs could result in more economical use of fissile material. In these talks Stassen defended the US proposal for an early cut-off date and argued that the UK could go on building up its stockpiles after that date by using fissile material which might be provided by the US – assuming of course that Congress was willing to change existing legislation.[38] After Ministers considered the issue, Noble told Stassen on 7 February 1957 that HMG accepted the US proposal on the understanding that the implications for the UK weapons programme were kept in the forefront of US thinking; and that, if the Russians accepted, the US would enter into negotiations with the UK with a view to ensuring that adoption of the US initiative would not prejudice the acquisition of the fissile material necessary for the UK's nuclear weapon requirements. Stassen agreed. This understanding was codified in an exchange of letters in February between Noble and Stassen to the effect that if the Russians looked interested in a cut-off, 'negotiations' should start with the US as to how the UK's own position was to be guaranteed. Stassen asked whether the UK would like these to start now

Requirements Note by the UKAEA ND (62) 14. For the Windscale fire see Lorna Arnold, *Windscale 1957: Anatomy of a Nuclear Accident* (London, 1992; 3rd edition 2007).

35 TNA CAB 134/2241, ND (64) 7, Cut-off and Fissile Material Supply for the UK Weapons Programme, Note by the Atomic Energy Authority, paragraph 15, 12 June 1964.

36 Glenn T. Seaborg with Benjamin S. Loeb, *Stemming the Tide: Arms Control in the Johnson Years* (Lexington, MA, 1987), p. 398.

37 TNA FO 371/171133, IAD 1036/21/G, the Stassen-Noble Agreement in Relation to Problems of Cut-off Production of Fissile Material, S.J. Barrett to Mr Shepherd, The Stassen-Noble Agreement, 20 November 1963.

38 TNA FO 371/129818, UN 1192/G/423, A.R.K. Mackenzie to Mr Hainworth, 9 May 1957.

and the MOD went off to consider.[39] This also included a clear agreement on timings:

> [I]t is proposed that all future production of fissionable materials shall be used exclusively for national and international non-weapons purposes under international supervision, beginning one month after the establishment of an effective inspection system to verify the commitment. A group of technicians representing the five powers ... shall meet on September 1st, 1957 to prepare an inspection system for these purposes; and the installation of agreed inspection system shall begin on March 1st, 1958 or as soon as possible thereafter upon ratification of the necessary agreements.[40]

The next significant cut-off exchange between the UK and US occurred in May 1957, when Stassen again assured Noble that the US would not submit any final language on a cut-off to the Russians until the UK was ready to accept it. (However, at the same time some US press statements also seemed to suggest a US policy at variance with the assurances given in January, which led to quiet remonstrations to Stassen.[41]) Despite this, Stassen went ahead and made disarmament proposals to the Russians without any prior consultation with the British. Macmillan and his colleagues were not best pleased:

> ... on the question of fissile material we were much more vulnerable. To our amazement the Stassen memorandum proposed an early end to the production of such material by any of the great powers. This plan came to be known as the cut-off. For us, this would be fatal. It would involve the abandonment of our nuclear ambitions ... Is this America's reply to our becoming a nuclear power – to sell us down the river before we have a stockpile sufficient for our needs. Some of my colleagues suspect this.[42]

London feared that the cut-off date proposed, if accepted, would have serious effects on the UK military programme unless satisfactory arrangements were

---

39 TNA FO 371/129820, UN 1192/473, Allan Noble to Secretary of State, Nuclear Disarmament, 31 May 1957.

40 TNA FO 371/129825, UN1192/560/G, D.S.L. Dodson, Foreign Office to J.M. Gibbon, MOD, 13 June 1957; see also OD (57) 10, Stassen Exchanges on Fissile Material – Missing Element from Noble-Stassen Agreement, penultimate sentence of 2nd paragraph, 8 March 1957.

41 TNA FO 371/129818, UN 1192/G/423, A.R.K. Mackenzie to Mr Hainworth, 9 May 1957; Hainworth to Dean, 13 May 1957; Noble to Hainworth, 15 May 1957.

42 Harold Macmillan, *Riding the Storm 1956–1959* (London, 1971), pp. 300–301.

made in advance for the supply of fissile material. A point picked up by Lord Cherwell who wrote to Macmillan to point out that compared with the holdings of the US and USSR, British stocks of fissile material were so trivial that the UK would be left practically defenceless.[43] However, Stassen had not cleared his lines in Washington either. Fortunately for the UK a subsequent conversation with Stassen indicated that the picture was not as bad as first feared and plans for the Embassy in Washington to probe further into US thinking were withdrawn.[44] Macmillan himself had spoken directly with Eisenhower and reminded him that the UK had already received from the American government what amounted to an undertaking that, whatever agreement might emerge, 'the development by the UK of nuclear weapon resources adequate to her needs should not be prejudiced'.[45] A subsequent Aide Memoire provided by the State Department did make it clear that the US would indeed be willing to discuss this issue of supply as soon as the Russians indicated that they could work with the US ideas. As the Russians did not do so, the UK was off the hook. As Macmillan noted, the Russians ignored the cut-off which 'saves us from the fatal cut-off'.[46]

Sir Richard Powell (Permanent Under Secretary, MOD) told the Cabinet Office's Working Group on Disarmament's first meeting in July 1957 that recent exchange of messages between the Prime Minister and the US Secretary of State had indicated that the US, while maintaining a benevolent attitude, was not prepared at that time to discuss UK detailed requirements for weapons or fissile material for weapons. There was thus a risk that disarmament negotiations might move to a stage when an agreement could be achieved and only then would the US be unwilling to meet UK requirements.[47] Powell suggested that, even if discussion about these requirements could not take place immediately, it would still be desirable to prepare, and to let the US have, a statement of what quantities and types of material the UK would need. At the same time he thought that the UK might also indicate what design information was needed in order to cover UK requirements for fissile material. Officials thought that it would be essential to give the US a clear sense of what the UK needed if it were to support US proposals for a cut-off. MOD and the UKAEA were tasked to go away and prepare a draft statement of what UK requirements would be according to

---

[43] TNA PREM 11/2168, Cherwell, Christ Church, Oxford to Macmillan, 30 June 1957.

[44] TNA FO 371/129820, UN 1192/470/G Foreign Office telegram 2440 to Washington, 30 May 1957.

[45] Macmillan, *Riding the Storm*, p. 302.

[46] Macmillan, *Riding the Storm*, p. 305.

[47] TNA CAB 130/130, GEN.604/1st Meeting, Item 3: Fissile Material for Weapons, 5 July 1957.

the date at which a cut-off might come into operation; it would also indicate what reductions in supplies could be made if warhead design information were provided. There is thus, at this time, a clear link between testing objectives and the UK's ability to support a cut-off in so far as more efficient designs would and did significantly reduce the need for fissile material.[48] This was one of the benefits from the Agreement for Co-operation on the Uses of Atomic Energy for Mutual Defence Purposes – henceforth the 1958 Mutual Defence Agreement.

The spring 1958 US proposal on a cut-off was unacceptable to the UK as it would stop production for civil as well as military purposes given the careless way in which it was drafted. It was therefore important that the UK should try to talk the US out of this before it committed itself to this position publicly.[49] The first task for the UK was to work out clearly what sort of cut-off it might be able to accept and then to specify the conditions that would govern UK acceptance. The Cabinet Office Working Group on Disarmament asked for a draft paper for Ministers outlining the present weapons programme and showing what effects on the programme would be if a cut-off came into force on different dates. This would indicate the UK deficiencies unless met by US supplies of material or complete weapons. While the UK might be able to acquire material for megaton weapons, the US would not be able to meet British requirements for defensive and tactical weapons. Unless the US provided these under the doctrine of interdependence, then the UK would have to do without. A paper was duly prepared for submission to the Prime Minister, Foreign Secretary and Defence Secretary, which the Working Group planned to submit towards the end of April 1958.[50] However, the 1958 and 1959 Mutual Defence Agreements set up the barter arrangements in which the UK exchanged plutonium from

---

48 We can obtain some idea of the advances in design and the reduction in quantities required from the following comment from TNA CAB 130/130, GEN.604/14 24 July 1959, The Technical Feasibility of Inspection and Control of the Future Production of Fissile Material Throughout the World: Summary of Conclusions, Note by the Atomic Energy Authority, 'The limitations of a control system need to be related to the effect of a given amount of material being diverted without detection. For example, the diversion of 50 kg of plutonium or 200 kg of HEU in any country that had not tested weapons would supply enough material to make a dozen kiloton bombs. In a country with test experience, about twice as many kiloton bombs could be made from the same material. Alternately, countries that had tested advanced thermonuclear bombs could make several megaton bombs from these quantities.' This suggests that the UK had designs using as little as 2.17 kg of plutonium per weapon.

49 TNA CAB 130/130, GEN.604/8th Meeting, Item 2: Suspension of Production of Fissile Material, 2 April 1958.

50 TNA CAB 130/130, GEN.604/9th Meeting, Item 3: Suspension of Production of Fissile Material, 24 April 1958.

its civil power stations for US U-235 and tritium and as such removed a major problem for the UK in fissile material supply.[51] In addition, after 1958 the issue of a fissile-material cut-off was relegated to a secondary position in disarmament discussions as UK, US and Soviet attention focused on negotiating a test ban treaty.[52] Concerns could thus subside for the moment, but would resurface again in 1962.[53] In terms of safeguards, the UK could not then agree to any measures that limited its own freedom to acquire supplies for the nuclear programme without having to give guarantees against military use. The UK was dependent on the import on uranium supplies, heavy water and to some extent graphite. Nor could there be any safeguards applied to the UK's civil reactors since some plutonium from these might be needed for military purposes.[54]

Previous forecasts in 1958 of fissile material requirements (these had changed between December 1957 and March 1958[55]) were based on the broad assumption that the UK should build up a stock of 100 megaton weapons and 1500 kiloton weapons by 1965.[56] Instead, with the then currently available resources and without allowing for any new requirements, the UK stock by

---

51 At that time the UK's military requirements for HEU were seen as lying between a minimum of five tonnes and maximum of 10 tonnes. TNA AB 48/197, St Clair C Hood, Economics and Programming Branch to Mr Cole, 8 November 1962. The exchange rate in the 1959 agreement was one tonne of plutonium for one tonne of HEU. In the event the UK agreed in 1959 to take 7.5 tonnes of US U-235, Defence Liabilities for U-235 Barter: Authority Case for Defence Taking Full Liability, 7 December 1962. Military demands were, however, subsequently reduced following the 1962 warhead review. The UK anticipated in the Special Materials for Defence Tables of 1959 that 13.5–15 tonnes of U-235 would be required to 1970, but by November 1960 the requirement to 1970 had gone up to 21 tonnes. These figures were based on estimates from the defence departments.

52 PPNN NPT Briefing Book Volume 1 (Southampton, 2004), p. 2. See also John R. Walker, *Britain and the Test Ban 1954–1973* (Farnham, 2010).

53 TNA CAB 134/1330, Cabinet Official Committee on Atomic Energy A. (O) (58) 11 Safeguards Against Diversion of Fissile Material from Civil to Military Purposes, 5 June 1958.

54 TNA CAB 134/1331, Cabinet Official Committee on Atomic Energy A. (O) (59) 1st meeting, 18 February 1959.

55 TNA CAB 130/130, Working Group on Disarmament GEN.604/3 Draft Submission to Ministers, Nuclear Disarmament, Foreign Office, 31 March 1958; AWRE also wanted to use about 400 grams of plutonium for the VIXEN tests in 1959 and anticipated needing another 1 kg in 1960. TNA AVIA 65/1111, N. Levin, AWRE to E.S. Jackson, DGAW, MOS, 19 May 1959. The MOD agreed that these tests were an essential part of the military programme and agreed to allocate 1.4 kg from defence stocks, F. Brundrett, MOD to E.S. Jackson, MOS, 26 May 1959.

56 See Richard Moore, *Nuclear Illusion, Nuclear Reality: Britain, the United States and Nuclear Weapons, 1958–64* (London, 2010) for detailed discussion of this period.

1965 would consist of 100 megaton weapons and about 200 defensive or tactical weapons. By 1970 the stock would be 100 megaton weapons and about 500 defensive or tactical weapons.[57] However, the provisional conclusions reached by the Cabinet Defence Committee in April 1962, together with the proposals made by the Minister of Defence for reductions in the programme for nuclear warheads had an important effect throughout the atomic energy field.[58] Indeed the UKAEA's estimates of the quantity of plutonium needed for 'weapons, etc.' in 1962 for the years out to 1971 were as follows: 1962–63 – 300 kg; 1963–64 – 150 kg; 1964–65 – 125 kg; 1965–66 – 75 kg; 1966–67 – 200 kg; 1967–68 – 150 kg; 1968–69 – 150 kg; 1969–70 – 125 kg; 1970–71 – 75 kg.[59] It is not clear whether this estimate was made before the Cabinet Defence Committee decided in 1962 to make major reductions in the weapons programme.

However, following the decision to build only four submarines, it appears that about one tonne of U-235 that would otherwise be needed for the 16 missile warheads for a fifth Polaris submarine – about 20.83 kg per weapon – would no longer be required for that purpose.[60] Military grade U-235 in the UK was derived partly from Capenhurst and partly from the barter agreement; the latter material could only be used for weapons purposes – R&D, production or use. 7,500 kg had already been provided under the barter agreement. This was sufficient for perhaps some 360 weapons.[61] SACEUR had also suggested that the UK might equip the new TSR 2 aircraft for the RAF with an additional 80

[57] TNA DEFE 7/2379, Safeguarding Plutonium Supplies, 7 March 1958. There was also a MOD review of weapon programme requirements underway at this time. This noted that if supplies for fissile material for making weapons were restricted to the output of the military reactors already built or under construction and of the diffusion plant at Capenhurst (30 per cent of which was for civil uses) then present plans for the production of nuclear weapons would not be completed until the mid 1970s at the earliest. TNA DEFE 7/2379, R.G. Powell to Minister Safeguarding Supplies of Plutonium, 14 March 1958.

[58] TNA AB 16/4189, Future of Capenhurst Effect on Atomic Energy Authority of the Proposed Revised Nuclear Warhead Programme, Memorandum by the Atomic Energy Authority, 29 May 1962 and TNA CAB 131/27, D (62) 7th Meeting, Item I, 18 April 1962. This later document is retained under Section 3 (4) of the Public Records Act 1967.

[59] TNA AB 48/195, Annex A to AEA (62) 61 Table 1, Plutonium Forecast – Supply and Demand (all isotopes rounded to nearest 25 kg).

[60] TNA CAB 134/2241, Nuclear Requirements for Defence Committee N.D. (64) 1, 24 February 1964, Requirements for Military Grade U-235 Memorandum by MOD and AEA.

[61] TNA CAB 21/6038, N.D. (64) 3 Military Grade U-235 Supplies, Note by the Secretaries Letter from Minister of Aviation to Minister of Defence; Note by Officials, 28 February, 1964.

WE177 tactical nuclear weapons.[62] However, it would appear that this would require an additional quantity of some 175 kg of plutonium and some 3,000 kg of U-235.[63] Therefore, the key question for the British now was what assurances should be sought from the US on guarantees for UK supplies.[64] The purchase option in the original 1959 agreement for U-235 could be reactivated. Officials made clear to the Americans that this was only being done from the possibility that a cut-off might take effect in the next few years and that, if implemented, HMG would want to be in a position to support it. The UK was paying the price for a policy of minimal production to be satisfied in the most economical way. It was now left in a much less favourable position to accept an early cut-off than was the case for either the US or USSR. London's view was that both superpowers had no shortage of such material.

Cut-off 1962–65: Continuing British Reservations

In April 1962 the US tabled a document in Geneva entitled 'Outline of Basic Provisions of a Treaty on General and Complete Disarmament (GCD) in a Peaceful World' in the new Eighteen Nation Disarmament Committee (ENDC), which had replaced the ill-fated Ten Power Committee.[65] The ENDC became the focus for multilateral disarmament negotiations for more or less the rest of the decade. The US plan included a halt in the production of fissile material for use in weapons; a transfer of U-235 from weapons to civil purposes; and a prohibition on the transfer of fissile materials for any military purpose. As such it resurrected the cut-off issue, which had been dormant since 1958. In May, however, the US suggested that cessation of production for military purposes could be introduced ahead of Stage One of the General and Complete Disarmament (GCD) plan. The UK, however, could only support a cut-off in

---

62 The WE177 A was a tactical nuclear weapon that replaced Red Beard. The high-yield variant, the WE177B (reportedly 500 kilotons), was, following a Cabinet Defence Committee decision, allocated to the Vulcan B2 bombers to act as an interim force until Polaris was fully deployed.

63 These figures suggest that roughly 2.1 kg of plutonium were used in the primary and 37.5 kg of U-235 in the secondary (or composite primary cores).

64 TNA FO 371/176370, IAD 1036/3/G, Memorandum [by the MOD] to the Defence and Overseas Policy (Official) Committee; A.D.F. Pemberton-Pigott to Mr A.D. Wilson, 10 February 1964.

65 The ENDC had the following members: from the West the UK, US, Canada, France and Italy; from the Warsaw Pact, the USSR, Poland, Bulgaria, Romania and Czechoslovakia and eight from the non-aligned: Brazil, Sweden, Mexico, Nigeria, India, Burma, Ethiopia and the United Arab Republic.

the context of Stage One of a GCD plan.[66] Fortunately, for the UK the Russians consistently refused to discuss the proposal on the grounds that it involved control without disarmament.[67] MOD officials told the Foreign Office in March 1963 that they could not support a cut-off as a pre Stage One measure without full Ministerial consideration. Penney made the same point to the Defence and Overseas Policy (Official) Committee.[68] The UKAEA would be able to make arrangements so that no civil plutonium would be used in the UK's military programme, but the cost of such modifications would be in the order of £2–3 million. In addition, they would also require re-affirmation from Washington that the US would continue to supply U-235 under the existing barter arrangements, despite a cut-off agreement.[69] Cut-off could only be considered by the UK in the context of general and complete disarmament; this was the view in mid 1963.[70] But Foreign Office officials worried that if the Russians and Americans agreed on a cut-off, the political pressures on HMG to follow suit would in all probability be irresistible. There was therefore a compelling argument for the UK to put itself in a position of being able to support the US proposals and for securing satisfactory guarantees on supply for the UK programme.[71] Echoes of 1957 can be clearly heard here. However, the Americans were warned in both Geneva and in Washington that a cut-off in advance of Stage One would cause difficulties for HMG and that the UK could thus not actively support the US position as it would cause the UK problems given the shortages of fissile material for the planned weapons programme.[72] One of the sub-plots here was that the US had suggested that Britain might provide warheads for the proposed Multilateral Force (MLF) involving surface ships armed with Polaris missiles and mixed-manned with personnel drawn from those NATO states that wanted to participate, which meant that UK plans for fissile material acquisition would

66 TNA CAB 148/50, OPD (O) (D) (64) 3, Arms Control and Disarmament Policy, Ministry of Defence Review Note by the Ministry of Defence, 24 November 1964.

67 TNA PREM 11/5172, Tam Bridges, Foreign Office to J.O. Wright, 10 Downing Street, 26 June 1964.

68 TNA CAB 148/4, D.O.(O)(64) 5th Meeting, Item 1, 10 March 1964.

69 TNA FO 371/171132, IAD 1036/10/G, Lord Norwich to Mr Pemberton-Pigott, 21 May 1963.

70 TNA FO 371/171132, IAD 1036/8 J.H. Lambert, UKDis Geneva to H.B. Shepherd, AEDD, Foreign Office, 6 May 1963 and J.K. Wright, MOD to H. Shepherd, Foreign Office, Nuclear Disarmament Item 5 (d), 30 April 1963.

71 TNA FO 371/171133, IAD 1036/24/G Cut-off of Fissile Material as pre-Stage I Measure, A.D.F. Pemberton-Pigott to Mr Wilson, 19 December 1963.

72 TNA FO 371/171132, IAD 1036/9/G UKDis Geneva telegram No. 12 Saving to Foreign Office, 14 May 1963; TNA PREM 11/5172, Tam Bridges, Foreign Office to J.O. Wright, 10 Downing Street, 26 June 1964.

need to be re-examined and this would be difficult to accept in the absence of any new assurances of supply from the US.[73] In August 1963 the US re-tabled the four major measures in Section C of Stage One of its outline GCD plan, but this omitted that part of the section dealing with the prohibition of the transfer of fissile material between states for other than peaceful purposes.[74] Although the US position was not without its ambiguities, which the US treaty language tabled in August 1963 did nothing to resolve, London thought it possible that the US did not rule out the continuation of the 1958 barter agreements should a cut-off and transfer requirements be implemented in advance of a Stage One GCD.[75] However, if as a result of any agreement on pre-Stage One GCD measures, a prohibition on transfer between states were put into effect in the next few years (1963–66), then serious problems would arise for the UK weapons programme.

The UK position on U-235 and plutonium supplies in relation to a cut-off was made clear in the Prime Minister's message to President Johnson on 17 March 1964.[76] Home suggested that British and American officials should have a closer look at the problems. Penney would lead for the UK. But this meeting's outcome, so far as the Americans were concerned, was that they only wanted to discuss the reductions in planned production that were to be the subject of simultaneous announcements that the three governments – namely the US, USSR and UK – might make. Shepherd (Foreign Office) and Dr Bob Press ACSA (N), MOD considered the interpretation of Article I of the 1958 Agreement against the background of the cut-off issue.[77] Britain did not envisage that the US would suspend the barter arrangements against strong UK objections, but the Foreign Office thought that, without threatening the suspension of supplies, the US could well plead Article I as a justification for asking the UK to negotiate new arrangements if it could be argued that these would facilitate the conclusion of an important disarmament agreement. Such anxieties about the risks of US political imperatives impacting adversely on UK nuclear defence interests were a recurring theme throughout our period. It applied in the test ban in the 1950s and would apply again during the SALT/ABM negotiations 1969 to 1972.[78]

---

73 TNA FO 371/171132, IAD 1036/10/G A.D.F. Pemberton-Pigott to Mr Wilson, 24 May 1963.

74 ENDC/PV.151, 14 August 1963, pp. 9–13.

75 TNA FO 371/171132, IAD 1036/209/G, The Control and Cut-Off of the Production of Fissile Material; US Paper ENDC/109.

76 TNA FO 371/171132, FO telno 3999, paragraphs 7 and 8.

77 TNA AB 48/642, H. Shepherd, Foreign Office to Dr R. Press, MOD, 20 April 1964.

78 For discussion on SALT/ABM and the UK see Chapter 10.

The US tabled a technical paper on verification of such an interim pre-Stage One GCD measure at the ENDC on 25 June 1964.[79] One of its main purposes was to show that such an arrangement would only require a lesser degree of control than would apply in the GCD context. Britain agreed, but reminded the US that since HMG could not participate in a cut-off agreement at that time, or even in the near future, the UK delegation to the ENDC would only be able to give the US paper a general welcome as a useful technical study of the kind in which London thought the ENDC could usefully engage itself. Dr Press (Cabinet Office) remarked that unlike the US, the UK could not start from a concept of 'plenty in the stockpile', therefore it could not commit itself to anything other than precise and detailed proposals that it could be sure would not be applicable until the UKAEA's 'mixed' production and processing plants (that is, that the same fuel cycle plants undertook civil and military work) were no longer required for the weapons programme, including production of material for the barter exchanges.[80] As the Russians had refused to discuss US proposals, the divergence in US and UK positions did not cause any embarrassment. However, if there were a chance that an agreement might be on the cards, the Foreign Office view was that there would be a strong political incentive for HMG to participate. If at that time UK fissile material stocks were less than adequate to meet the weapons programme requirements, then there would be a painful choice. However, there had to be a chance that the US might be prepared to do as much as possible to help the UK out of its dilemma. Although the US might be able to provide U-235, plutonium was another matter and the UK would have to accelerate its own production at Calder Hall and Chapelcross. Supplies of U-235 alone would thus not solve the UK problem; it was essential that the US understood this.[81]

When proposing a cut-off outside a three-stage GCD plan, the Americans had never specifically proposed that it should include a ban on transfers between states. If such a ban were to be introduced, it would conflict with US obligations under the 1959 barter agreements. Reliance on this exchange was a determining factor in the UK decision to cease production of weapons grade U-235 at

---

[79] ENDC/134, 25 June 1964, United States of America Working Paper on Inspection of a Fissionable Material Cut-off.

[80] TNA AB 48/642, R. Press ACSA (N) to DS 12 Inspection of a Fissionable Material Cut-Off, 15 June 1964; about 40 per cent of the fissile material content of returned weapons was recycled at AWRE. The remainder was sent to Windscale for recovery. TNA AB 48/196, St Clair Hood to Mr Gillams, Civilianisation of Windscale, 13 October 1964.

[81] TNA PREM 11/5172, Tam Bridges, Foreign Office to J.O. Wright, 10 Downing Street, 26 June 1964.

Capenhurst in 1963.[82] Consequently the UK was now critically dependent on this exchange for the completion of its defence programme as then planned. Change in the weapons programme since HEU production ceased meant that U-235 supplies were only just sufficient to meet the approved weapons programme up to 1970.[83] Its seems that in order to make the plutonium needed for Polaris, the UKAEA had to resort to a more economical production route, which meant a much longer supply period. This may well have stemmed from the wider economic pressures facing the Chancellor that resulted in a major reduction in the planned weapons programme in July 1962 and appears to have been related to a requirement for the financial arrangements that were then nearing completion to relieve the defence budget of 'historic costs' for UKAEA production of military material.[84] Vic Macklen, then at the UKAEA's Reactor Group, noted in 1964,

> ... in my days at defence we were always concerned to acquire the stock of fissile material that we considered to be necessary for defence purposes as soon as possible. Thus the accountants and economists were kept out of the business to the maximum extent possible. The net result of all the interference over the last five years has hardly been to save money but merely worsen the UK situation and make harder and probably more costly to get the desired stocks of fissile material at short notice.[85]

President Johnson's Proposals on Cut-back: Spring 1964

When President Johnson shared his ideas for some steps on a cut-off with the UK in the shape of a draft letter to Khrushchev on a cut-off in mid February 1964, the UK was quick to respond. London would be glad to be associated with such a proposal provided it was made clear that the UK would not commit itself to

---

82 See TNA DEFE 7/1888. As early as July 1961, the Treasury was placing heavy pressure on MOD to cut back on U-235 production at Capenhurst, but this was being resisted as the present rate of production up to 1966 was fully committed for the Phase I warhead requirements (strategic weapons). DRPC Sub-committee on Atomic Energy Brief for Meeting 12 July 1961, R. Press, 12 July 1961.

83 TNA DEFE 19/125, Availability of Fissile Material, COSC Confidential Annex COS 33rd Meeting/64, Part I, 5 May 1964, Availability of Fissile Material. These orders were for Polaris ET 317, WE177 A and WE177B warheads.

84 TNA CAB 21/6038, NRDC Brief for Meeting on Thursday 28 March, Item ND (63) 4, Provision of Fissile Material, 25 March 1963.

85 TNA AB 65/186, Item 3: Cut-off AEX (64) 54 Annex 8 to X (64) 9th meeting.

anything that it could not undertake.[86] Although the Johnson Administration succeeded to some extent in surrounding its initiative with an aura of arms control, the real reason was probably the excess of materials production built up in the 1950s.[87] The UK's own 1964 Defence White Paper had stated that production of military U-235 at Capenhurst ceased in 1963. Military plutonium production was also continuing and civil plutonium was required for the UK's barter exchange obligation on U-235 for defence purposes. Without the barter U-235, the UK would have been unable to close the high-level enrichment stages at Capenhurst. The UK took the chance too to make suggested amendments to Johnson's draft message to Khrushchev. Although the proposed changes, which were accepted, safeguarded the UK position, the Embassy in Washington was instructed to make it abundantly clear to the Americans that the UK had no room for manoeuvre on the need to continue plutonium production and the continuation of planned deliveries from the US of U-235 for military purposes in exchange for plutonium production in the UK. A revised Presidential letter to Khrushchev went on 25 February.

The Soviet reply was encouraging as it indicated that the USSR would be prepared to announce that two production reactors would not be completed and Khrushchev had no objection to the UK being drawn into the agreement, but criticised the UK position on fissile material. Although UK civil plutonium was exchanged for U-235, and although the US could use this material for civil purposes, it directly served a military purpose. In addition, under the then arrangement for reprocessing, all spent fuel from military and civil reactors was intermingled in the same plant at Windscale, so the UK could not say that no plutonium derived from the civil power stations would ever go into its military stockpiles.[88] The Foreign Office felt that it would be helpful if the UK could rid itself of this complication; and the MOD agreed that this issue should be taken up at the Cabinet Office's Defence and Overseas Policy (Official) Committee meeting soon.[89] In the meantime, a telegram was sent to Washington to set out UK comments on the US draft reply to the Soviet premier.[90] Given the UK civil plutonium barter arrangement and co-processing (where fuel from the civil and military reactors was mixed together during reprocessing, at this stage there were no separate campaigns), London thought it would be better to avoid making any premature statement about UK plutonium production for weapons

---

86 TNA PREM 11/5181, Foreign Office telno 2837 to Washington.

87 Seaborg, *Stemming the Tide*, p. 398.

88 TNA PREM 11/5181, H.B. Shepherd, AEDD to Lord Hood, 5 March 1964.

89 TNA FO 371/176370, IAD 1036/8/G, H.B. Shepherd to Lord Hood, 5 March 1964.

90 TNA PREM 11/5181, Foreign Office telno 3489 to Washington, 6 March 1964.

use. It would thus be better to state in Johnson's reply: 'I am taking this up with the British Prime Minister and I feel sure that he would be prepared, at the right time, to amplify the recent announcement about United Kingdom reductions as part of a joint action by our three countries.'

Johnson attached very considerable importance to his exchanges with Khrushchev on this issue and treated the matter with great urgency. In fact, the US hoped that an announcement could be made publicly before the end of March 1964 and preferably a good deal sooner.[91] Embassy officials were satisfied that the Americans accepted that the UK should be fully associated with this initiative, but the Americans were a little puzzled by the UK's apparent reticence regarding the initiative.[92] The Administration knew of no other area in which there seemed to so much hope of an early agreement with the Soviet Union in order to maintain the momentum resulting from the Partial Test Ban Treaty agreed in July 1963. Washington's reply was to be handed to the Soviet Ambassador on 9 March. London meanwhile was anxious to assure the Americans that there was in fact no reticence on the UK's part and consequently the Embassy in Washington was tasked to provide the necessary assurance. Britian fully recognised the importance and urgency of this issue and that the UK would want to be associated fully.[93]

Rab Butler (the Foreign Secretary) and Sir Alec Douglas-Home (Prime Minister) discussed the messages between Khrushchev and Johnson on 9 March. Butler felt that it was important that the UK should be associated with the proposals under which the Soviet Union and the US would announce independently the shutdown of some facilities for production of fissile material for weapon purposes.[94] However, the UK had already made all the reductions it could in fissile material production and any further reductions would impact severely on its defence programme. In his message to Khrushchev of 22 February, Johnson included at UK request the statement: 'You will have noticed that the UK has announced the cessation of production of U-235 for military purposes and I would hope that the UK would want to join in any discussion we may have.' Khrushchev in his reply drew attention to 'large new plutonium plants which were being put into operation in England'. Johnson included in his reply to this letter, again at UK request, the following:

> ... the UK has already announced that it has stopped the output of U-235 ... and is gradually bringing to an end the production of plutonium for weapons

91 TNA PREM 11/5181, Washington telno 922 to Foreign Office, 7 March 1964.

92 TNA PREM 11/5181, Washington telno 923 to Foreign Office, 7 March 1964.

93 TNA PREM 11/5182, Foreign Office telno 3568 to Washington, 9 March 1964.

94 TNA PREM 11/5172, Butler to Prime Minister, 9 March 1964.

> purposes. This is in contrast with your concern that new plutonium plants were being put into operation ... I have been assured ... that no expansion of plutonium production is in fact being undertaken.

Almost, but not quite, all the plutonium for military purposes was produced at Calder Hall and Chapelcross. The remaining civil reactors were designed for electricity production, although they made a small proportion of plutonium that was suitable for military purposes. The then intention was to use this for such purposes. The co-processing problem, where the civil and military plutonium were mixed in the same plant, was to be addressed so that it would be possible to say that none of the plutonium derived from the civil power stations would be used in the weapons programme. This would be a clear response to Khrushchev's claim that the UK was increasing its production. It would, however, call for some re-arrangement of production that would cost the UKAEA an estimated £2–3 million. Butler did not think, in response to Lord Harlech's (UK Ambassador in Washington) report, that the UK was being reticent, but it had been necessary to consider very carefully what might be said in public and what new arrangements might be required. Harlech was asked to reassure the Americans on this point.[95] The Foreign Secretary was quite clear that the UK should be associated with this development.[96] Once officials had discussed the matter in the Defence Overseas Policy (Official) Committee on 10 March it emerged that the costs of making sure that no civil plutonium was intermingled with military material were much less than the UKAEA had originally thought. Butler then informed the Prime Minister (Alec Douglas-Home) that as a result the UK would now be willing to make new arrangements under which no civil material would be used in the military programme and that this should be relayed to President Johnson.[97]

The Minister of Defence, Peter Thorneycroft, was sufficiently worried about the cut-off and its implications to write to the Foreign Secretary on 12 March 1964.[98] Thorneycroft stressed that the UK had to continue to produce warheads for some five more years in order to maintain the deterrent and if the UK had to produce U-235 and tritium domestically again this would be at the cost of considerable trouble and expense. Thorneycroft felt that it would be bizarre indeed if the UK ended up without essential fissile material as a result of stopping production when the Russians were halting no more than the construction of

[95] TNA PREM 11/5172, Reduction of Production of Fissile Material for Weapons Purposes, Foreign Office, 9 March 1964.

[96] TNA FO 371/176370, IAD 1036/13/G, Butler to Prime Minister, 9 March 1964.

[97] TNA FO 371/176370, IAD 1036/16/G, R.A. Butler to Prime Minister, 13 March 1964; TNA PREM 11/5182, Burke Trend to Prime Minister, 16 March 1964.

[98] TNA PREM 11/5182, Peter Thorneycroft to Foreign Secretary, 12 March 1964.

two new reactors, which was all that Khrushchev had indicated was on the cards. Foreign Office officials, however, felt that it was most important politically that the UK should play its full part in the expected agreement between Johnson and Khrushchev.[99] Home sent a personal message to President Johnson on the issue of reduction and a possible cut-off on 17 March. He recalled that he had agreed that it would be appropriate to indicate to Khrushchev that he would be ready to amplify the recent UK announcement about reductions as part of a joint action. London was bothered about the Soviet reference to new large plutonium plants in the UK, clearly an allusion to the Central Electricity Generating Board's (CEGB) Magnox stations. The co-processing problem was highlighted as an issue that the UK planned to address so that the UK could truthfully state that no civil plutonium was involved in the military programme.[100] In view of the UK U-235 problem Home asked President Johnson to advance US deliveries of U-235 so that the total amount due to the UK up to 1969 was transferred immediately. The other aspect of the problem was that the barter agreement envisaged interest payments for U-235 if its supply exceeded that in any one year from the UK plutonium exchange; London hoped that the Americans would not make this an issue. Payment would be as per the barter agreement for the plutonium as the UK could not afford to pay cash for this.[101] It would certainly help the cause if the US could state that all such UK plutonium would only be for civil purposes, which was its intention anyway.

Ministers (Home – Prime Minister, Butler – Foreign Secretary, Hogg – Education and Science Secretary, Maudling – Chancellor and Thorneycroft – Defence Secretary) and officials (Burrows –Foreign Office, Zuckerman – MOD, Peck – Foreign Office and Penney – UKAEA) met on 16 March 1964 to review the question of UK fissile material supplies and the US interest in cutting back on production.[102] The Minister of Defence noted that the political aspects of the situation were more disturbing. Although the existence of the barter agreement with the US was publicly avowed, its implications were not generally understood. HMG's task in defending the independence of the deterrent would be further complicated if Ministers exposed themselves to the criticism that the UK was not only dependent on the US for delivery systems, but also for the fissile material for the warheads. He even suggested that, in trying to

99 TNA FO 371/176371, IAD 1036/23/G, A.D.F. Pemberton-Pigott to Sir B. Burrows, 16 March 1964.

100 TNA PREM 11/5182, Prime Minister's Personal Telegram Serial No. T99A/64, Foreign Office telno 3999 to Washington, 17 March 1964.

101 TNA PREM 11/5182, Foreign Office telno 4000 to Washington, 17 March 1964.

102 TNA CAB 130/197, GEN.848/1st Meeting Supplies of Fissile Material, 16 March 1964.

maintain the illusion of independence, the UK would be sacrificing a promising opportunity of making progress towards general disarmament. However, the UK's position would be greatly improved if it already had acquired the U-235 it needed from the US in bulk, but there was still a need to obtain tritium from the US in exchange for military plutonium produced in the UK.

As a separate matter, Home also raised the prospect of future Russian support for a cut-off as proposed by the US in Geneva at the ENDC. If a cut-off were to take effect before the UK received the balance due of U-235, then the UK could not support such a cut-off. And since plutonium production for weapons was scheduled to continue until 1969, a cut-off before that date would be an obstacle to UK adherence. To deal with this, the Prime Minister suggested that British and American officials should meet to consider the political implications for the Western posture in various circumstances and whether it would be worth examining some standby arrangements that might enable the UK to give active support to the US cut-off proposals. Johnson replied swiftly on 20 March.[103] UK experts were welcome to come to Washington to discuss the problems raised by the Prime Minister. The US impression was that the existing arrangements did not in fact impose any real obstacle to the proposed concurrent announcements that the US was hoping to make with the Soviet Union; nor did the US see any difficulties in meeting both sides' barter commitments under a cut-off. However, the specific proposal for early delivery of U-235 gave the US some problems, but Johnson did ensure that Glenn Seaborg (Chairman, US Atomic Energy Commission), John Foster (Defense Advanced Research Projects Agency) and Harold Brown Director of Defense Research and Engineering, Department of Defense), were aware of the problem as expressed in Home's message and would be prepared to discuss the issues at any time. Penney headed off to Washington on 23 March for talks on the 24th.

At this meeting, it became clear that there was no chance of any advance delivery of U-235.[104] The Americans argued that since what they were proposing was a cut-back and not a cut-off, the barter arrangements were not affected. Moreover, while the US stated that it would not use UK plutonium for weapons purposes; to say so publicly would probably present difficulties. Since the barter agreement was presented to and accepted by Congress as a military and not a civil agreement, in this context it would be hardly possible to set aside the 'equivalent quantity' modification as earlier suggested by the UK.[105] In short, the US position was that it would not agree to early delivery of barter U-235;

103 TNA PREM 11/5182, Foreign Office telno 4142 to Washington, 20 March 1964.

104 TNA FO 371/176371, IAD 1036/32/G, Washington telegrams Nos. 1149 and 1150, H.B. Shepherd to Sir B. Burrows, 26 March 1964.

105 TNA PREM 11/5182, Washington telno 1149 to Foreign Office, 25 March 1964.

Washington would honour the barter agreement as it then stood; and support, as necessary, any public statement made by the UK to the effect that UK civil plutonium would not be used for weapons purposes, but would not go as far as to say that no equivalent quantity would be set aside for military purposes.[106] William Penney (now Chairman of the UKAEA) thought that the best way of avoiding difficult questions on the statement that the Prime Minister was to make to the House of Commons on the UK position was for him to say that none of the plutonium being made in the civil reactors would be used in UK weapons. He also suggested several possible ways of responding to questions on the uses of UK civil origin plutonium in the US.[107] Drafts of the statement for the House were cleared by relevant Whitehall departments, but given the outcome of Penney's talks in Washington the statement could not be as forthcoming as the UK would have wished.[108] In particular, Home would have liked to say that UK civil plutonium would be used only for peaceful purposes.[109] However, the Minister of Aviation, Julian Amery, told Home that he hoped that the draft was a statement of fact and intention and not a commitment. He felt that it was important that the UK should remain free to resume production of U-235 for military purposes and to step up production of military plutonium if either course were to become necessary for the maintenance of the UK deterrent. Recent studies had shown that the UK was operating on a dangerously narrow margin and was already uncomfortably dependent on the exchange agreements with the US.[110] This was the reality of the Conservatives' nuclear policy, having won the great prize in 1958 with the restoration of Anglo-American nuclear weapons co-operation, the UK had put itself in hock to the US in order to remain in the game.

Home's proposed draft parliamentary statement was, however, shared with the US and the Americans were advised that the UK had tried to make it as helpful as possible to the US initiative.[111] Lord Harlech saw Dean Rusk on 30 March and despite the earlier indications, Rusk told him that the Administration would now have no objection to the Prime Minister including in his statement an assurance that the plutonium sent from the UK would not be used for

---

106 TNA FO 371/176371, IAD 1036/34/G, A.D. Peck, MOD to Sir Bernard Burrows, Foreign Office, Proposals for Reducing Supplies of Fissile Material, 23 March 1964.

107 TNA PREM 11/5182, Washington telno 1150 to Foreign Office, 25 March 1964.

108 TNA PREM 11/5182, Butler to Prime Minister, 26 March 1964.

109 TNA PREM 11/5182, Prime Minister's Personal Telegram Serial No. T122/64, Foreign Office telno 4407 to Washington, 27 March 1964.

110 TNA PREM 11/5182, Minister of Aviation to Prime Minister, 26 March 1964.

111 TNA PREM 11/5182, Prime Minister's Personal Telegram Serial No. T122/64, Foreign Office telno 4407 to Washington, 27 March 1964.

weapons purposes.[112] Rusk went on to say that he thought that the UK draft was a bit disingenuous as it made no reference to the UK's intention of using US U-235 for weapons purposes. Perhaps it would not be wise to conceal this and it would make more sense to put all the cards on the table and thereby avoid any accusation of attempts to conceal the facts. Harlech thought that the point was covered in the first sentence of paragraph 2 of the draft which read: 'For their part HMG have already adjusted their supplies of fissile material to the minimum necessary to attain our independent nuclear deterrent and to meet all our defence requirements for the foreseeable future.'[113] London obviously agreed with Rusk because there was a sentence that expressly referred to the exchange of U-235 for plutonium for UK weapons purposes in the Prime Minister's eventual statement to the House of Commons on 21 April 1964. Officials had started drafting a brief in early April for possible discussions with the Americans on a cut-off as proposed by Home in his message to the President on 17 March. As a basis for recommendations officials thought it necessary to make some assumptions on the likelihood that a cut-off would put an end to transfers for weapons purposes and the likelihood of a cut-off. US views were sought on the assumptions.[114] The main feature here in UK thinking was the fear that the US–UK barter arrangements might become too embarrassing politically for the Americans to continue. Worrying about contingencies was a recurring feature of this particular story; it was also to be same story on SALT/ABM, but that issue is for chapter ten.

On 16 April Johnson wrote to Home to tell him that he had decided to make his announcement of further reductions in US fissile material production on 20 April, without waiting for a parallel announcement from Khrushchev. He would, however, make one more try to persuade the Soviet leader to join in.[115] Home decided to send his own message in support and emphasised that he intended to make clear the UK's position on policies already adopted and was continuing to pursue to contribute fully to the President's initiative on fissile material.[116] The extent of any UK statement, however, would depend on whether Khrushchev

112 TNA PREM 11/5182, Washington telno 12151 to Foreign Office, 1 April 1964.

113 TNA PREM 11/5182, Prime Minister's Personal Telegram Serial No. T122/64, Foreign Office telno 4407 to Washington, 27 March 1964.

114 TNA PREM 11/5182, Foreign Office telno 4857 to Washington, 9 April 1964.

115 TNA PREM 11/5182, A.D.F Pemberton-Piggott to Private Secretary, Reduction of Production of Fissile Material for Military Purposes, 17 April 1964.

116 TNA PREM 11/5182, Prime Minister's Personal Telegram Serial No. T155/64, Foreign Office Telno 1171 to Moscow, 17 April 1964.

actually joined in.[117] He did and so Home made the following statement to the House of Commons on 21 April after a last minute effort by the Defence Secretary to limit its scope, which was opposed by the Foreign Secretary:[118]

> HMG welcomes the announcements ... Production of fissile material in the UK ... has always been very much smaller scale than in the US or USSR. ... HMG have already adjusted their supplies of fissile material to the minimum necessary to maintain our independent nuclear deterrent and to meet all our defence requirements for the foreseeable future. Military plutonium production is being gradually terminated. Our plans do not envisage the use of any plutonium produced by our civil reactors in the UK weapons programme and I am informed by the United States Government that they have no intention of using the plutonium received from us for weapons purposes.[119]

In practice the situation was that forward planning for the UK weapons programme was calculated on the quantities of plutonium expected to be produced by the military reactors and did not take account of any possible bonus of military grade material that might be forthcoming from CEGB or South of Scotland Electricity Board (SSEB) nuclear reactors. Both MOD and the UKAEA intended to ensure that this continued.[120] The full amount of plutonium produced from the civil reactors and subsequently processed would be held for civil use only. In practice, however, neither the MOD nor UKAEA could ensure that plutonium atoms from civil power stations never changed places with atoms from Calder Hall and Chapelcross reactors during the co-processing stage at Windscale.[121]

---

117 TNA PREM 11/5182, A.D.F. Pemberton-Pigott to Private Secretary, Reduction of Production of Fissile Material for Military Purposes, 17 April 1964.

118 TNA FO 371/176371, IAD1036/35/G, Butler to Prime Minister, Reduction of Production of Fissile Material, 15 April 1964; Peter Thorneycroft to Prime Minister, Reduction of Production of Fissile Material, 10 April 1964.

119 Hansard Vol. 693, session 1963–64, cols 1097–8.

120 TNA FO 371/176372, IAD 1036/75/G, E.H. St G. Moss, DS 3, MOD to H.B. Shepherd, Foreign Office, 28 May 1964.

121 TNA FO 371/176373, IAD 1036/103/G, A.D.F. Pemberton-Pigott to Sir B. Burrows, 12 October 1964.

Capenhurst: Cut-off Inspection or Resumption of HEU Production?

Another issue to the fore at this stage was the possibility of verifying the cessation of U-235 production for military purposes. This had implications for Capenhurst as US officials were studying the technical problems in confirming that a gaseous diffusion plant was in fact shut down. These, however, would be greater for Capenhurst than for some of the US plants since the low enrichment section was still working on a 'tick-over' basis for civil purposes, so that inspection from the perimeter could not provide conclusive evidence that military production had ceased. UKAEA experts were reviewing the issue and preparing a paper for consideration in Whitehall, but they did not wish to hand this over to the US until such times as American views were known. In the event the UKAEA disagreed with the US assessment that all that was required for effective monitoring of a diffusion plant was to do so from the perimeter; diversion could still take place in such circumstances, in particular, the UKAEA's Production Group thought that it was unlikely that the mass and isotopic balances could be calculated to the desired accuracy from information obtained at the main gate unless there was also some knowledge of what was happening in the plant.[122] They did not want to commit to any particular position.[123] Subsequently AWRE's Safeguards Team visited Capenhurst and possible inspection techniques were formulated on a shrinking perimeter basis. This started with inspection at the site perimeter and proceeded in stages via limited site access to limited diffusion plant access and finally a complete knowledge of all plant details. At each stage the effectiveness of inspection was correlated with the required release of classified information. Research work here led to a general conclusion that effective inspection was not possible without access to the plant. However, if subterranean pipelines for example could be eliminated from consideration, then inner perimeter inspection at the diffusion plant building's perimeter might be seriously considered as an acceptable inspection technique.[124] Meanwhile the UKAEA's Cut-Off and Safeguards Working Party accepted the conclusions and agreed that this study had provided sufficient information to allow the UK to take part

[122] TNA AB 65/186, The Inspection of Gaseous Diffusion Plants, UKAEA Production Group, 9 February 1966.

[123] TNA PREM 11/5182, Quentin Hogg to Foreign Secretary, Reduction of Production of Fissile Material, 20 March 1960.

[124] TNA AB 9/2430, Minutes of the Seventh Meeting, UKAEA Item 6: Diffusion Plants, 24 April 1967; Minutes of the Ninth Meeting, Item 3: Control of Gaseous Diffusion Plants, reference A Technical Assessment of Perimeter Control of Production at Gaseous Diffusion Plants Producing All Concentrations of U-235 (COSWP) (67) p. 8, 29 September 1967.

in any future technical discussion on perimeter control of diffusion plants. In general, there was no need to carry the study any further.[125]

There was another important reason for not taking any initiative. Since the UK could not support a cut-off at that time given the shortage of U-235, and since the US might not want to provide satisfactory assurances over U-235 supplies to the UK, it could not be completely ruled out that the high-level enrichment stages at Capenhurst might have to be reopened.[126] So when the prospect of safeguarding Capenhurst arose again in early 1964, following Johnson's interest in reduction of fissile material production, the UKAEA was still reluctant to commit itself wholly to international inspection, since to do so might reveal technical details and might also give some indication of UK stockpile figures.[127] MOD shared this view: it was unwilling to explore with the US the verification problems of confirming closure until all discussion of the barter agreement was out of the way and until it was quite clear that there was no question of re-opening the plant.[128] Nor could the UKAEA accept any inspection of Calder Hall or Chapelcross and certainly not the option to follow material through from reactor to reprocessing plant.[129] Nothing could be considered here until the UK–US barter arrangements had run their course in 1969. But these were initial views: once some detailed work had been done in the UKAEA's Production Group it became clear to officials in the Authority and Whitehall that some international inspection of the high separation plant at Capenhurst would in fact be possible to confirm that it was non-operational. It would even be possible to establish that the low separation stages were only being used to produce low enriched uranium (LEU) for civil reactors by permitting an inspector to withdraw appropriate samples of uranium hexafluoride.[130] But since there was no segregation of material going to or from civil or weapons purposes at Springfields or Windscale, then the Production Group would find it extremely difficult to effect any such physical segregation. The Authority's London office was not so sanguine as they feared that there was a risk that the Foreign Office

125 TNA AB 9/2430, Item 3: Control of Gaseous Diffusion Plants, Minutes of the Ninth Meeting, 29 September 1967.

126 TNA EG 1/495, Lord President to Foreign Secretary, Reduction of Production of Fissile Material, 20 March 1964.

127 TNA AB 48/196, J. Gaunt to Dr J.M. Hill, Production Group Risley.

128 TNA FO 371/176371, IAD 1036/24/G, A.D.F. Pemberton-Pigott to Sir B. Burrows, 18 March 1964.

129 TNA AB 48/196, J. Gaunt to Mr Gillams, 29 January 1964; J.L. Gillams to Dr J. Gaunt, 7 February 1964.

130 TNA AB 48/196, J.M. Hill, Production Group to Dr J. Gaunt, UKAEA, London, 11 February 1964.

would jump on any suggestion that Capenhurst might be inspectable without noting that any offer would have to be limited. Instead any offer ought to be that a walk-round inspection was all that would be possible at that time because of the danger of disclosing vital technological information that was classified both for commercial and military reasons. Detailed inspection was out of the question unless there was a really serious possibility of disarmament. It would only be possible to suggest that a CEGB station could be subject to inspection. Bradwell was subsequently offered in order to match the US proposal that one of its civil stations be placed under safeguards. (An agreement was concluded on 20 June 1966 with the IAEA for safeguarding Bradwell for a period of five years to expire on 1 September 1971. It was not renewed but subsumed within the UK's 1967 voluntary safeguards offer.[131]) In order to ensure that the IAEA inspectors could not then follow the reactor fuel through the reprocessing stage at Windscale, which had to be excluded from safeguards because of its military work (up until 1969 it would be predominately and almost continuously engaged in military work), a substitution arrangement was agreed that enabled the UKAEA to substitute an equivalent amount of plutonium existing in other facilities (ZEBRA – an Authority 100 MW research reactor at Winfrith intended primarily to investigate the physics of large fast reactors[132]) that could be inspected. The substitution concept was devised to avoid admittance of IAEA inspectors to reprocessing plants and other facilities that were not yet subject to safeguards in themselves, but became so only because they temporarily contained a quantity of safeguarded materials from other facilities.[133] The UKAEA were satisfied that such an offer would not result in supplies for defence needs being directly affected. Given this assurance the MOD accepted that Bradwell could be opened to IAEA inspection.[134]

There were two separate aspects of the Authority's inability to accept inspection at Windscale: first, it had to conceal the specification of military material (its isotopic composition) and to ensure delivery to time of quantities

[131] IAEA INFCIRC/86, 23 September 1966, *The Texts of the Agreements for the Application of Agency Safeguards to the Bradwell Reactor Facility in the United Kingdom.*

[132] UKAEA, *Annual Eleventh Annual Report 1964–65* (London, 1965), Appendix III, p. 88.

[133] TNA EG 1/495, The Effects of Safeguards Upon Authority Operations, Note by the Chairman, 29 January 1965.

[134] TNA CAB 148/20/40, OPD (65) 67, Inspection by the IAEA of the CEGB Nuclear Power Station at Bradwell, Memorandum by the Minister of State for Foreign Affairs, 7 April 1965.

of military material.[135] Windscale was designed for continuous operation and any batching of safeguarded civil material specifically to meet IAEA inspection and control arrangements would require shutting down the plant for up to six days for washing out. This imposed unacceptable costs in terms of loss of production capacity and delay to important programmes.[136] Until then there had been no reason to maintain military and civil activities in separate watertight compartments; facilities had been designed and constructed to meet the combined requirements of both programmes in the most efficient and economical manner and this had led to an extremely close interweaving of military and civil activities at the same facilities.[137] Separating the facilities into exclusively civil and military establishments would present major problems of reorganisation and would seriously affect the efficiency of the UKAEA's operations. However, from 1970 or 1971 onwards Windscale could be opened to and made available for inspection whenever spent fuel came forward for processing from a civil station as the earlier difficulties had been overcome. If the plant were required for military activities, then it would have to be temporarily withdrawn from safeguards. For these amongst other reasons the UK was not prepared to accept US proposals for safeguards on civil materials provided by the US. (The US later wanted to propose that all transfers of fissile material be subject to NPT safeguards; the UK could not agree to this without first completing its current study of the implications of safeguards applied to material transferred to the UK.[138]) Instead UK officials thought that it would make more sense to offer a progressive application of safeguards, on the basis of making specific offers of facilities from time to time. The Cabinet Overseas and Defence Policy Committee agreed this in October 1965.

Although it was thought unlikely by UKAEA officials that the high diffusion stages at Capenhurst would be needed again for defence purposes, the Ministry of Aviation appears to have been anxious about the reliability of US HEU supplies fearing that the Treasury would no longer agree that the 'tick over'

---

135 TNA AB 48/196, Atomic Energy Official Committee Brief on 'Revised Safeguards System of the IAEA' A (O) (65) 3 for the Meeting of the A (O) Committee on 17 February 1965.

136 TNA AB 48/196, Safeguards and the New UK/US Agreement on Civil Atomic Collaboration, Note by the Ministry of Technology, Annex A, 5 August 1965; Safeguards Working Party SGWP (65) P.33 (revise 5/1/66) Safeguards Working Party Possible Questions and Answers Dealing with the UK's Difficulties in Accepting Safeguards under the New Bilateral.

137 TNA CAB 148/23/16, O.P.D. (65) 156, Nuclear Safeguards and the New UK/US Agreement, Note by the Ministry of Technology, 28 October 1965.

138 TNA CAB 165/746, M. Davis to Mr E.F. Newley, Brief for the Committee on Safeguards for Nuclear Materials Meeting on 22 July 1966.

arrangements could be sustained after autumn 1964, so it suited the Authority to keep defence interests alive; this was another key reason for not making any offer to open Capenhurst to international inspection.[139] And in fact this point was made clear in a draft paper from UKAEA officials to the Lord President, in his capacity as responsible Minister, namely that the possibility could not be wholly ruled-out that in certain circumstances it might be necessary to re-open Capenhurst for the production of HEU for military purposes.[140] It would therefore not be prudent to agree to a Foreign Office request to give the US a paper on the possibilities of inspection at this site. In fact the UK started to think again in the mid to late 1960s that Capenhurst might need to be re-opened if supplies could not be secured from the US in the event that a cut-off might enter into force before the end of 1969. But they were most anxious to avoid this because of the expense and the earlier public statement that closure of the plant would not prejudice the independent deterrent.

Nuclear Requirements for Defence Committee (NRDC): 1964[141]

As of early 1964 military grade plutonium production for the then current approved warhead programme requirements would not be fulfilled until about the end of 1969.[142] About one third of the UK's U-235 needs had been met from Capenhurst's production. Deliveries of U-235 and tritium from the US were scheduled to continue until 1969. There would be enough material by then to satisfy both weapons and naval propulsion programmes until about 1980. However, there would be no reserve of plutonium other than that recovered from old weapons withdrawn by 1972–73, which was expected to amount to 0.42 tonnes. This would leave a balance of about 0.7 tonnes of U-235 by 1980. On this basis, there would be no material for a NATO Multilateral Force or for TSR2 WE177s as proposed by SACEUR or for new national requirements. If there were to be a permanent cut-off, then perhaps the UK would need to

139 TNA AB 48/196, J.L. Gillams to Mr Pierson, 18 February 1964.

140 TNA AB 48/196, Draft Submission to the Lord President, Reduction of Production of Fissile Material.

141 The NRDC was a Cabinet Office committee consisting of MOD, UKAEA, FO (and later the FCO), Ministry of Aviation and Cabinet Office officials dealing with planning of UK nuclear requirements for the weapons and propulsion programme. It was the direct successor of the Nuclear Tests Policy Committee that had been established by Macmillan in 1959. It was chaired by the MOD Permanent Under-Secretary.

142 TNA CAB 130/197, GEN.848/1st Meeting, Supplies of Fissile Material, 16 March 1964.

acquire a reserve stockpile amounting to 50 per cent of the stock as then planned – about 2.1 tonnes of military grade material by 1969. This entailed carefully phased small off-takes of military grade plutonium from Calder Hall and Chapelcross.[143] If a cut-off began in 1965, it would only be possible to accelerate production by about 0.1 tonnes. And this would mean that the UK would only have half the planned requirements for Red Snow – the warhead for the Yellow Sun Mark II free-fall bomb and the Blue Steel stand-off bomb, WE177A, Polaris and WE177B. There would not be enough even to complete the Polaris programme alone.[144] Things would be better if 1967 were the target date since about 1.7 tonnes could be produced by that stage, giving a total of 2.0 tonnes by mid 1967; an additional 0.2 tonnes could be produced, but this would be at the expense of adverse publicity and greater costs. In extremis, material could be made in CEGB stations, but this too would be costly both financially and politically. The Prime Minister had said in 1959 that only one CEGB station would have modifications suitable for military production and that his April 1964 statement to the House of Commons on cut-off had made any change in policy politically costly. UK U-235 production had led to about 4 tonnes, rising to 11.5 with barter supplies during the next five years until 1969. Capenhurst's high enrichment stages could be reopened too, but that would take about two years to come to fruition.

The UK's weapons programme requirements had been intentionally geared to the most economic production rate so that it would not be until the end of 1969 that there would be sufficient material even for the programme then envisaged. U-235 shortages were even more critical than plutonium. When the decision to close Capenhurst was taken in 1962, it was estimated that by 1969 the UK would have a reserve of about 4.5 tonnes in addition to the requirements then foreseen. The position was drastically changed when the Skybolt air-launched ballistic missile was cancelled and even then the warhead envisaged for Skybolt, although lighter, still used more fissile material than the Red Snow warhead in Yellow Sun Mark II.[145] Polaris required a very large increase in the requirements for U-235, so too did the high yield WE177B of which 53 were eventually built.[146] By the time that these changes had been approved the UK had lost the option under the barter agreement to purchase an additional five

143 TNA T225/2500, P.H.F. Dodd to Mr Jay, D.O. (O) 10: Cut-Off of Fissile Material, 9 March 1964.

144 TNA AB 48/336, Aide Memoire, History of Cut-Off, Undated Chronology Covering 10 March 1964 to 21 April 1965.

145 TNA DEFE 7/2217, BND (SG) (60) 20, Retarded/Laydown Bomb for the TSR 2 in the Deterrent Role, Report by the Ministry of Aviation, 3 January 1961.

146 TNA AIR 2/18209, E. Plumtree, D Air Plans to D of Ops (S) RAF, 16 July 1969.

tonnes from the US. Thus if the UK were to accede to a cut-off agreement in mid 1965, it would actually have little more than half of its planned production. Recycling of old weapons would release some material, but this would not cover the approved build of new warheads for Polaris, the RAF and Navy. Without additional supplies the UK could not complete the Polaris programme, but a cut-off taking effect in mid 1967 would improve things somewhat, but even then there would be a shortfall of 2.1 tonnes of plutonium and 11.5 tonnes of U-235 without any reserve.

Officials identified several options that could be taken singly or in combination to overcome these problems. These ranged from accepting a cut-off in principle, but delaying UK accession until stocks were built up; buying more material from the US; exercising the barter options; resuming UK HEU production; and accelerating plutonium production. As ever so much depended on what the US would be willing to provide, so it was necessary to ascertain as soon as possible exactly what Washington might be able to offer and/or sell. Until such times as answers were forthcoming, it would be difficult to advise Ministers on the best course of action. The need for urgency arose partly from the fact that, if the UKAEA was to accelerate plutonium production, then action was required without delay, and partly from the political desirability of taking an early initiative in this field, if it were possible to take one at all.[147]

As matters stood in mid March 1964, when the barter arrangements expired in 1969, the UK would still have the option of purchasing further supplies of fissile material from the US or of producing it in the UK. But if agreement was reached on a complete cut-off, the UK defence programme would be liable to be crippled in the period after 1969. Moreover, since the then present estimate of requirements contained no reserve to meet additional contingencies between 1964 and 1969, a cut-off would prevent the UK from meeting any additional requirements for weapons developments before 1969. Home wanted the UK to be associated with the US as far as possible on its disarmament proposal, but in return the UK would need to have U-235 transferred to the UK in bulk, which preferably would not involve any additional financial liability.

Fortunately for the UK the Soviet Union had yet again shown no disposition to accept a cut-off proposal – so often on this issue Soviet reluctance had saved UK bacon, but the Foreign Office still thought it possible that in a few years time a prohibition might be agreed. Why it thought this is far from clear. Although it would not automatically prevent transfers of fissile material between allies,

---

147 TNA CAB 134/2241, N.D. (64) 7, Existing plutonium plans required a total of 754 kg by 1969–70: 120, 160, 190, 200 and 84 kg in each year 1964–65 to 1969–70. It took about a year to 18 months to produce a billet of military grade plutonium starting with natural uranium.

the US might find it difficult in such circumstances to agree new arrangements for supplying the UK with military grade U-235.[148] NRDC members discussed military grade plutonium production in March 1964 and were informed that, if there was agreement on a cut-off, it could well affect Britain's production programme in its later stages, but since the Soviet Union did not agree, the problem was not immediate. Nevertheless, the UKAEA was then authorised to proceed with the production of a further 160 kg of military grade plutonium. Ministers had considered Johnson's ideas on cut-off on 16 March and the extent to which the UK could afford to be associated with such an agreement. They had concluded that, while Britain would wish to participate as far as possible, it would in return expect arrangements to be made by the US to meet British defence requirements, especially for U-235 transfers.[149] Penney told the meeting that, if plutonium from civil reactors could in future only be used for civil purposes, it might be necessary, in order to meet UK requirements for military grade plutonium, to produce an additional 150 kg of this material from Calder Hall and Chapelcross at an extra cost of £2–3 million. If further domestic supplies of U-235 were needed, then Capenhurst could be re-opened, but at a cost on both financial and presentational terms. Tritium was provided from the US in exchange for military grade plutonium from Calder Hall and Chapelcross. Unless there was an unexpected early cut-off agreement, UK defence needs were fully covered until the end of the decade.

Although a cut-off agreement was not on the horizon, if such an agreement did materialise, officials thought that it would be politically awkward if the UK were not in a position to subscribe to it. Moreover, once there had been an agreement to reduce the production of fissile material, the Foreign Office thought that Ministers might wish to advocate more strongly than hitherto the adoption of a complete cut-off. The implications of such a decision would require considerable further study, apart from the supply of tritium that was a special problem in its own right because of its short-half-life and that it was also part of the barter scheme.[150] It seemed unlikely in the summer of 1964 that a cut-off agreement would have any serious effect on the weapons programme up to 1969. However, thereafter the position was obscure and the then current estimates of fissile material requirements made no provision for nuclear surface

148 TNA CAB 134/2241 N.D. (64) 2nd Meeting, Item 1: Requirements for Military Grade U-235, 3 March 1964.

149 TNA CAB 134/2241, N.D. (64) 3rd Meeting, Supplies of Fissile Material, 17 March 1964.

150 The stockpile of tritium bearing weapons in the UK stockpile was only planned to increase relatively slowly as of May 1960. TNA AB 16/3708, E.F. Newley to Chairman (UKAEA), 12 May 1960.

ship propulsion or any other contingencies that could only be met either by reactivating Capenhurst or, preferably, by purchasing the necessary material in the US. Sir Henry Hardman, NRDC Chairman, then arranged as a matter of urgency for the creation of a working party to examine the implications of a reduction and a cut-off in fissile material production for military purposes in preparation for forthcoming discussions with the US on the question of supplies under the barter agreement.[151]

The NRDC Working Party report outlined the main implications for the UK of a cut-off agreement and of UK support for it.[152] US ideas only envisaged a cut-off affecting material for use in weapons; it did not apply to propulsion or civil uses. Neither did they apply to transfers for weapons purposes between governments that had agreed to cessation. However, the Working Group felt that such a ban on transfers could not be ruled out entirely. In any case, such transfers might be difficult to sustain politically even if there were no prohibition in an agreement. This report assumed two scenarios for planning purposes: one in which a cut-off would come into force in mid 1965; the second that it would not apply until mid 1967. The prospect that the latter date might come about was sufficiently real to make it prudent to examine the consequences. Most of the working group's members believed that if a cut-off were to come to pass, it would be difficult for HMG to stand aside and not associate itself with the US and Soviet Union. The 1964 Defence White Paper had made clear publicly that the reduction in the production of fissile material would not affect in any way the UK's ability to maintain its deterrent and that production, if necessary, could be resumed. Officials therefore did not want to take any steps that would place the deterrent in jeopardy or make it appear even more dependent on the Americans. Even if the US were prepared to meet any UK shortfall, Ministers would need to consider whether they could rely on such assurances and also whether this would be a long-term solution, bearing in mind any future increase in requirements. The Working Group also thought that some practical work should be undertaken on verification; the various UKAEA groups suggested that this might include a trial inspection of plutonium production at Windscale, a trial inspection of the U-235 cycle at Dounreay and development of instrumentation at AWRE.[153]

---

151 TNA CAB 134/2241, N.D. (64) 3rd Meeting, 17 March 1964; TNA FO 371/176373, IAD 1036/84, A.D.F. Pemberton-Pigott to Mr Palliser, Working Party on the Cut-off, 10 July 1964.

152 TNA CAB 134/2241, N.D. (64) 5, Implications of a Possible Cut-Off of Supplies of Fissile Material, Report by a Working Party of the Committee, 27 April 1964.

153 TNA AB 65/186, M. Davis to Chairman, Cut-Off: Practical Investigation of Verification, 1 July 1965.

Penney approved Phase 1 of this work in order to develop detailed proposals.[154] This eventually went ahead under the codename Operation Libator in 1966.[155]

The NRDC discussed the Working Party's report on 5 May 1964.[156] Members quickly recognised that it would be unjustifiable for the UK to initiate costly technical action at that stage to meet the unlikely hypothesis of there being an effective cut-off by mid-1965. Any action that the UK might need to take, on the assumption that a cut-off agreement became effective by mid 1967, would depend largely on the extent to which the US would be prepared to help the UK to meet its defence requirements. Given the latent criticism in the US of the barter agreement for U-235 and tritium, officials certainly felt that it would be unwise to reopen these very lucrative deals. However, the military barter agreement had an option for the UK to purchase military grade U-235 and it was therefore open to the UK to buy additional material – up to 10 tonnes – as a safeguard against an early cut-off. This material would be both for weapons and for naval propulsion. Plutonium was different. It would be cheaper to make any additional supplies in the UK and as a safeguard against a 1967 cut-off the UKAEA might need to accelerate production for military purposes. The Chiefs of Staff were informed earlier that day that there would be a zero balance in 1970 between supply and demand, but that a small surplus would be recovered from obsolete weapons in the next three years. This would amount to about 15 per cent of total plutonium investment in weapons.[157]

Since the US was believed to want to negotiate a cut-off in association with other disarmament measures and to insist upon adequate inspection, it was unlikely that any agreement would be reached between the US and the USSR on a complete cut-off that would be effective by mid 1965. If, however, this eventuality arose, Britain would have to rely on US help to provide additional supplies of fissile material to enable the UK to subscribe to such an agreement. Alternatively, if this were not the case, the UK would have to stay outside such an agreement until such times as it had produced sufficient supplies for its defence programme. Penney advised the Committee at its next meeting in June 1964 that since both the US and USSR had large stockpiles, they might decide to

---

154 TNA AB 65/186, M. Davis to C. Allday, Production Group, Risley, V.H.B. Macklen, Reactor Group and Dr F. Morgan, AWRE, 28 July 1965.

155 TNA AB 65/186, D.B.B. Janisch to Dr F. Morgan et al., 30 November 1965. Dr Fred Brown was the Chief Inspector assisted by Ken Johnston and Dr Monk – all from AWRE.

156 TNA CAB 134/2241, N.D. (64) 4th Meeting, Implications of a Possible Cut-off of Supplies of Fissile Material, 5 May 1964.

157 TNA DEFE 19/125, COS Availability of Fissile Material, 33rd meeting/64 Part I, 5 May 1964.

take unilateral action instead of negotiating an agreement and thereby avoiding the technical difficulties of inspection and control and undesirable information disclosures inseparable from any negotiations. Unilateral actions of this nature might be taken at any time. Therefore the Committee felt that it was of some urgency for the UK to prepare a further technical study in greater depth on the technical implications of a cut-off. US criticisms of UK hesitations underlined the need for further study to formulate a British position quickly.[158]

The UK had made some enquiries in the US about the possibility of reinstating an option to purchase U-235 and the possibility of buying military grade plutonium. However, although the U-235 option was possible, Dr Harold Brown (Director of Defense Research and Engineering, Department of Defense) had told Zuckerman that there was little likelihood of the UK obtaining an option to purchase plutonium.[159] When the Committee considered this report, it thought that the US should be left in no doubt that an agreement to reinstate the UK option to buy military grade U-235 from the US would not remove all the difficulties facing the British in a cut-off, since UK supplies of military grade plutonium were still inadequate. However, it would be possible to give the US a general reassurance that provided UK plans for an accelerated plutonium production plan developed satisfactorily and the UK was permitted to exercise the U-235 purchase option, the UK would in due course be able to support the US in any cut-off negotiations. As a result, the UK decided to exercise the buy-back option for up to 10 tonnes of military grade U-235 at any time up to the end of the barter agreement in December 1969. Penney reminded the Committee too that he had previously made clear that in order to avoid interference with the weapons programme in the event of a cut-off by 1967, it was necessary that any decision about accelerated production of military grade plutonium should be made by 1 July 1964. The Minister of Defence advised the Prime Minister in July 1964 that if there were a cut-off agreement before 1969, UK worries about plutonium supply would remain. Until these were assured, the UK would be unable to subscribe fully to implementation of a cut-off.[160] MOD officials believed that since the UK could not start from a concept of 'plenty in stockpile', it could not commit itself to anything until the UKAEA's mixed production and processing plants were no longer required for the weapons programme, including production of material for barter exchange.[161] Home was invited to agree to the NRDC's recommendations that the US should be asked for an

158 TNA CAB 134/2241, N.D. (64) 5th Meeting, 25 June 1964.

159 TNA CAB 134/2241, N.D. (64) 5th Meeting, 25 June 1964.

160 TNA CAB 134/2241, Letter from Thorneycroft to Prime Minister, 7 July 1964.

161 TNA FO 371/176372, IAD 1036/81/G, R. Press, ACSA (N), MOD to DS 12, 15 June 1964.

option to purchase up to ten tonnes of U-235, that an accelerated programme of plutonium production that would advance the completion date to mid 1967 should be adopted for planning purposes and that fuel cans should be purchased for this forthwith. Home agreed.[162] Thus Ministers agreed in July 1964, subject to confirmation by 1 October, that the UKAEA should be authorised to purchase forthwith fuel element cans worth £200,000. In addition, it was agreed that the UK adopt for planning purposes an accelerated plutonium production programme that would advance the completion date to mid 1967. However, the October 1964 General Election intervened and the decision was deferred for the new government, but the same recommendation remained extant in a new MOD draft paper.[163] Despite this, MOD felt that there was no need to put this recommendation to new Ministers at that stage (late 1964) or in advance of more general decisions by the Labour government about the future UK weapons production programme. The paper was kept in reserve, but seems to have been overtaken by events: election of a new government.[164] However, Penney had been tasked to set up a working party, which was chaired by Dr Bob Press, MOD, to study the implications of effective verification of a cut-off in the production of fissile material for weapons purposes. Its main function was to prepare a statement to be used as an authoritative reference whenever a cut-off matter arose, with particular reference to the inter-play of technical considerations with negotiating positions that might be taken up by the UK, public statements that might be made and confidential exchanges with the US.[165]

This group concluded in mid December and noted that mid 1967 or possibly mid 1968 would be, in the group's view, the earliest date that the UK could implement a cut-off agreement as a pre-Stage One disarmament measure unless an alternative supply of military plutonium could be arranged, or unless the political need to adhere to that agreement superseded completion of the UK nuclear weapons programme. This conclusion assumed that three considerations still applied. First, fissile material requirements for the weapons programme as at mid October 1964 and approved by the previous government would still stand;

---

162 TNA PREM 11/5172, Home to Secretary of State for Defence, Ref. MO 18/1 July 7 1964 'Supplies of Fissile Material', 9 July 1964.

163 TNA FO 371/176373, IAD 1036/106/G, E.H. St G. Moss, D.S.3, MOD to P. Nicholls, HM Treasury, 13 October 1964.

164 TNA FO 371/176373, IAD 1036/112/G, B.A.B. Burrows to Mr Shepherd, 4 November 1964.

165 TNA CAB 134/2241, Nuclear Requirements for Defence Committee CO/P (64) 24, The Implications of a Cut-off in the Production of Fissile Material for Weapons Purposes with Particular Reference to Verification Considerations, Ministry of Defence, 18 December 1964.

second, that there would be accelerated plutonium production in the UK; and third, that accelerated delivery of U-235 from the US under barter arrangements would ensue. The purpose of a cut-off would be to inhibit the proliferation of weapons and, as would accord with the West's general approach to disarmament measures, to ensure verification would be introduced to deter possible diversions of fissile material to weapons purposes. The group concluded that the most comprehensive system of verification could not provide absolute assurance against cheating, either in the use of declared facilities or by the construction of clandestine facilities. Verification should therefore aim primarily to detect significant diversions in production and to deter smaller diversions. Verification effectiveness would depend on the rights of access to declared facilities and to suspected areas or facilities. Given the rights of access envisaged at that time and which the UK might accept, it might be possible to achieve control to an accuracy of five per cent for allowed uses of fissile material and ten per cent for declared production of such material. Such control would have an effective deterrent value against diversion of much less than five to ten per cent of throughput. Continuing weapon facilities, defence but non-weapon applications of fissile material and some declared production facilities for allowed uses would all impose some conflict between military security and verification requirements. In such facilities, a risk of diversion of some fissile material would either have to be accepted or military security and the policy of non-proliferation of nuclear weapon information would be prejudiced. Finally, the group recommended that practical investigations should be undertaken in the UK to test the verification proposals envisaged in their report.

The Labour Government and NRDC Deliberations: 1965

In January 1965 the NRDC determined that there was now a good case to rescind the confirmation decision on accelerated plutonium production. There were other weapons materials issues too; but recoveries and recycling of material from old weapons plus existing supplies of lithium would be sufficient to meet weapons programme requirements as then envisaged.[166] The NRDC noted later

---

[166] TNA PREM 13/2004, S. Zuckerman, MOD to Prime Minister, 13 January 1965. Post 1968 the UK would also become totally dependent on US for tritium and lithium. TNA PREM 13/2004, Extract from Zuckerman's Meetings with PM, 5 and 6 January 1968. For discussion on early UKAEA production problems see Donald McIntyre, *UK Nuclear History Working Paper Number 5: Project Crystal: Lithium 6 for Thermonuclear Weapons* at http://www.mcis.soton.ac.uk/Site_Files/pdf/nuclear_history/Working_Paper_No_5.pdf, accessed 31 March 2011.

in the year that this proposal had never been submitted to Ministers. Factors involved here were the current review of overall defence expenditure and the Polaris programme, where there was now a good chance that there would now be only four boats, possibly even three.[167] However, if the Atlantic Nuclear Force (ANF) – the Labour government's alternative to the Multilateral Force – were to proceed, there would still be a shortfall of plutonium by mid 1967 and a lesser one by 1968. (The ANF was a factor in the considerations in so far as the Foreign Office believed that the UK should not prejudice a future ANF force by agreeing to a cut-off at a level that would prevent any UK warhead contribution.[168]) Foreign Office officials believed that this fact should be made clear in any submission to Ministers. Sir Bernard Burrows (Foreign Office) told the NRDC that British spokesmen on disarmament had been considerably embarrassed by the UK's inability to support the US proposal on a cut-off. Ministerial embarrassment remained a Foreign Office concern; its Ministers did not like not being able to welcome such an important US disarmament initiative. This was one of the main reasons why the Foreign Office was so keen on securing assurances from the US on supply to meet MOD concerns.[169]

Although the US had been asked formally about fissile material supplies, no reply had been received by June 1965. The issue was raised again in June in the context of a draft US cut-off resolution for the United Nations Disarmament Commission (UNDC). One of the features of this text was approval for a cut-off. However, Whitehall departments considered that it would be very difficult for HMG either to abstain on this resolution, thus exposing disagreement with the US, or to support it and thus appearing publicly to commit the UK to supporting the concept. MOD officials were of the view that, given the fissile material shortages, defence considerations should be the over-riding factor governing the UK position and that nothing should be done that would prejudice UK future defence interests.[170] Ministers eventually agreed that the UK would support the text provided that the US were also asked to give some sort of advance assurance that in the event of a cut-off they would look after UK fissile material needs. MOD calculated that if a cut-off were to become effective

167 TNA CAB 134/2241, N.D. (65) 1st Meeting, Item 1: Nuclear Policy and Underground Tests 1965/66, 22 January 1965.

168 TNA FO 371/181374, IAD 1036/8/G, E.J.W. Barnes to Lord Hood, Destruction of Nuclear Weapons, 13 May 1965; IAD 1036/17/G, J.E.D. Street to Mr Stuart, WOCD, 2 July 1965.

169 TNA FO 371/187445, IAD 1036/18/G, J.E.D. Street to PUS, Fissile Material for Weapons Purposes, 8 September 1966.

170 TNA DEFE 13/871, R.J. Andrew, Head of DS 12, MOD to PS/S of S, 3 June 1965; R.J. Andrew to PS/S of S, 9 June 1965.

in April 1967, the UK would need 400 kg of plutonium.[171] As it transpired the US withdrew its text; but in reply to the UK approach for an assurance over supply the Americans made it clear that this would not be possible without a full study of the UK supply situation. A US Aide Memoire of 14 June 1965 recognised the UK's problems should a cut-off agreement be reached and agreed that the US would consider what assurances could be given to meet the UK's difficulties. It was then agreed that the UK would review the long-term UK requirement of fissile material for weapons purposes and would then discuss the matter with the US.[172]

Even though the US was not currently pressing for a cut-off, there was a chance that the Americans might revert to it at any time. The Russians might be prepared too to contemplate support for a cut-off in 1968, which the Joint Intelligence Committee (JIC) thought would be the earliest date by which Soviet stocks might be sufficient.[173] During discussion of the last US cut-off resolution for the UNDC in 1964, the UK had sought assurances about accelerated supplies under the current barter agreement. The US had said that they could not arrange this without a full study of the UK position, information about which they would require as soon as possible. Sir Patrick Dean (Permanent Under-Secretary, Foreign Office) advised that this did not betoken any new negative attitude to Britain's deterrent even though there were differences in Washington on the desirability of a continued UK capability.

However, the longer the delay between 1965 and 1967, the easier it would be for the UK as fissile material stocks increased to meet weapon requirements.[174] Nevertheless, the Foreign Office now felt that it was time to invite Ministers to reconsider what steps could and should be taken to make it possible for the UK to accept a cut-off, in the unlikely event that one came into being before UK stockpiling was complete. This would have the advantage that if suitable arrangements could be made with the Americans on a contingency basis, the UK would then announce its complete support for the cut-off proposal and Britain would thus be spared the ever-present risk of friction with the

[171] TNA DEFE 13/871, R.J. Andrew, Head of DS 12, MOD to PS/S of S, 10 June 1965.

[172] TNA FCO 10/162, Arms Control Measures (General) Cut-off of Fissile Material for Weapon Purposes, Supply of Fissile Material Draft Aide Memoire to State Department Minute to Mr Andrew and Mr Street.

[173] TNA CAB 134/2241, N.D. (65) 2nd Meeting, 24 November 1965.

[174] TNA CAB 134/2241, CO/(64) 24 The Implications of a Cut-off in the Production of Fissile Material for Weapons Purposes with Particular Reference to Verification Considerations, Ministry of Defence, 18 December 1964.

Americans at disarmament discussions.[175] Foreign Office officials were keen that a further approach should be made to the US to seek clear assurances on supply, but neither the MOD nor UKAEA were convinced that the time was ripe.[176] Neither was Sir Solly Zuckerman. Sir Patrick Dean's brief for the NRDC meeting on 24 November 1965 noted that the UK was obliged to say something to the US; that HMG wished to be able to support US proposals on the cut-off; and that the UK was on strong ground in asking the Americans to put the UK in a position where it could do so. This was set out in a Foreign Office paper for the NRDC; Lord Chalfont, Minister of State, felt strongly that the situation in which the UK was unable to support the US could not be allowed to continue irrespective of Russian support for a cut-off.[177] If an approach were to be made, then the US had made clear in June that it would have to be supported by detailed figures and explanations as the US could not give the UK a blanket assurance in general terms. NRDC members agreed that before an approach to the US could be properly considered, it was necessary to explore further with the US certain aspects of its current cut-off proposals.[178] In particular, would the cut-off apply to weapons purposes only or extend to research and propulsion and was it the US's intention that a cut-off should be verified or simply left as a gentlemen's agreement? If the latter, it might appeal to the Russians and thus enter into force much more quickly. Technical talks were held in Washington on 16 and 17 December 1965 to find out. The US response was that in answer to the first concern, propulsion should be excluded and as for the second, some sort of verification would be needed to deter cheating, but the regime for the existing nuclear weapon states would be less onerous. In fact US ideas for what would be required for the non-weapon states were much laxer than the UK's.[179] Other detailed technical verification issues were also examined.[180] The Americans had also made clear that they would be emphasising their initiative at the next session of the ENDC, which for the Foreign Office only underlined the need for

---

175 TNA CAB 134/2241, N.D.(65) 6 Cessation of Fissile Material Production for Weapons Purposes (The Cut-off), Memorandum by the Foreign Office, 9 November 1965.

176 TNA FO 371/181375, IAD 1036/31/G, J.E.D. Street to Sir B. Burrows, 8 November 1965.

177 TNA FO 371/181375, IAD 1036/34/G, The Cut-Off Item 1, AEDD, 22 November 1965.

178 TNA FO 371/181375, IAD 1036/42/G, J.E.D. Street to Sir B. Burrows, 11 January 1966.

179 TNA FO 371/181375, IAD 1036/42/G, J.E.D. Street to Sir B. Burrows, 22 December 1965.

180 TNA FO 371/181375, IAD 1036/42/G, Draft Memo for NRDC Cessation of Fissile Material Production for Weapons Purposes (The Cut-Off).

the UK to secure a supply assurance from Washington. Increasing US emphasis would only bring into greater prominence the fact that the UK was not giving full support to it and this would cause increasing embarrassment to HMG. The NRDC should endorse the need for another approach to the US.

Britain's initial decision to increase plutonium production rested on an arbitrary assumption that a cut-off treaty might be agreed and enter into force in 1967. Weapons material production would then have to be completed by then instead of 1969 as planned. If there were four Polaris boats and a cut-off on 1 April 1968 and then there was no acceleration in plutonium production, there would be a shortfall of 160 kg instead of 475 kg.[181] This was one of the reasons why the second Polaris Economy underground nuclear test had been so important given the failure of the first shot. A saving of 100 kg meant a saving to the tune of £2.5 million and would also enable the UK to carry out somewhat earlier any disarmament agreement involving a cut-off.[182] The Foreign Office thought that it would be very dangerous to assume that the US would meet UK needs if required. Political contexts that might apply in future could not be foreseen, but it was probable that the Russians would not agree to a cut-off unless the UK participated as well. Indeed, it could be fairly assumed too that the US would put the UK under a fair amount of pressure to join in. Moreover, there was a danger that at the time the US might be unwilling to make good UK stockpile needs unless they could justify this action by reference to a binding obligation assumed previously. In these circumstances, the Foreign Office thought that Ministers might feel that they could not refuse to co-operate in a cut-off, in spite of the effect it might have on the defence programme. In working out what to do about this, the UK had to think about both current requirements and any possible future needs as present plans envisaged no reserve. Future contingencies might include additional warheads for the ANF, but it was difficult to see beyond the current needs. As for the barter agreement, perhaps the best option would be to persuade the US to supply all the rest of the scheduled deliveries of U-235 at once to be paid for by annual deliveries of plutonium.

There were two alternatives for plutonium. If a cut-off were delayed, the UK could embark on an accelerated production schedule. This of course had been considered in 1964 and a decision to do so was overturned by the Labour government as part of its wider defence review. Alternatively, the UK could ask the US if there were any possibility of supplies from them, but this had been

---

181 TNA DEFE 24/291, (*Draft*) Brief for PUS on Nuclear Weapons Policy, 12 January 1965.

182 TNA CAB 21/6037, Nuclear Tests (General) Draft Paper for Cabinet Polaris Economy Test, Memorandum by the Secretary of State for Defence. (See also TNA CAB 21/6038 for more on cut-off.)

rejected the last time that the UK had enquired. To do so again might provoke the US into reconsidering the terms of the Barter agreement – something to be avoided if possible. The choice between accelerating UK production and buying some plutonium from the US would in practice depend on financial considerations as well as on the degree of risk at the time and whether public knowledge of such acceleration would then be politically acceptable. Foreign Office advice was that Ministers should be told that an approach to the US should be made as soon as possible for assurances of supplies in the event of a cut-off. The NRDC agreed and its chairman noted that in light of the discussion a submission to Ministers would need to be prepared explaining the need to clarify US proposals and recommending a technical discussion on the nature, scope and implementation of their cut-off proposals. At the same time, the measures to increase availability of UK plutonium should also be recommended to Ministers as a prudent precaution against an early cut-off.[183]

Penney told the NRDC at the end of 1965 that UK plutonium supplies for weapons purposes could be readily improved, without additional cost, by drawing on a small 'civil reserve', delaying certain civil work and using a contingency reserve held against delays in obtaining plutonium to meet UK obligations under the barter agreement. He would welcome discussions with the US on the scope and nature of their cut-off proposals, in particular whether they envisaged that the agreements would be backed up by inspection, and whether they would apply only to weapons or to all military uses including naval propulsion. If they were to apply to weapons only, it was not clear how practical arrangements for inspection could be made. If inspections were involved, Penney thought that discussions on a cut-off would go on for years.

Verification and Fissile Material

Penney went to Washington in December 1965 to discuss the UK Working Party's report on verification of a cut-off.[184] The upshot of this meeting was that it remained the policy of the US that fissile material for nuclear propulsion should be excluded from a cut-off, whilst accepting that there would be a consequent risk of diversion. Verification would be needed for other facilities, but the burden on the existing weapon states would not need to be as extensive

183 TNA CAB 134/2241, N.D. (65) 2nd Meeting, 24 November 1965.

184 TNA CAB 134/2241, N.D. (66) 1, Cessation of Fissile Material Production for Weapons Purposes (The Cut-off), Memorandum by the Foreign Office, 15 March 1966.

as that which might be needed for non-nuclear weapon states.[185] Foreign Office officials believed that the UK would have to be able to declare full support for an early cut-off, even if there was little likelihood that such an agreement would be concluded in the near future. Increasing US emphasis on its cut-off proposal would bring into greater prominence the fact that the UK was not giving full support to it and would cause ever-increasing embarrassment to HMG. It was also clear that the US wanted to use the IAEA safeguards system for any cut-off verification regime. However, the US did agree with the British view that there would need to be much greater comparability in treatment of nuclear weapon states and non-weapon powers. Therefore IAEA procedures would have to be further evolved to supersede the Agency's then current safeguards regime as set out in the Agency document INFCIRC 66.[186] Until then the UK had considered the two verification requirements separately – cut-off and IAEA safeguards.[187]

Penney thought, following recommendations from UK officials who attended the December talks, that the UK should set up a new standing committee with representatives from MOD, the Foreign Office and Ministry of Technology to deal with the technical aspects of both issues in a single place given the connections between the issues.[188] (Con Allday, one of the two UKAEA officials who went to Washington, thought that if the US were left to make the running the result would probably be less practicable and more expensive, practical work in the UK could also be used to influence IAEA work.[189]) Penney recommended to the Foreign Office that such a body should be created. The UK would need to decide whether it should follow the US approach in seeking to apply IAEA safeguards for all purposes. If this were so, then it would have to be amended considerably to make it compatible with what an effective cut-off agreement would require, and thus those primarily concerned with cut-off should be given the opportunity to influence the more immediate issues that arose in the IAEA context. Dr Bob Press (Assistant Chief Scientific Adviser (ACSA (N)), MOD) also agreed; he thought that since the US now accepted that IAEA safeguards for a cut-off agreement had yet to be evolved, it was clearly in the UK's interests to influence that evolution, as far as possible through practical experience

---

185 TNA FO 371/187445, IAD 1036/3, US–UK Meeting on Verification of a Cut-Off of Production of Fissile Materials for Weapons Purposes.

186 TNA CAB 165/745, Henry Hardman, PUS, MOD to Sir Maurice Dean, Ministry of Technology, 14 March 1966.

187 TNA CAB 165/745, W.G. Penney, UKAEA to Sir Bernard Burrows, Foreign Office, 21 January 1966.

188 TNA AB 48/642, Dr M. Davis to Chairman, 22 December 1965.

189 TNA AB 65/186, Note of a Meeting Held in the Chairman's Office on 20 January 1966.

gained in the UK.[190] The Foreign Office shared Penney's view and suggested that a new interdepartmental committee should indeed be established, chaired by the UKAEA and which would report to both the senior level bodies – the NRDC and Atomic Energy Official Committee (AE (O) C) – as required.[191] This was agreed at a meeting chaired by the Cabinet Secretary on 16 March 1966 with a remit to advise both Committees on technical questions relating to the production, acquisition or control of nuclear materials, including the application of international safeguards to such materials and proposals for a cut-off in the production of fissile material for military purposes. Ted Newley, Director AWRE, chaired it.[192] In addition Penney asked the UKAEA to establish its own body – the Cut-Off and Safeguards Working Party, which included representatives from the MOD, Ministry of Technology and the CEGB.[193] This group's main task was to work out and consider inspection trials, in particular those that were already started as verification trials in aid of cut-off, but which would henceforth also be in aid of safeguards.

This new body was the Committee on Safeguards for Nuclear Materials and it held its first meeting on 25 May 1966 and one of the first points discussed was that the conversion of the UK's existing continuous reprocessing plant to batch-working would be costly and would have a disruptive effect on the UK civil and military programmes. However, its members thought that it would probably be practicable to put Windscale under safeguards on a campaign basis, as this would make it possible to reprocess military material at times when the plant was not subject to inspection. The UKAEA was tasked to look at both the technical and economic aspects of such a proposition for future consideration.[194] Authority experts concluded in July that there was no technical advantage to the IAEA in being allowed to inspect a campaign containing safeguarded material, but if the UK felt there was some political merit in allowing access, it could be re-examined in greater detail. It would still not be possible, given the then present

190 TNA AB 65/186, R. Press, ACSA (N), MOD to Sir William Penney, UKAEA, 22 December 1965.

191 TNA CAB 165/745, B.A. Burrows, Foreign Office to Sir William Penney, UKAEA, 1 March 1966.

192 TNA CAB 165/745, Note of a Meeting Held in Sir Burke Trend's Room, 16 March 1966.

193 TNA AB 9/2430, Minutes of the First Meeting of the Cut-Off and Safeguards Working Party, 24 February 1966.

194 TNA CAB 134/3014, Committee on Safeguards for Nuclear Materials: Meetings 1–6; Papers 1–17, 1967 NM (66) 1st Meeting, 25 May 1966.

military commitments, to place the plant itself under safeguards.[195] It would cost about £20m to build a separate reprocessing plant and this would not come into operation for about four years.

The UK could still not accept safeguards in early 1966 on the whole of its civil programme for two basic reasons. First, for historical reasons, and reason of economy, much of the fissile material for the UK civil and military programmes was still processed in the same plant – Windscale. Moreover, the civil and military research and development programmes were to a certain extent intermingled at AWRE. If the material in these installations, which was destined for civil use, was subject to safeguards, then considerable security problems would arise if inspections were permitted. Moreover, because of the method of reprocessing, it would mean a loss of time and efficiency to separate the two programmes. Second, the impact on the UK's research and development effort would not be matched by its main competitors and the UK would consequently lay itself open to commercial espionage.[196]

The NRDC examined cut-off issues again at its first meeting in 1966.[197] As in previous years the same basic problems remained, an early cut-off would catch the UK short of the supplies needed for its defence needs. US assurances of supply were required to avoid continuing anxieties over shortfalls and being out of step with US disarmament policy imperatives. Glenn Seaborg remarked that US efforts on behalf of a cut-off and transfer to civil purposes reached their peak in 1965 and early 1966.[198] Although it seemed that an immediate cut-off was an unlikely prospect, UK interests were best served by working out its own positions on the technical issues involved. In technical discussions, the UK might hope to influence US policy in much the same way as it had done over nuclear testing.

In further discussion on the fissile material supply problem in September 1966, the NRDC agreed that Ministers should be asked to endorse the view that a cut-off should not apply to materials for nuclear propulsion and that the UK should earmark the national stock of uranium ore specifically for military purposes. Furthermore, it was also agreed that a recommendation for a minimum reserve of 2,000 kg of U-235 should be included; one of the objectives for this reserve was to keep open the possibility of manufacturing

---

195 TNA CAB 134/3014, The Safeguarding of the Reprocessing Plant at Windscale, Note by the Atomic Energy Authority, 13 July 1966.

196 TNA CAB 134/3014, Political Aspects of Safeguards, Note by the Foreign Office, 17 May 1966.

197 TNA CAB 134/2241, N.D. (66) 1st Meeting, Item 3: Cessation of Fissile Material Production for Weapons Purposes, 22 April 1966.

198 Seaborg, *Stemming the Tide*, p. 400.

nuclear warheads in the future. However, Ministers would not be asked to commit themselves to such a course of action, only to keep the option open. It is not clear yet what action Ministers took on these recommendations; the NRDC memorandum that provided the basis for them has been retained under Section 3(4) of the Public Records Act 1967. Denis Healey suggested to his Nuclear Policy Committee colleagues in mid December 1966 that the UK's fissile material supplies could be solved by asking the US to accelerate their deliveries under the barter agreement and by building up a reserve for weapons purposes taken from stocks earmarked for naval propulsion.[199] Ministers appear to have decided that a fissile material reserve should be created.[200] We do know, however, that the MOD sought an assurance that enriched uranium produced in the future enrichment organisation (URENCO) could be used in the UK weapons programme, which implies that anxieties over future assured supply on the long term were still a cause for concern even in 1969.[201]

Cut-off and Weapons Requirements 1967

The Ministerial Committee on Nuclear Policy agreed on 9 January 1967 that assurances should be sought from the US that, in the event of an international agreement to halt fissile material production for weapons purposes, the US would supply the UK with sufficient fissile material to complete the UK programme. This would then enable the UK to support fully in pubic this important US arms control proposal.[202] It also decided that 2,000 kg of available U-235 should be set aside as a weapons reserve and that the future requirements for fissile material for naval propulsion should be met by purchases from the US or by domestic production.[203] The Foreign Secretary wanted to confirm that Wilson was still content that such an approach, in the form of an Aide Memoire, should be made in Washington, but Wilson asked George Brown (Foreign Secretary) to

---

199 TNA FCO 10/162, J.E.D. Street to Mr Rennie, Fissile Material Supply Problems, 16 December 1966.

200 TNA FCO 10/162, V.H.B. Macklen, ACSA (N), MOD to J.E.D. Street, Foreign Office, 10 March 1967.

201 TNA PREM 13/2569, Note for the Record, Sir Solly Zuckerman's Call on the PM, 28 July 1969.

202 TNA FCO 10/162, Arms Control Measures (General) Cut-Off of Fissile Material for Weapon Purposes, R.E. Parsons, Foreign Office to A.M. Palliser, 10 Downing Street, 25 April 1967.

203 TNA DEFE 19/125, D.C. Fakley D Sc 3 to AUS (R&D), 16 February 1967.

raise the matter orally at the 18 May Nuclear Policy Committee meeting.[204] The Foreign Office's Atomic Energy and Disarmament Department recommended that the Foreign Secretary should emphasise *inter alia* that the UK had long wanted to give full support to the US cut-off initiative, but in order to do so the US had to provide assurances that UK defence needs would be met and they had specifically asked for the details of British outstanding requirements; hence the need to approach Washington with a formal note setting all of this out. Ministers appear to have agreed as the Foreign Office instructed the Embassy on 19 May 1967 to hand over the Aide Memoire. The Embassy however sought further clarification on a number of points, including the comment on the final sentence of the proposed Aide Memoire, which they felt offered too much of a hostage to fortune.[205] Ian Smart (Washington) argued that the Aide Memoire handed over in August 1964 said that UK support for a cut-off would have to be in proportion to the extent to which UK requirements were satisfied; the next one in June 1965 was even less dogmatic and said only that it would greatly help HMG to have an assurance that UK fissile material needs would be safeguarded. Neither document suggested that such an assurance would be essential in determining UK attitudes; such an inadvertent threat might provoke an adverse US reaction. It would be better to soften the line or better still omit the offending phrase, 'assurances of this nature are essential if we are to play our full part in promoting and supporting the measures of nuclear disarmament which the West has put forward'. London accepted that the last sentence should be changed, but made clear that there should be no substantive discussion when the Aide Memoire was handed over – a team would come to Washington once the Americans had had time to consider the UK request.[206] Smart handed the document over on 16 June, but by mid July there was still no US response.[207] However, initial reactions a few days later from the State Department were positive, but further inter-agency deliberations were needed before an affirmative response could be given. Furthermore, it was already clear that the US would make a distinction between U-235, already covered by the barter agreement and plutonium that might need a new agreement. The Ambassador was called in to see Foy Kohler (Deputy Under Secretary in the State Department) to discuss the Aide Memoire on 25

[204] TNA FCO 10/162, J.E.D. Street to Private Secretary, Ministerial Committee on Nuclear Policy the 'Cut-Off', 16 May 1967.

[205] TNA FCO 10/162, I.M.H. Smart, Washington to J.E.D. Street, AEDD, Foreign Office, 29 May 1967.

[206] TNA FCO 10/162, J.E.D. Street, AEDD to I.M.H. Smart, Washington, 9 June 1967.

[207] TNA FCO 10/162, I.M.H. Smart, Washington to R. McAndrew, AEDD, Foreign Office, 13 July 1967.

July. The US disposition towards the request was favourable. However, a great deal of work remained to be done, including consultations with the United States Atomic Energy Commission (USAEC) and Joint Congressional Committee of Atomic Energy (JCAE).[208] But the UK request for an assurance on plutonium was more complicated than that for U-235 and if it became necessary to sell such material to the UK, then a new agreement rather than a modification to the existing barter arrangements would be required. Thus there would still be some delay whilst all this was looked into before a formal reply could be given. However, by November 1967 no such response had been provided; this was raised by Sir William Cook (MOD) during a visit of Dr Gerald Tape, USAEC Commissioner, to London for a meeting on HEU supplies for naval reactors. Cook told Tape that the UK now regarded the enquiry on assurances of fissile material supply to be increasingly of academic interest since the prospects for a cut-off before the UK had acquired the stocks required for its weapons programme were very remote. Tape replied that this was the US attitude too and he thought it unlikely that the US would make any formal response to the Aide Memoire.[209] And so it proved, but not before the Foreign Office instructed Washington to have one more go at eliciting a formal response, but Smart thought that there was no need to press the matter at that stage.[210] The State Department confirmed on 2 February 1968 that the Americans did not intend to provide a formal reply to the Aide Memoire.[211] It had become clear to them that the only reply that they would be able to give would be so vague as to serve little purpose. Tape had reported his conversation with Cook and he had, quite reasonably, taken from this that the UK itself no longer expected to receive a formal reply. As an afterword to this part of the UK–US debate on cut-off we should note a comment that Zuckerman made to Wilson in July 1968. He told the Prime Minister that the UK had been at odds with the Americans for years about the possibility of a cut-off. The US line had been to promote the idea: 'our people have been blocking it.' He went on to recall that the Americans had been irritated in the past by the UK's failure to give them the public support that

---

208 TNA FCO 10/162, I.M.H. Smart, Washington to R. McAndrew, AEDD, Foreign Office, 25 July 1967.

209 TNA FCO 10/162, Record of a Meeting between Commissioner Gerald Tape, USAEC and Sir William Cook, CA (P) MOD, 2 November 1967 to Discuss Enriched Uranium Supplies and Other Matters, paragraph 7 D Sc 3, 3 November, 1967.

210 TNA FCO 10/162, R.C. Hope-Jones to I.M.H. Smart, Washington, 21 December 1967, I.M.H .Smart, Washington to R.C. Hope-Jones, AEDD, Foreign Office, 4 January 1968.

211 TNA FCO 10/162, I.M.H. Smart, Washington to R.C. Hope-Jones, AEDD, Fissile Material Supplies, 4 February 1968.

they had hoped for. In fact they refused or avoided giving the UK the assurances about supplies in the event of a cut-off being negotiated before 1969. There was no 'goodwill' in this particular area of disarmament. The implication of a special relationship was, in Zuckerman's opinion, vastly overemphasised.[212] UK defence of its nuclear interests at the expense of arms control imperatives had thus not gone unnoticed, but as we have seen, officials and Ministers had not enjoyed being impaled on the horns of this particular defence and diplomatic dilemma and had tried to unhook themselves.

The UK was now more relaxed about a cut-off since officials had concluded in early 1967 that it would not now need any additional U-235 for weapons purposes above the amounts to be provided under the barter agreement, instead the UK intended to ask the US to provide the remaining balance before any cut-off agreement entered into force. Plutonium shortages were also an issue and the UK planned to ask the US to sell the UK the necessary quantities should the need arise. All of this was precautionary as the UK did not think that the Russians were likely to change their attitudes, so the chances of an agreement entering into force before 1969 were slim indeed. However, HMG could not prudently assume that such a delay would in fact occur and in keeping with the general Western line that the UK must be prepared to accept immediately for itself disarmament proposals that it had supported in principle, US assurances were essential if the UK were to play its full part in supporting Western disarmament proposals. From the MOD perspective an inability to obtain further supplies of fissile material for weapons after the end of 1969, either because of a cut-off or because of a ban on transfers, would not prevent completion of the then current weapons programme; nor would it unduly restrict UK freedom to re-arrange the composition of the weapon stockpile in future, but it would entail inevitably some loss of flexibility in defence planning.[213] We must assume that defence planning included both Polaris and the WE177A for the RAF. Although the first order for the 44 kiloton weapon the WE177A was not placed until early 1969, it had been the intention since the early 1960s that this weapon should be introduced.[214] No new supplies of fissile material were required as stocks

---

212 TNA CAB 168/125, Zuckerman to Prime Minister, Further Measures of Disarmament (OPD 968) 54, 25 July 1968.

213 TNA FCO 10/162, Mr D.C. Fakley, D Sc 3, MOD to R. McAndrew, AEDD, Foreign Office, 14 August 1967.

214 TNA AIR 20/12080, J.G. Matthews DD ops(S) RAF to DD Mech Eng 3 (RAF), Future Deployment of Nuclear Weapons – Availability of Existing SSAs for Conversion to Other Purposes; British Nuclear Weapons in Germany, D.C. Humphreys, Head of DS 9, 27 September 1968; J.D. Thirlwell A/D of Ops (B&R) (RAF) to D Air Plans WE177A – Weapon Requirement, 21 November 1967; see also TNA DEFE 19/125.

were already available from stockpile and from obsolescent high level delivery weapons (Red Beard) as these were phased out.[215] It had been decided in 1965 to defer the WE177A given the need to introduce the high-yield WE177B first. The Cabinet Defence Committee had decided on 23 January 1963 that the high-yield version should be developed in order to enable the RAF to have a low-level capability in order to defeat Soviet air defences.[216] Cancellation of Skybolt meant that in 1963, the WE177B variant took priority and this was intended initially for the TSR 2, which could fly under air defences and then later for 48 V-Bombers until such time as Polaris took over the deterrent role.[217] Plutonium would be recovered from obsolescent Red Beards for use in the WE177A.[218]

The UK Voluntary NPT Safeguards Offer 1967–71: Military Exemptions

On 4 December 1967 the UK made a voluntary offer to accept NPT safeguards. This aim was to persuade the non-nuclear weapons states that the UK would be willing to suffer similar inconveniences, economic or otherwise, which safeguards procedures might impose on them for the sake of the NPT and to persuade these states to vote for the draft Treaty in the UN as well as sign and ratify it.[219] Nevertheless, this offer was not intended, and could not be allowed, to hinder or reduce the availability of material or facilities for UK nuclear weapons

[215] TNA AIR 2/18209, WE177A – Initial RAF Production Order, J.M. Nelson F.6 (AIR) to M.G.F. Hall, Treasury, January 1969; TNA AIR 20/12198, Deployment of Nuclear Weapons to Germany, Note by the MOD in D.C. Humphreys, Head DS9 to DUS (P) Nuclear Weapons, 12 May 1969; TNA DEFE 24/304, Draft Minute from Healy to Prime Minister, Nuclear Weapons, 30 May 1969 under cover of minute from R. Haynes AUS (AS) to DUS (P) Nuclear Weapons; this refers to Red Beard as high level.

[216] TNA AIR 8/2393, Weapons for Bomber Command VCAS to Secretary of State, 5 April 1963, Cabinet Defence Committee Reference D (63) 1st Meeting, 23 January 1963. This would take about three years and its development would cost up to £2 million in addition to the costs of developing the corresponding low-yield tactical weapons.

[217] TNA CAB 21/5728, G. Plowden to Mr Rogers, V-Bombers and Nuclear Weapons, 1 July 1965.

[218] A total of 87 WE177As appear to have been built. One source notes that that 120 WE117As were to be built by April 1975, 2–3 per month – aim for Plan R was 230; 53 WE177Bs were be retained in service as an insurance throughout the LTC, September 1970 – September 1979. This does not appear to have been the case. Some of the WE177A orders were changed into orders for the WE177C of which some 125 appear to have been produced in total.

[219] TNA CAB 134/3015, NM (68) 15, Principles for Safeguards Agreement between the IAEA and the UK to Cover the UK Offer in the Context of a NPT Note by the MOD, 24 October, 1968.

and nuclear military activities.[220] If it were so allowed, the UK would find itself drifting into an undeclared cut-off and it was in no position to accept such a thing at that time. The problem therefore was to devise a 'safeguards' system for nuclear weapon states in the context of the NPT that would impose the full severity of safeguards procedures on their civil programmes, but which would not hinder their legitimate military programmes (by excluding military R&D activities at civil sites, Windscale and Capenhurst or civil work at military sites such as material used in plutonium fuel manufacture at AWRE).[221] The work at Capenhurst was so inter-mixed that physical separation was not a feasible proposition.[222] In the UK context this would mean that inspections would only be applied to nuclear materials inside declared areas, or in transit between one declared area and another. There could be no right of pursuit of nuclear material outside such declared areas, nor could the origin of such material be stated when it entered such areas. At Springfields for instance, there would be no objection to occasional inspection, but the UK could not allow continuous quantitative monitoring of throughput of fuel elements because these would include those designated for the military reactors.[223] This limit on following through of safeguarded material would, however, enable the UK to consider extending its voluntary offer to some mixed facilities without having to withdraw such facilities every time material intended for weapon purposes was being processed. However, the UKAEA and Ministry of Power noted that the transition from a safeguarded condition to an unsafeguarded one and back again, possibly on several occasions, would cause administrative difficulties for operators.[224] At that time the IAEA system allowed a choice of attaching safeguards to material or facilities. In the case of Bradwell, the safeguards were on the facility and it would be possible to extend the Bradwell arrangement by putting additional power reactors under safeguards. That said, UK defence interests could be severely impaired if safeguards were thus applied since material would automatically attract safeguards – at this stage the UK envisaged the possibility of using

[220] TNA DEFE 19/61, V.H.B. Macklen, MOD to M.I. Michaels, Ministry of Technology, 1 July 1968; V.H.B. Macklen, MOD to Mr E.F. Newley, Director, AWRE, 10 June 1968.

[221] TNA CAB 134/3014, NM (67) 8, International Safeguards on Nuclear Material in the United Kingdom, Elaboration of the Offer Made in NATO, Note by the Secretaries, 7 June 1967.

[222] TNA DEFE 19/61, D. Healy to Foreign Secretary, 13 May 1968.

[223] TNA CAB 134/3014, NM (67) 4th Meeting, Item 2: International Safeguards on Nuclear Materials in the UK: Elaboration of Offer Made to NATO, 29 June 1967.

[224] TNA CAB 134/3015, NM (68) 17, The Effect on the UK of IAEA Safeguards, Note by the UKAEA and Ministry of Power, 11 November 1968.

plutonium as a fuel for naval propulsion and this was one of the main areas to be shielded from safeguards. Officials thought that it might be necessary in future to replace production of plutonium for such purposes in the Calder Hall and Chapelcross reactors with production in a civil power station.[225] By contrast, only those materials that the UK received by import with attached safeguards, or those materials voluntarily offered were retained in this regime. For this reason officials studying the problem strongly recommended that any elaboration of the UK voluntary offer made it clear that it referred to material and not facilities.[226]

The Newley Committee eventually decided that it would be premature to consider the effect of safeguards on the UK until it was known what safeguards were to be applied to the non-nuclear weapons states ratifying the NPT.[227] By September 1969 a great deal of work had been done on the practical considerations of the problems of inspecting nuclear facilities and this had led to the development of a philosophy of safeguards that would provide for an efficient, economic and non-intrusive system. A solid core of practical work on the development of inspection techniques and measurement instruments had arisen from these studies and currently formed a large part of the safeguards' investigational programme. An important role had been played in the international sphere where the UK had been able to influence individual nations and the IAEA both in practical techniques and in the general philosophy of safeguards. This role could not have been performed without the investigational programme.[228]

As for the potential impact on UK facilities the MOD was tasked to prepare a paper on the problems likely to arise in the implementation of the UK's voluntary offer on safeguards, but this was not produced before the Newley Committee was dissolved on 3 September 1970 on creation of the General Sub-Committee of the Official Committee on Nuclear Policy.[229] In the event the

---

225 TNA CAB 134/3014, NM (67), Elaboration of Offer Made in NATO, Note by the UK Atomic Energy Authority, 23 June 1967.

226 TNA CAB 134/3014, NM (67) 8, International Safeguards on Nuclear Material in the United Kingdom, Elaboration of the Offer Made in NATO, Note by the Secretaries, 7 June 1967.

227 TNA CAB 134/3015, NM (68) 8th Meeting, Committee on Safeguards for Nuclear Materials, Item 2, 13 November 1968.

228 TNA AB 9/2430, Item 4: A Review of the UK Present Effort on Safeguards, Minutes of the Fifteenth Meeting, 30 September 1969. The UK effort cost some £100,000 per year and involved six to seven individuals full-time.

229 TNA CAB 134/3017, NM (70) 11, Dissolution of the Committee Note by the Cabinet Secretary, 3 September 1970; TNA CAB 134/3017, NM (70) 2nd Meeting, 26 June 1970, Item 2; NM (70) 1st Meeting, Item 1, 22 April 1970.

paper was not ready until October. The new committee's secretary circulated a draft in advance of its first meeting.[230] One of the key principles was that the UK voluntary offer should not hinder or restrict in any way the UK's legitimate military activities and in particular should not constitute an unofficial cut-off; neither should safeguards increase costs to defence. AWRE, the Royal Ordnance Factory at Burghfield and the Admiralty Reactor Test Establishment (ARTE) Dounreay, would all have to fall outside any safeguards inspection system at all times and under all circumstances. Other UKAEA sites, which had, or might have an element of defence work within them, would have to be excluded temporarily while any sensitive military work was underway. MOD accepted too that safeguards should apply to fissile material and not primarily the facilities in which it happened to be held. Nevertheless for security reasons MOD insisted that there were establishments and enclaves within establishments to which no access could ever be permitted and the inspections would need to be excluded from other establishments at certain times when sensitive defence work was in progress. The UK would also need to be able to withdraw material from safeguards for any military purpose, but when it ceased to be required, it could be returned. Military material would never attract safeguards even if it passed through predominately civil sites and neither would civil material if it entered exclusively military sites. All this would need to be negotiated in a separate bilateral agreement with the IAEA.

The Committee considered the MOD paper at its first meeting in 1971 and agreed that it embodied acceptable principles to govern the implementation of the UK voluntary offer.[231] MOD and UKAEA were tasked to identify the establishments to be covered by these agreed principles and prepare a further document for consideration. This appeared at the end of May and the places to be subject to total exclusions were: AWRE, Foulness, Burghfield, ARTE, Rolls Royce and Associates Derby, Rolls Royce (Nuclear) Derby and Source Material Stores (ores) at Inskip. Enclaves within civil establishments wholly excluded from inspection were Windscale (the submarine spent reactor core pond, the residue pre-treatment plant and associated store); the North Group (enriched uranium recovery and residue recovery plant (B203); the Capenhurst CANDLE plant (tritium extraction); Springfields ore rafts; at Harwell, parts of Buildings 220, 338 and 393 and 459 naval programme, naval store house, the DIDO reactor; and the Calder Hall and Chapelcross reactors. Establishments where safeguards inspection would need to be occasionally suspended were

[230] TNA CAB 134/3020, NPO (G) (70) 6 Implementation of UK Offer on Safeguards, Note by the Secretary, 5 October 1970.

[231] TNA CAB 134/3420, NPO (G) (71) 1st Meeting, Item 1: Implementation of UK Offer on Safeguards, 18 February 1971.

Springfields, Winfrith, Windscale, Harwell (parts of including possibly the LIDO reactor) and Dounreay's caves for post-irradiation examination and reprocessing facilities.[232] This served as a supplement to the brief for the UK delegation's negotiations with the IAEA on the UK's voluntary offer that started in Vienna on 9 June 1971.[233] At a subsequent meeting of the Nuclear Policy (Official) Committee on 2 June the high separation diffusion and low separation diffusion plants at Capenhurst were also added to the list of exclusions.[234] UK went into these tasks with the objective of relating the safeguards agreement to facilities rather than nuclear material since at certain vital facilities, chiefly Windscale and Springfields, IAEA safeguards could at times be unwelcome from a defence point of view as they could disclose information about military stocks even if some of the material being processed was civil.[235] A further meeting was held with the IAEA on 28 September and it, like the 9 June session, was taken up largely by an examination of the text of Part I of the model IAEA safeguards agreement in INFCIRC/153, with a view to identifying points where different formulations would be required in order to meet the UK's national security requirements.[236] In the event no agreement was concluded with the IAEA until 6 September 1976 and it only came into force on 14 August 1978.[237] Article 1 of the Basic Undertaking reflected MOD requirements; safeguards would be applied to materials in facilities or parts thereof subject to exclusions for national security only. The delay was in part explained by the UK accession to EURATOM in 1973, which meant that the agreement had to be a trilateral one. And under EURATOM safeguards the UK insisted that certain facilities would

232 TNA CAB 134/3420, NPO (G) (71) 14 1971 Implementation of the UK Offer on Safeguards, Note by the Secretary, 27 May 1971.

233 TNA CAB 134/3420, NPO (G) (71) 2nd Meeting, 2 June 1971, Item 2; TNA EG 8/84, F.H. Jackson, Vienna to P. J. Kelly, AE Division DTI Negotiations on the UK Offer, 21 July 1971.

234 TNA EG 8/84, NPO (G) Meeting Item 2, Discussions with the IAEA about Safeguards under the UK Offer, 2 June 1971.

235 TNA EG 8/84, NPO (G) (71), Draft Official Committee on Nuclear Policy General Sub-committee Discussion with the IAEA about Safeguards under the UK Offer, Joint Memorandum by the FCO and DTI.

236 TNA EG 8/84, Official Committee on Nuclear Policy General Sub-committee Discussions with the IAEA about Safeguards under the UK Offer, Memorandum by the FCO and DTI Draft, 9 November 1971.

237 IAEA Information Circular INFCIRC/263, October 1978, *The Text of the Agreement of 6 September 1976 between the United Kingdom of Great Britain and Northern Ireland, the European Atomic Energy Community and the Agency in Connection with the Treaty on the Non-Proliferation of Nuclear Weapons*; see also Darryl A. Howlett, *EURATOM and Nuclear Safeguards* (Basingstoke, 1990), pp. 181–6.

need to be excluded from safeguards and other civil facilities would have from time to time to see exclusions too.[238]

The Decline of Cut-off: 1968–69

The US indicated at a NATO Disarmament Experts' meeting in early 1968 that a cut-off might be the main priority after the conclusion of the Non-Proliferation Treaty. London was worried that this might make life difficult before UK supply problems for the weapons programme were overcome in 1970. In fact the UK thought that a CTBT ought to be a higher priority and that was likely to be the view of the non-nuclear states.[239] MOD confirmed in March that the UK ought to be able to go along with any new US proposal for a cut-off provided that any agreement did not become operative before the end of 1970.[240] This suggests that somewhere along the line the earlier target date of 1969 had slipped. However, after the debacle over nuclear propulsion in the NPT negotiations when some such as Italy wanted to retain the right to use HEU for naval reactors, MOD thought this would make it much more difficult for it to go along with US cut-off proposals if they were resurrected. Until the landscape could be seen more clearly, Dr Frank Panton (ACSA (N), MOD) asked that the earlier statement that MOD could accept a cut-off after 1970 should be placed in abeyance.[241] So by the end of 1968 the MOD had misgiving about a cut-off similar to those that it held in the case of the CTBT and for the same reasons: there was no established UK nuclear policy given the uncertainties over Polaris modernisation and the deterrent remained vital to the West's strength. We should not forget the Cold War context of the times: the Soviet Union had invaded Czechoslovakia and the Vietcong had launched the Tet Offensive in the Vietnam War – both in 1968. Meanwhile, Washington was also losing interest in a cut-off after the high water mark in 1965–66. The NPT

[238] TNA AB 48/1073, Implementation of EURATOM Control in the UK Record of Informal UK/Commission Discussions in Luxembourg, 17–18 October 1972, Annex A, Conditions Necessary to Meet UK National Security Requirements; see also Implementation of EURATOM Control in the UK Record of Second Informal UK/Commission Discussion in Luxembourg, Thursday 7 and Friday 8 December 1972.

[239] TNA DEFE 19/169, M.H.C. Warner, Fissile Material Supplies, Draft Letter to Ian Smart, British Embassy, Washington Cut-Off and Disarmament Negotiations, 18 April 1968.

[240] TNA FCO 10/162, M.H.C. Warner, DS 12, MOD to R. McAndrew, AEDD, Foreign Office, 12 March 1968.

[241] TNA FCO 10/162, Dr F.H. Panton, to Mr R. McAndrew, AEDD, Foreign Office, 20 March 1968.

had become the main priority and from the military point of view a cut-off was probably becoming less advantageous to the US given the prospect that Soviet U-235 stockpiles might exceed those of the US.[242] ABM warhead requirements were also a factor – both in Washington and Moscow. But the cut-off issue had not quite gone away. There was another attempt, as the US raised the cut-off once again at the ENDC in April 1969. This now included the suggestion that specified quantities of material should be transferred from dismantled weapons into safeguarded civil programmes. Although the acute problem of UK supplies would soon be overcome by the end of the year, there were still some defence drawbacks for the UK as a cut-off would hinder expansion of the weapons stockpile or the hardening of Polaris warheads. Foreign Office officials thought that the arms control advantages outweighed the disadvantages, which were not insurmountable in its view, from the defence perspective. They recommended that Ministers support the new US position.[243] A draft paper for the Defence and Overseas Policy Committee was prepared by Ronald Hope Jones, Head of the FCO's Atomic Energy and Disarmament Department (AEDD). However, MOD, after some reflection and in light of various difficulties that they were facing with the US – renewal of the 1959 barter deal had not yet been concluded and the US was asking whether the UK could pay in cash instead of plutonium, arguments over the centrifuge (UK plans to collaborate with Germany and the Netherlands on uranium enrichment and US concerns that the British were using US technology) and the Enrico Fermi affair (the proposed Italian naval supply ship, which would have required HEU to be withdrawn from safeguards) – took the view that it could not support any recommendation that Ministers commit themselves to support a cut-off before the barter transactions were satisfactorily concluded.[244] Moreover, in the continuing absence of definitive decisions about nuclear weapons policy – the Polaris Improvement Programme – MOD found it difficult to be completely sure of what the UK's own nuclear needs were likely to be over the next five years, both for Polaris and indeed for anything else.[245] Senior FCO officials then decided that there would be no point in pushing the MOD further on this topic and felt that Ministers would come

242 Seaborg, *Stemming the Tide*, pp. 400–401.

243 TNA FCO 66/116, Draft Paper for OPDO (D), Cut-off of Production of Fissile Material for Weapons Purposes, 5 May 1969.

244 TNA FCO 66/116, D.C.R. Heyhoe, DS22 to R.C. Fisher, Disarmament Department FCO, Cut-off of Production of Fissile Material for Weapons Purposes, 16 June 1969.

245 TNA FCO 66/116, R.A. Sykes, Defence Policy Dept. to Sir T. Brimelow, Cut-Off Production of Fissile Material for Weapons Purposes, 24 June 1969. Sykes was reporting Peter Nailor, Head DS 22, MOD's views.

down in favour of the defence side of the argument if pushed. No proposals were thus put to the Defence and Overseas Policy Committee.

MOD had made clear to the FCO that there was now no way in which a cut-off agreement would work to UK security advantage.[246] Under certain circumstances it might well be seriously harmful to UK defence interests. This would have been the case if the negotiations under the 1959 Barter Agreement had not been satisfactorily concluded. MOD believed that it would not be wise to commit to a more public support of this idea until the barter supplies had been delivered for 1969 and the current plutonium production programme was satisfactorily concluded. Nor could MOD agree to the transfer of any material from the defence programme to civil stockpiles. In short, MOD found it difficult to regard a cut-off as being of any real disarmament significance and, moreover, thought it likely to bite on the UK more than anyone else.[247] The primacy of US–Soviet SALT negotiations removed cut-off as both a threat to the UK nuclear weapons programme and as a mainstream arms control issue. Indeed as an issue, cut-off would not be revived again as a serious prospect until the 1990s. Despite the fact that a cut-off was a recurring worry to the UK until it had settled on and brought into service Polaris and the WE177 systems, and while much expense and time was occasioned by AWRE, the UKAEA and MOD, in retrospect the cut-off did not have a major influence on the direction and speed of the UK's nuclear weapons research, development nor indeed its production programme.

Conclusions

A fissile material cut-off treaty hung like a Sword of Damocles over the UK nuclear weapons programme in the 1950s and 1960s. The UK faced continual shortages of fissile material throughout this period and developing improved warhead designs that made more efficient use of what material was available was a critical driver in weapons development work. Fending off a cut-off treaty until such times as the UK had acquired more economical designs and fissile material to support the planned weapons production programmes was thus a major arms control objective. However, like the test ban treaty, the UK did not want to be seen domestically or internationally as obstructing, or hostile to, arms control and disarmament initiatives and therefore its policy was essentially a juggling act. In this the UK succeeded. This was the same for both Conservative and

[246] TNA FCO 46/304, Peter Nailor, Head of DS 22, MOD to R.A. Sykes, Defence Policy Department, FCO, 26 June 1969.

[247] TNA FCO 66/116, W.F. Mumford, DS 22, MOD to R.C. Hope-Jones, Disarmament Department, 14 October 1969.

Labour governments. Much of what we see in the mid 1960s was contingency planning against the prospect of a cut-off treaty coming into force – and a good deal of defence planning comes into this category. In retrospect it is clear that a treaty was never really on the cards; and indeed, some 40 years later at the time of writing, it is still as elusive as ever. UK officials, however, were not prepared to take the risk. There is another parallel with the test ban treaty. The UK conducted a good deal of practical work on the technical aspects of verification, in this case drawing on its own nuclear fuel-cycle experiences. If such a treaty were to be negotiated and enter into force it was essential to understand beforehand the problems in verification and to develop effective measures to deter cheating; it was also critically important to establish the levels of access that could be accepted at the UK's own facilities without compromising classified information – facility design, throughput and production levels for instance. Once again it was a question of balance – just how much access could be provided to international inspectors to enable them to conduct an effective inspection without compromising classified information. It seems that the UK was content that a balance could be achieved and much of its work evolved into support programmes for the IAEA safeguards systems that would apply to civil nuclear facilities in non-nuclear weapons states under the NPT. The UK role in the NPT negotiations was considerable and this proposed treaty also raised issues that could impact on the UK's weapons programmes and its collaboration with the US. We now turn to look at some aspects of this negotiation and related issues in detail: plans for an Atlantic Nuclear Force and the Treaty's Articles I and II and nuclear sharing arrangements in NATO.

Chapter 9

UK Nuclear Weapons, the Atlantic Nuclear Force (ANF) and the NPT 1962–1968

Introduction

This chapter provides an account of UK thinking on Atlantic Nuclear Force (ANF) and Multilateral Nuclear Force (MLF) issues – essentially differing and competing approaches to the vexed question of nuclear sharing and nuclear decision making within the NATO alliance. MLF was originally promoted by the Kennedy Administration, whilst the ANF was a UK alternative favoured by the Labour Government of Harold Wilson from 1964 onwards. Ultimately both schemes came to naught. We will look in particular at how these concepts could impact on the UK's own warhead programme and its and international efforts to negotiate a nuclear non-proliferation treaty (NPT), which had begun in the ENDC in 1965. We also see for the first time some detailed estimates of the actual costs of nuclear warhead production. This chapter will also summarise the extent to which UK–US nuclear collaboration under the 1958 UK–US Mutual Defence Agreement was taken into account during the formative stages of the NPT negotiations in 1965 and 1966, which in many ways was the key multilateral arms control agreement negotiated in the 1960s. Its main provisions called for the then nuclear weapons states (namely those possessing nuclear weapons as of 1 January 1968 – US, USSR, UK. France and China would not accede to the Treaty until much later). not to transfer nuclear weapons to any other state and for the non-nuclear weapons states to foreswear nuclear weapons.

Atlantic Nuclear Force (ANF): Origins

At the Nassau meeting in December 1962 Kennedy and Macmillan agreed to use their best endeavours to develop a NATO MLF to which both the US and UK could contribute national forces. In February 1963 the US outlined in NATO's North Atlantic Council proposals for a MLF of surface ships open to all members who wished to contribute. These ships would be armed with Polaris missiles and would be jointly owned and mixed-manned. These proposals were warmly welcomed by Germany, but the UK had strong reservations, doubting whether there was military justification for any substantial addition to the Alliance's strategic nuclear forces. Furthermore, the MLF would have involved considerable additional expenditure for all participants and a force without the UK would be dominated by the US and Germany.

On the Thursday after the election in October 1964 the new Foreign Secretary – Patrick Gordon Walker and the Prime Minister – met in the latter's house and Wilson proposed that the UK might think of committing the UK's Polaris submarines absolutely to NATO. Development of what was to become the ANF – became a major foreign policy initiative of the new Labour government.[1] The idea appeared the following morning in a piece in *The Times* newspaper by Gwynne Jones, the paper's military correspondent. Wilson always said that this was a coincidence as he and the Foreign Secretary were first to discuss the idea, but in Gordon Walker's view it must have been written and set before he and Harold Wilson had spoken. He could not believe in such coincidences; Wilson, in his view, must have discussed the idea with George Wigg who was close to Jones. The idea was tentatively outlined in Washington to Dean Rusk in December 1964 and Walker proposed the name ANF as well as the concept that it should be under an authority in which all participating European powers should be equal.

Later in December 1964 the UK duly proposed an alternative to the MLF to its allies – the ANF – as a means of internationalising UK submarines by committing them irrevocably to NATO for the duration of the alliance. The difference between the ANF and MLF was that the former would have consisted of existing or planned forces (US and UK Polaris submarines and the UK V-bomber force and possibly some Minuteman missiles in the US). It would not, however, include mixed-manned UK Polaris submarines, as the US was also unwilling to contemplate such arrangements, which meant that the UK could not accept it too because of its use of US know-how and the restrictions under the 1958 US Atomic Energy Act. Even if the US were to relax its position, the UK was reluctant to follow suit since it did not believe that anything more than mixed-manning was feasible under operational conditions in Polaris submarines.[2] This

---

1 TNA FO 800/951, Secretary of State's Tentative Comments on the ANF Minute, 26 December 1964; see also Mohamed I. Shaker, *The Nuclear Non-Proliferation Treaty: Origin and Implementation 1959–1979, Volume 1* (London, 1980), pp. 174–8.

2 TNA FO 371/190665, WUN 1193/49, Non-Proliferation and Nuclear Sharing; TNA PREM 13/220, Nick Fenn, Foreign Office to J.O. Wright 10 Downing Street, 8 October 1965.

left open the possibility that some weapons systems such as bombers or missile sites could be mixed-manned. However, the UK did not intend to commit all of its nuclear forces to the ANF: there were tactical nuclear weapons – land-based and carrier borne that fell outside the proposal.[3] If, as was intended, the UK effectively relinquished the 'supreme national interests' exemption clause, then it would not be easy to refuse a request to install Permissive Action Links (PALs – electronic and/or mechanical safety and security controls on the use of weapons) in British weapons when they became available without casting doubt on the UK's sincerity of its renunciation of national control. MOD pointed out that if UK nuclear weapons were assigned irrevocably, without any strings, a period of months might elapse before full national control could be resumed, if indeed it ever could.[4]

ANF Command and Control: An Irrevocable Commitment?

The Prime Minister, Foreign Secretary and Defence Secretary discussed an officials' paper on ANF considerations on 11 November 1964.[5] Ministers thought that in the first instance the UK might offer to make available planned or existing UK nuclear forces, which would be juridically committed but without any form of physical restraint through a permissive link or electronic lock. However, such a device, which might be produced by the US in another year or so, could in any event be circumvented and its acceptance in subsequent negotiation would not mean that the UK could not regain independent control of its nuclear forces in the event that NATO dissolved. Under existing plans for the control of the UK Polaris force the intention was to rely on careful drills and tight discipline to ensure that the commander of a submarine did not launch without proper authority or be misled by spurious instructions.[6] For AWRE to develop a PAL would require, as of 1965, up to four years of R&D effort, followed by extensive tests and trials, and it was thought unlikely that all the UK's Polaris boats could be fitted with British PALS before 1972. Further study of the technical aspects by AWRE and the Ministry of Aviation was required,

---

3 TNA DEFE 25/33, MISC.17/4, Atlantic Nuclear Force, A Report by the Defence and Overseas Policy (Official) Committee, Cabinet Office, 18 November 1964.

4 TNA FO 371/184411, WUN 1192/243, F.W. Mottershead, MOD to The Viscount Hood, Foreign Office, 6 April 1965.

5 TNA CAB 130/212, 1 MISC. 16/1st Meeting, 11 November 1964.

6 TNA CAB 21/6048, Atlantic Nuclear Force, 1965, The Mechanism of Command and Control: Permissive Action Link, Memorandum by the Ministry of Defence, August 1965.

but it was already clear that if the function of the PAL was extended to provide a means of exercising multilateral political control over the firing of weapons then the technical problems became very much more severe, especially when it came to communications. A revised ANF proposal for NATO was prepared in January 1965 and this noted that PALs designed to make operational command more effective could be installed in all elements on the force, provided that they did not prevent reversion to national control if the alliance came to an end.[7]

Command and control and PALs presented an apparently irreconcilable dilemma. Although the UK had ostensibly promised to commit its nuclear forces to ANF, it also thought that it might need to reserve the right to take a subjective decision unilaterally in the event of a crisis in which there were fundamental differences in the alliance on the possible use of nuclear weapons. There was also the question of whether the UK should maintain the right to withdraw its forces in order to meet a threat outside the NATO area. Ministers needed to decide whether they could accept command and control arrangements including PALs, that would in practice prevent the UK firing the weapons independently, or that the UK should press for arrangements that would allow HMG to recover its weapons in an emergency. In the latter case this would become apparent in the command and control arrangements and would indicate to the allies that the UK was not serious about committing its forces to the ANF, as it would retain a special status.[8] MOD's view was that an effective physical system of unilateral control over the weapons of the ANF was incompatible with the UK's need to be able to withdraw its contingent and was likely to detract from the credibility of the force. In such circumstances the best course was to emphasise to the allies the technical difficulties of such a system and to seek to dissuade them from insisting on PALs, using the argument that this would cause the deterrent to lose its credibility.[9] Ministers had to decide whether the UK attached more importance to the ability to resume immediate national control of the Polaris submarines in a emergency and the maintenance of the credibility of the force; or convince its allies by this means that the UK had sincerely decided to abandon independent control by committing its warheads to NATO for the duration of the alliance.[10]

---

[7] TNA CAB 148/19/5, O.P.D. (65) 5, Note by the Chairman of the Defence and Overseas Policy (Official) Committee, 12 January 1965.

[8] TNA FO 371/184412, 1965 WUN 1192/271/G, The Mechanism of Command and Control: Permissive Acton Links.

[9] TNA FO 371/184412, WUN 1192/274/G, R.J. Andrew, MOD to E.J.W. Barnes, WOPD, Foreign Office, Draft Paper by the MOD for the DOPC Sub-committee on the ANF, the Mechanisms of Command and Control: Permissive Action Links, 1 June 1965.

[10] TNA CAB 21/6048, The Mechanism of Command and Control: Permissive Action Link, Memorandum by the Ministry of Defence, (Revised Draft), 20 August 1965

British Warheads for the ANF

The Foreign Secretary (Patrick Gordon Walker) raised the question of UK warheads for the ANF with officials during a meeting on 15 December 1964. He thought that it might only be necessary to produce rather more warheads than were already planned and that the financial arrangements might be such that the UK could sell these warheads to the force. He commissioned a paper on the options and costings.[11] In his view there were political advantages, apart from responding to an oft-expressed US wish that a UK contribution would mean that the mixed-manned element was not exclusively armed with US warheads. One possibility was that the UK might make warheads more cheaply by buying U-235 from the US, a fact linked to the possibility of a cut-off. UK and US warheads were basically interchangeable, but the differences were significant enough to make it desirable on technical and security grounds, that any one ship should have US or UK warheads, but not both.[12] Under the then plans, warheads for a fifth Polaris boat would have been ready by early 1969. Acceleration of plutonium production would have involved an extra charge to the defence budget of some £5.5 million in 1965–66 and £8 million in 1966–67, though savings would have balanced nearly all of this in later years. Despite this, there was still strong opposition within the MOD and the Navy Department to the idea of making UK warheads available for the ANF; both wanted the idea to be expressly rejected in the draft paper for the Defence and Overseas Policy Committee (DOPC) sub-committee on the ANF.[13]

Partly as a result of this, and because there was little prospect of this issue actually arsing in the on-going discussions about the ANF in the NATO Paris Working Group that was considering MLF/ANF options for collective nuclear forces decision making, the Foreign Office thought it best to put the paper to one side for the present. The paper noted that the rate of new production together with stocks and material recovered from old weapons would have sufficed to complete the existing weapons programme and leave a margin that

---

under Cover of Letter from R.J. Andrew to E.J.W. Barnes, Foreign Office ANF – Permissive Action Links, 20 August 1965.

11 TNA FO 371/179080, WUN 11948/112/G, E.J.W. Barnes to Mr Hugh-Jones, 16 December 1964.

12 TNA FO 371/179080, WUN 11948/112/G, Suspended Draft Paper ANF, The Mixed-manned Element and British Warheads, under Cover of Compliment Slip MOD to W.N. Hugh-Jones, Foreign Office, December 1964.

13 TNA FO 371/184414, WUN 1192/327/G, D.C. Thomas to Mr Barnes, Provision of British Warheads, 29 October 1965; R.J. Andrew, MOD to E.J.W. Barnes, Foreign Office, British Warheads for the ANF, 14 September 1965.

would have served for the extra 12 warheads required for the fifth submarine. There was probably enough even for 20 warheads, although the need to provide for working stocks would have left no room for contingencies of any sort.[14] Nevertheless the paper, which appears to have been the last word on the subject, concluded – reluctantly for some in MOD – that the UK should only be prepared to offer to supply warheads to the mixed-manned surface component of the ANF as a last resort, if HMG thought it essential in spite of the difficulties to meet a demand for a visible UK participation in the mixed-manned component and if there was no other way of satisfying their demand. An offer to supply British warheads would also have pre-supposed a decision about the UK's future weapons programme, which would have far-reaching financial and other effects. The UK could contribute only the type of warhead then being made for its own Polaris force. Even if a 25-ship force were accepted, their manufacture in the UK would not delay commissioning. Depending on other considerations, the UK might have been able to provide enough British plutonium to meet such a programme, but it would have to obtain the necessary U-235 and possibly other special materials from the US. The key was getting German support, but the UK could not discuss details affecting its Polaris missiles or submarines without US approval, otherwise the 1958 exchange agreement and co-operation arrangements could be endangered, which would be bad news as the UK depended on these to sustain its weapons programme. If exchanges with allies on the non-proliferation provisions should prevent acquisition of control over nuclear weapons by a group of non-nuclear states, this would go on to cause trouble with the NPT.[15]

Officials looked again at the question of UK warheads for the ANF in a meeting in the MOD on 17 June 1965. This revealed that there were all sorts of snags to the idea and that none of the MOD, Ministry of Aviation and Navy Department experts liked it.[16] However, they concluded that a revised paper on this topic should not conclude that the idea should be abandoned and that it would be for the Defence and Overseas Policy (Official) Committee to make recommendations to Ministers. The Ministry of Aviation produced an estimate

[14] TNA FO 371/184414, WUN 1192/327/G, R.J. Andrew, MOD to E.J.W. Barnes, Foreign Office, British Warheads for the ANF, 14 September 1965; The Provision of British Warheads for the ANF, Draft Paper by the MOD for the DOPC Sub-Committee on the ANF, 15 September 1965.

[15] TNA CAB 21/6047, E.J.W. Barnes, Foreign Office to F.W. Mottershead, MOD, 1 February 1965.

[16] TNA FO 371/184413, WUN 1192/279, D. Thomas to E. Barnes, 21 June 1965.

of the potential costs of a UK warhead contribution towards the end of June.[17] This noted that the manufacturing cost excluding fissile material consisted of work done and parts made in the UK and parts made in the US; costs of the former would come in at £60,000 per warhead, but US costs could only be guessed at. If parts were bought immediately, then they would cost £44,000 each, but since normal production would end in early 1966, orders placed after that date could cost £180,000 to £280,000 on the best information available to the Ministry. Historic costs – a bit of a black art – consisted of recovery of R&D by AWRE, recovery of development expenditure by the Ministry of Aviation, recovery or pre-production expenditure by the Royal Ordnance Factories, amortisation of capital employed in production facilities and all other overheads attributed to normal accounting principles to the Polaris programme. If 16 warheads were produced for the ANF, each warhead would attract £200,000 for these charges. The best estimate for fissile material at this stage was £600,000 per warhead. The cost was most unlikely to fall below this sum, but the effects of a complete or partial international agreement on a cut-off in the production of fissile material might well raise it.

In addition to these costs, there were other charges that had to be considered: UK overheads and warhead functioning demonstration costs. Provision of warheads to an ANF imposed heavy overheads on the UK that would have to be recovered. It would be reasonable to expect the non-nuclear ANF participants to desire proof of the functioning of the UK warhead and if missile flight trials were needed then these might amount to between £2 and £3.5 million, roughly £130,000 and £220,00 per warhead procured. Total cost would thus range from £945,000 to £1,145,000 million depending on the aforementioned variables. UK supplies of fissile and other special materials had been carefully tailored to a minimum UK programme for weapons and propulsion.[18] Unless some new arrangements were negotiated with the US, any UK warheads for ANF could only be at the expense of the domestic programme. On the most favourable assumption regarding the re-supply and maintenance of the ANF ships the UK could supply 16 warheads and continue to meet its own naval requirements until about 1975–76. If 20 warheads were needed, including reserves, the effect on the domestic programme would have been felt as early as 1971–72. These calculations were based on the assumption that the production and procurement

[17] TNA FO 371/184413, WUN 1192/294/G, J.A. Battersby, Ministry of Aviation to R.J. Andrew, Head of DS 12, MOD, 25 June 1965.

[18] TNA FO 371/184413, WUN 1192/295/G, The Provision of British Warheads for the Atlantic Nuclear Force, Draft Paper by the Ministry of Defence for the DOPC Sub-committee on the ANF, July 1965; R.J. Andrew, MOD to E.J.W. Barnes, Foreign Office, 13 July 1965.

of fissile and special materials continued as then planned. The then programme to meet UK weapons and propulsion needs was not going to be fulfilled until 1969. If a cut-off in the production of fissile material for weapons purposes entered into force before then, the UK would be seriously embarrassed. Provided fissile material were made available, the UK could provide the warheads for the surface ships built in the latter stages of the proposed programme – to be completed about springtime 1969. However, given the costs both for the ANF and the defence budget, the MOD would have much preferred to make any UK offer to supply the warheads as a last resort. The US could provide warheads much more cheaply than the UK could.

In order to ensure that the proposed ANF would not result in, or be accused of leading to, proliferation, and in particular that the UK could not be said to be bringing a German finger nearer to the nuclear trigger, Ministers and officials thought it essential to incorporate into the ANF Charter clauses whereby the nuclear members would undertake not to transfer nuclear weapons and the non-nuclear members would undertake not to acquire them. Germany would undertake not to own or control nuclear weapons and in order to fulfil non-proliferation principles, there would also be a prohibition of nuclear weapons passing into the ownership or control of a group of countries.[19]

Working on the assumption that some of the allies decided to proceed with a mixed-manned surface fleet within the ANF and the UK was pressed to participate, one possible form of contribution would have been to supply some UK warheads.[20] The warheads then planned for Polaris contained some US material and components. British warheads were thus defined to mean warheads partially manufactured and wholly assembled in the UK. The US had privately suggested that the UK consider contributing some warheads to the MLF. As to the number of warheads that might be involved, for planning purposes it was assumed that the UK contribution to the ANF would be four Polaris submarines; the total number of missiles was about 200; a mixed-manned fleet would include a third of this – nine ships each of eight missiles and if the production of fissile material and special materials continued as then planned, then the UK might supply warheads for 16 missiles – roughly 20 per cent of the warheads for the mixed-manned component. The UK could thus supply warheads for 16 missiles from the existing programme provided that there was no requirement for additional reserve missiles and warheads to be maintained and that the nuclear warheads were supplied from UK's existing supply facilities.

[19] TNA DEFE 25/33, 65, Atlantic Nuclear Force, Outline of HMG's Proposal (Original as Proposed to US Authorities in Washington), 7 December 1964.

[20] TNA DEFE 25/33, Atlantic Nuclear Force, British Warheads. A.M. Mackintosh Head DS12, MOD to E.J.W. Barnes, Foreign Office, 11 February 1965.

Provision of warheads for the mixed-manned component meant keeping the warhead programme at about its then present level. The UK would have to supply personnel to operate and control the warheads. Warheads for a fifth UK submarine would have been ready by early 1970. (The new money cost of a British warhead, excluding fissile and special material was £150,000. The comparable price to the defence budget in which historic costs were taken into account was £400,000. If fissile material cost, using US prices, then the fissile material for each warhead was about £600,000.) MOD concluded that the UK should undertake to supply additional warheads to the mixed-manned component only if HMG considered it essential for political reasons to meet a demand by NATO allies for a visible British participation in that commitment. In fact, the immediate effect on the defence budget of supplying warheads would be to somewhat reduce the savings which were estimated to accrue from cancellation of the fifth Polaris submarine.[21]

The Treasury noted that once the UK incurred a commitment to supply warheads initially, it would have to service them but also keep pace with the US in modernisation. They thought too that it would be difficult to envisage a situation where the ANF would accept a situation where UK warheads, including re-entry bodies, would progressively fall below the standards of their US counterparts.[22] Officials thought that unless the UK was prepared to abandon national control of at least some part of the strategic nuclear force, the ANF proposals would not be acceptable to the allies.[23] However, by the end of April 1965 officials concluded that until the completion of the study on the possible UK contribution of warheads to the force, it was not certain that it was in British interests to offer such a contribution, especially financially.[24] It seems as if the results of this study are still closed to historians since two meeting minutes, one from May and the other from August 1965, are retained under section 3(4) of the Public Records Act 1967.[25] At the end of the day of course all this became academic. Although UK proposals were still on the table in

---

21 TNA FO 371/184406, WUN 1192/44/G, A.M. Mackintosh, Head DS 12, MOD to E.J.W. Barnes, Foreign Office, 15 January 1965; TNA FO 371/184407, WUN 1192/117/G, Atlantic Nuclear Force, British Warheads, 11 February 1965.

22 TNA FO 371/184406, WUN 1192/43/G, P.H.F. Dodd, Treasury Chambers to A.M. Mackintosh, MOD, 8 January 1965.

23 TNA CAB 21/6047, D.S. Laskey to Sir Burke Trend, 5 November 1964.

24 TNA CAB 148/41/11, O.P.D. (O) (65) 10th Meeting, Item 3: ANF: Brief for Multilateral Discussions, 30 April 1965.

25 TNA CAB 148/48, OPD (O) ANF (65) 6th Meeting; OPD (O) (ANF) (65) 5th Meeting, 12 August 1965.

NATO in September 1966, they were no longer under active consideration.[26] Discussions on McNamara's proposals – 'the McNamara Committee', what was to become the Nuclear Defence Affairs Committee and Nuclear Planning Group in NATO – had essentially superseded both the MLF and ANF.[27]

The NPT and Nuclear Sharing 1965–66

The identification of the NPT as the most feasible and desirable measure at that stage in the disarmament field coincided largely with the appointment in October 1964 of a Minister of State with special responsibility for disarmament.[28] In January 1965 the Foreign Office's Atomic Energy and Disarmament Department (AEDD) revised an earlier working draft of a possible non-dissemination treaty (the initial term for non-proliferation) whose main obligation would have required that the 'governments of France, the UK, USSR and USA, hereinafter called the "nuclear states", undertake not to transfer national control of nuclear weapons to states not now possessing control of nuclear weapons to states or to any association in which power to prevent the use of such weapons is not retained by the nuclear states which are members of the association concerned.'[29] This was later revised:

> Each of the nuclear states party to this Treaty undertakes not to transfer control of nuclear weapons to any non-nuclear state or to any association if the association in question does not include among its members one or more nuclear states who retain the power to prevent the use of the nuclear weapons supplied to the association concerned.

There was of course nothing in this that posed a threat to existing patterns of UK–US nuclear cooperation; if there had been, then the MOD would have

---

26 TNA FO 371/190666, 1966 WUN 11913/88, Secretary of State's Visit to the USA, October 1966, NATO Nuclear Sharing and Non-proliferation, WOCD, 26 September 1966.

27 Glenn T. Seaborg with Benjamin S. Loeb, *Stemming the Tide: Arms Control in the Johnson Years* (Lexington, MA, 1987), pp. 126–9; Paul Buteux, *The Politics of Nuclear Consultation in NATO 1965–1980* (Cambridge, 1983), pp. 48, 55–7; Shaker, *Nuclear Non-Proliferation Treaty* (London, 1980), pp. 188 and 286.

28 TNA CAB 151/138, The Non-Proliferation Treaty Background Notes, 3 December 1968.

29 TNA FO 371/181386, Non-dissemination of Nuclear Weapons, 1965, IAD 1052/19, UK Draft, Non-dissemination Agreement, S.J. Barrett to Mr Shepherd, 7 January 1965.

been quick to object.[30] MOD was more concerned with the need to ensure that China was included in the definition of a nuclear power and the need to prevent withdrawal from any treaty.[31] Officials proposed that this draft might then be shown to the US with the view to tabling a Western draft in the next session of the ENDC.[32]

This text was passed to ACDA on 23 February 1966.[33] Initial US reactions were not given to the British Embassy in Washington until 5 March, these noted that the main provisions in Article I seemed to rule out the European option, which would cause difficulty for certain NATO governments, but as it explicitly sanctioned MLF/ANF it would be unlikely to be acceptable to the Russians.[34] London's intention all along of course was to ensure that ANF type associations were compatible with the principles of non-proliferation; the draft had been deliberately created in order to be consistent with UK's ANF proposals. However, the Americans did accept that a draft should be placed on the table in Geneva after necessary consultations within the Alliance.[35] London felt that it was important that the West should come forward with specific proposals when the ENDC reconvened in the summer of 1966.

The US response was thus seen as cautious, but not discouraging.[36] The British view remained, however, that creation of an ANF had to be linked with provisions for non-proliferation of nuclear weapons by the nuclear weapon states and non-acquisition or control by the non-nuclear weapon states.[37] Consequently a revised text was shared with three other Western (US, Canadian and Italian) delegations in the UN Disarmament Commission in New York on 2

30 TNA FO 371/181386, IAD 1052/33, MOD Comments on Draft Treaty on Non-dissemination, B.M. Day, MOD to H.B. Shepherd, Foreign Office, 27 January 1965.

31 TNA FO 371/181388, IAD 1052/68, B.M. Day, MOD to Andrew Stuart, Foreign Office, 25 May 1965.

32 TNA FO 371/181386, IAD 1052/38, H.B. Shepherd to Lord Hood, 17 February 1965.

33 TNA FO 371/181386, IAD 1052/42, R.S. Faber, Washington to H.B. Shepherd, AEDD, 23 February 1965.

34 TNA FO 371/181386, IAD 1052/45, Washington telegram 529 to Foreign Office, 5 March 1965.

35 TNA FO 371/181386, IAD 1052/45, J.E.D. Street, AEDD to A.M. Mackintosh, MOD, 17 March 1965.

36 TNA FO 371/181386, IAD 1052/50, S.J. Barrett to Mr Shepherd, 11 March 1965.

37 TNA FO 371/184411, WUN 1192/240/G, Atlantic Nuclear Force Brief for Multilateral Discussions, Summary of Negotiating Positions [Draft by the Foreign Office], 12 March 1965.

June 1965.[38] The consultation and tabling process was to be step by step and these three agreed that the next one would be to share the draft with the Germans, as long as it was made clear that this was a UK draft, prior to presenting this to the North Atlantic Council (NAC). London continued to grapple with the MLF/ANF problems and sought solutions on how to make clear the non-proliferation nature of the MLF/ANF proposals.[39] Sir Evelyn Shuckburgh (UK Ambassador to NATO) introduced the UK draft to the NAC along with an explanatory memorandum on 7 July, but detailed consideration would not take place in the Council until 26 July 1965.[40]

Prior to that meeting on 26 July the US suggested some amendments to the UK draft particularly on Article I.[41] These appeared to leave open the door to the establishment of a nuclear organisation in Western Europe or elsewhere operating by majority vote without the veto of its nuclear members.[42] As such HMG felt that it would be extremely difficult to take responsibility for putting forward a proposal that envisaged a European option totally at variance with its own proposals for the ANF and with its public statements on the latter.[43] US proposals were therefore rejected. For the UK at this time the dual objective was to increase the nuclear interdependence and cohesion of NATO whilst at the same time avoiding any proliferation of nuclear weapons. It was with these twin purposes in mind that the UK had offered to internationalise its nuclear weapons and commit them to the ANF for the duration of the NATO alliance.[44] There was no contradiction in UK eyes: both policies were complementary.

Following the 26 July NATO meeting, Western countries in Geneva agreed to establish a working party to draft a treaty text that they could all sign up to.

---

38 TNA FO 371/181388, IAD 1052/67, New York telegram No. 1339 to Foreign Office, 3 June 1965.

39 TNA FO 371/181388, IAD 1052/79, J.E.D. Street, Foreign Office to A.A.S. Stark, Bonn, 15 June 1965.

40 TNA FO 371/181388, IAD 1052/100, UKDel NATO telegram 181 to Foreign Office, 7 July 1965.

41 TNA FO 371/181389, IAD 1052//123, Foreign Office telegrams 126 and 127 to UK Delegation to Disarmament Conference, 19 July 1965.

42 TNA FO 371/181389, IAD 1052//123, Geneva telegram 12 to Foreign Office, 20 July 1965.

43 TNA FO 371/181389, IAD 1052//129, Foreign Office telegram 5741 to Washington, 21 July 1965.

44 TNA FO 371/181389, IAD 1052//123/129, The British ANF Proposals and the Non-Dissemination Question, Speaking Notes for Lord Chalfont, 23 July 1965.

The UK delegation gave some revised Articles I and II to the Americans first.[45] Article I in the UK draft now read:

> Each of the nuclear states party to this treaty undertakes not to transfer any nuclear weapons into the national control of any non-nuclear state, either directly, or indirectly through a military alliance; and each undertakes not to take any other action which would result in the acquisition, by any State or Organization not now possessing it, of an independent power to use nuclear weapons.

The Americans did not like this and went ahead with their own version.[46] Canada and Italy preferred the US text, thus isolating the UK.[47] Britain's draft would only permit a European nuclear association if either an existing nuclear power was a member of it and continued to hold a veto on the use of nuclear weapons, or a European federal state were created as a legal successor of at least one of the existing nuclear powers.[48]

The Foreign Secretary wrote to Dean Rusk on 6 August 1966 saying that the UK could not honestly support any proposal that left open the prospect of proliferation, however theoretical. He made clear that the UK approach would permit the creation of the ANF. Whatever happened in Geneva, the UK would defend this proposition most vigorously and would emphasise that it wanted to press on with discussion and conclusion of an ANF without unnecessary delay.[49] The dual aim of UK policy was a fundamental aspect of its proposals: arrangements for the multilateral organisation of nuclear weapons within NATO would have had little point if it did not serve to prevent proliferation of nuclear weapons.[50] The UK decided not to pursue its own draft in the face of such opposition, but the four Western powers were able to agree the text of a US draft that could be tabled in the ENDC.[51] The British agreed with the US that the Americans would table a draft treaty on 17 August 1965 supported

---

45 TNA FO 371/181389, IAD 1052//140, Geneva telegram No. 14 to Foreign Office, 27 July 1965.

46 TNA FO 371/181389, IAD 1052//140, Foreign Office telegram No. 148 to Geneva, 29 July 1965.

47 TNA FO 371/181390, IAD 1052//149, Geneva telegram No. 28, 3 August 1965.

48 TNA FO 371/181390, IAD 1052/165, C.M. MacLehose, Foreign Secretary to J.O. Wright, 10 Downing Street, 6 August 1965.

49 TNA FO 371/181390, IAD 1052/166, Foreign Office telegram No. 6122 to Washington, 6 August 1965.

50 TNA FO 371/181392, IAD 1052/254, H.B. Shepherd to Mr Street, 23 July 1965.

51 TNA FO 371/181391, IAD 1052/174, Geneva telegram No. 40 to Foreign Office, 11 August 1965; Seaborg, *Stemming the Tide*, p. 164.

by the Canadians and Italians, but which the UK would support as a basis for negotiations and with reservations on Articles I and II.[52]

Ministers and officials were satisfied that the conclusion of a treaty was entirely compatible with nuclear sharing arrangements of the type foreshadowed in the UK ANF proposals into which the UK intended to merge its nuclear capacity.[53] As the Prime Minister told Lord Chalfont after he had spoken at the ENDC, it had been most valuable that he had reserved the UK position fully on any possible development of a European nuclear grouping and thereby preserving the essence of the UK's ANF proposals with their anti-proliferation emphasis.[54] Despite this at least one senior Foreign Office official was of the view that creation of an ANF quickly was contrary to the wish of some Ministers and of influential people in the MOD.[55] An alternative Soviet draft treaty appeared at the UN on 24 September, but its provisions would have had no impact on existing UK–US nuclear arrangements – its sole aim, as Andrei Gromyko had intimated to the UK – was to prevent any form of access by Germany to nuclear weapons.[56] [57]

In a review of the UK deterrent and non-proliferation policy commissioned by Lord Chalfont in the autumn of 1965, AEDD concluded that an offer to drop the deterrent would not significantly increase the chances of a non-proliferation treaty. However, Labour's stated policy was to abandon the independent capability as soon as it could be integrated into a suitable regional security system and the enunciation of this policy was thought to be helpful in working for the freezing and reduction of existing nuclear forces.[58]

In the opening NPT negotiation phases the major area of disagreement among the Western powers was clearly the question of reconciling nuclear sharing

---

[52] TNA FO 371/181391, IAD 1052/192, Foreign Office telegram No. 343 to Certain Missions, 13 August 1965; Seaborg, *Stemming the Tide*, pp. 164–5.

[53] TNA FO 371/181391, IAD 1052/201, A.C. Stuart to Mr Darwin, 20 July 1965.

[54] TNA PREM 13/652, Prime Minister's Personal Telegram Serial No. T 302/65, Foreign Office telno No. 216 to Geneva, 24 August 1965.

[55] TNA FO 371/181391, IAD 1052/205/G, E.J.W. Barnes to Sir P. Gore-Booth, 9 August 1965. Most complete statement on ANF was Wilson's statement to House of Commons on 16 December 1964 – Hansard, Vol. 704, cols 430–35.

[56] TNA FO 371/181392, IAD 1052/264, Notes on the Soviet Draft, Non-dissemination Treaty Tabled at the UN, 24 September 1965.

[57] TNA FO 371/190666, WUN 11913/88, Secretary of State's Visit to the USA, October 1966, NATO Nuclear Sharing and Non-proliferation, WOCD, 26 September 1966.

[58] TNA FO 371/181393, IAD 1052/299, J.E.D. Street to Lord Hood, 15 November 1965.

in NATO with a non-proliferation treaty.[59] Three options were on the table: an arrangement corresponding fairly closely to the ANF proposal and providing that a existing nuclear power retained a veto; an arrangement loose enough to permit at some time in the future majority control, which would make possible the out-voting of the existing powers; and finally, a European federation, which would be the legal successor of the sovereignty of its component states and could therefore inherit from one of them the status of a nuclear state. Britain thought that the second option would represent proliferation, but the US was not ready to reject this idea. In the event the UK decided to drop its draft because it had little support in NATO and it let the US take the lead with a draft of its own. Again there appears to have been no consideration of any potential impact of the draft on UK–US nuclear collaboration.[60] British and American officials met in Washington in January 1966 to look at the drafting of Article I and II of an NPT; this had focused exclusively on text dealing with prohibitions on transfers to non-nuclear weapons states, thereby excluding by default transfer between the UK and US under the 1958 Mutual Defence Agreement and 1963 Polaris Sales Agreement.[61] This went some of the way to dealing with UK objections to the notion that transfers might be permitted to a mixed association of states in which the existing nuclear weapon states did not have a veto. In March 1966 after discussion with Chalfont, the US came part way to meet the UK view by accepting an amendment that banned the non-nuclear association, though not the majority vote option. When the US and USSR joint NPT text appeared in June 1967, Article I was effectively in its final form as it would appear in the text opened for signature the following year.

There is no indication that either the UK or the US thought about the implications for UK nuclear testing in Nevada as the transfer of nuclear weapons would be illegal; perhaps since there were no UK plans for any more testing, the question did not arise and in any case the procedures used for the tests between 1962 and 1965 did not involve any loss of UK control. Experimental test devices underwent final assembly at the test site by AWRE experts. However, in December 1966 the Foreign Office noted that there was an asymmetry between the two halves of the US Article I text drafted in light of recent exchanges with the Russians, since the first referred to 'any recipient whatsoever' and the second to 'any non-nuclear state'. But if 'any recipient whatsoever' appeared in both

59 TNA CAB 148/51, O.P.D. (O) DO (65) 2nd meeting Sub-committee on Disarmament, 11 August 1965.

60 TNA DEFE 13/871, R.J. Andrew, Head of DS 12 to PS/Secretary of State, 13 August 1965.

61 TNA FO 115/4612, Dec 31, R.S. Faber, Washington to J.E.D. Street, AEDD, 21 January 1966.

places, then the US would be prevented from assisting the UK to manufacture nuclear weapons and the effect of this would presumably be to rule out cooperation under the 1958 Mutual Defence Agreement.[62] The final version of Article I reflected UK requirements with the phrase 'non-nuclear weapon state.'

Discussions between the USSR and US leading to final agreement on Articles I and II took place between November 1966 and January 1967. US officials made clear to the Russians that the deal did not apply to the transfer of nuclear delivery vehicles, provided that there was no transfer of nuclear weapons or control over them.[63] NATO nuclear consultative arrangements were resolved at a meeting in Rome 1966 and were adopted by the NAC in December. An accepted difference between the US–Soviet draft of January 1967 and the previous US–UK drafts was that the former would have banned the transfer of nuclear warheads (as distinct from nuclear delivery vehicles), not only between nuclear weapon states and non-nuclear weapon states, but also between the nuclear-weapon states themselves. The difference had hitherto been accepted by the MOD on the grounds that the text removed all difficulties associated with the non-nuclear association option and the majority vote option. It was at the same time accepted that in the context of possible supplies by the US to the UK, the undertaking reflected the actual position, since the US legislation already prohibited the delivery of warheads to the UK. The undertaking was therefore considered by both the Foreign Office and the MOD as acceptable, provided that the Article did not prohibit, as it specifically did not, assistance between nuclear weapon states other than delivery of nuclear warheads.[64]

NATO allies asked the US what could and could not be transferred under the draft NPT. For the Americans the position was clear: it dealt only with what was prohibited, not with what was permitted. It prohibited transfer of nuclear weapons, not delivery systems or control over them so long as this did not

62 TNA FO 115/4618, Foreign Office telegram 11484 to Washington, 19 December 1966. The Article draft read as follows: 'Each nuclear weapon state party to this treaty undertakes not to transfer to any recipient whatsoever nuclear weapons or other nuclear explosive devices or control over such weapons or explosive devices, directly or indirectly: and not in any way to assist encourage, or induce any non-nuclear weapon state to manufacturer *or* otherwise acquire nuclear weapons or other nuclear explosive devices, or control over such weapons or explosive devices.' Foreign Office telegram 2650 to Washington, 22 December 1966.

63 TNA PREM 13/2442, Fred Mulley to Secretary of State for Defence, 6 February 1968.

64 TNA CAB 168/16, Minister of State for Foreign Affairs to Sir Solly Zuckerman; the same point was put in a minute from Fred Mulley to the Secretary of State for Defence on 6 February 1968.

involve transfer of bombs or warheads.[65] The US in short said that they intended to interpret 'transfer' as the surrender of custody of, or any ownership interest in, nuclear weapons while 'control' meant the independent power to use nuclear weapons.[66] It was an integral part of the agreement with the Russians that the words 'transfer' and 'control' would not be defined, either in the text or outside it, but the application of these words in relation to nuclear sharing arrangements should be covered by an understanding between the US and Soviet Union in the form of 'interpretations', which were later handed to the Russians for private use and the US later made them public during the Treaty ratification hearings in the Senate.[67]

At the second Overseas Policy and Defence Committee meeting in 1968 it had been questioned whether Articles I and II of the draft NPT were compatible with NATO strategy.[68] The views of the Law Officers were sought on this point, and their comments bore out the view that there was no inconsistency. They also agreed that, failing public Soviet approval of UK–US interpretations, the next best course would be a public declaration of these interpretations by the UK.[69] The meeting decided that the Foreign Secretary should keep under review the timing and form of any UK public declaration of its interpretation of these Articles; it is instructive that no one worried that they might be incompatible with the 1958 Mutual Defence Agreement or the 1963 Polaris Sales Agreement. The Foreign Secretary eventually made UK interpretations clear in a response to an inspired parliamentary question on 8 July 1968.[70]

Conclusions

It seems then that neither the NPT nor ANF/MLF proposals had a concrete impact on the UK nuclear weapons programme, although it is clear that if either the MLF or ANF plans had come to fruition, then there could well have been a significant impact in terms of costs and other UK plans, especially for

[65] TNA CAB 168/16, IMV (67) 4, Visit of Italian PM to London, June 1967, Non-proliferation Brief by the Foreign Office, Annex 2, US Interpretation of the NPT, 19 June 1967.

[66] TNA CAB 168/16, FCO Guidance No. 215, Commentary on the Draft Non-proliferation Treaty, 24 August 1967.

[67] TNA CAB 168/16, NATO Nuclear Sharing Arrangements and the Non-Proliferation Treaty, Minister of State for Foreign Affairs to Sir Solly Zuckerman.

[68] TNA CAB 148/35, OPD (68) 2nd Meeting, Item 3, 30 January 1968.

[69] TNA CAB 148/35, OPD (68) 3rd Meeting, 1968, Item 1, 14 February.

[70] TNA DEFE 19/169, Michael Stewart to Prime Minister, 4 July 1968.

the tactical nuclear weapons programme – WE177A warheads for the Royal Navy and RAF. Although the UK programme would be subsumed in a sharing arrangement – ultimate UK control would not be surrendered. The key issue was NATO nuclear sharing arrangements in both the ANF/MLF and NPT contexts, and care was taken to ensure that the NPT did not hobble UK–US nuclear weapons collaboration. However, since the UK had stated that it would not be acquiring any successor to the Polaris system and had ceased underground nuclear testing in 1965, the flow of nuclear weapons information from the US under the 1958 Mutual Defence Agreement was drying up to a trickle by the time the NPT was concluded in 1968.[71] It was only following the decisions to begin what became known as the Polaris Improvement Programme in the late 1960s that requirements for renewed close collaboration with the US emerged. By that time another arms control negotiation was about to begin and this could also contain provisions that might constrain the UK nuclear weapons programme, and the UK was not even a party to these negotiations. It is to the UK approach to the bilateral US–USSR strategic arms limitation talks that we now turn.

[71] See John R. Walker, *British Nuclear Weapons and the Test Ban 1954–1973* (Farnham, 2010), chapter 11.

Chapter 10

UK Nuclear Weapons and the Strategic Arms Limitations Talks 1969–1973

Introduction

There were two interrelated issues of fundamental importance for the UK nuclear weapons programme during the US and Soviet strategic nuclear arms control negotiations that eventually began in 1969: anti-ballistic missile (ABM) controls and no-transfer of strategic delivery systems, materials or components. These may be defined as provisions in a US–Soviet bilateral agreement that either prohibited Anglo-American nuclear co-operation under the 1958 Mutual Defence Agreement and/or the 1963 Polaris Sales Agreement or imposed serious constraints on such co-operation; and ABM deployments that might inhibit the ability of UK missiles to reach their targets in the Soviet Union. The main issue that appears to have been at stake for the UK was continued US freedom to transfer nuclear delivery systems. However, we now know that the considerations being given to the Polaris Improvement Programme in the same time frame as SALT meant that the UK also required and received other technical assistance from the US that was critical to the success of this programme. At various times during SALT I the Soviet Union tried to limit, constrain or prevent both the quantitative and qualitative expansion of the British deterrent. Failing that, Soviet negotiators tried implicitly or explicitly to have UK forces taken into account in the numerical ceilings on Strategic Nuclear Delivery Vehicles (SNDVs) negotiated with the US. It is also clear that although the US kept the rest of its NATO allies briefed through the North Atlantic Council in Brussels, the UK had preferential treatment in both the detail and quality of briefings on the progress of the SALT negotiations and insights into US negotiating objectives. Nor at any time did the UK have any discussions with the French on SALT/ABM issues even though they shared similar concerns – ABM effectiveness and unsolicited inclusions of their force in the numerical ceilings set out in a bilateral agreement between the super-powers.

British policy objectives throughout this period can be characterised as a continuing attempt to ensure that Polaris was not in any way limited or otherwise taken into account and to ensure that nothing in SALT foreclosed the options of future US transfers of delivery systems or other technical assistance essential to sustain the UK nuclear weapons programme. The 1963 Polaris Sales Agreement was the hallowed text here. British officials worried constantly about how the Americans proposed to handle ABM controls and no-transfer issues with the Russians. Washington considered various formulations, and some of these were thought to be contrary to UK long-term interests. British officials throughout our period were concerned that UK interests would be marginalised by any wide-ranging bilateral agreements.[1] It should be noted that in the FCO at least, initial reactions focused on the political and military implications for NATO and allied cohesion rather than the no-transfer issue – that was to come later, although continued anxieties about the impact on US Forward Based Systems (FBS) continued to gnaw away throughout this period.[2] In a paper circulated to the North Atlantic Council at the end of May 1969, the UK argued that the principal European interest needing protection was the maintenance of the US nuclear deterrent and that the deterrent should continue to offer the same degree of security to Europe as it had hitherto.[3] Nevertheless, FCO officials certainly thought that the UK needed to think collectively about the wider impact of SALT as it could give rise to a situation where decisions might be forced on the UK at very short notice, or changes in US policy arising from SALT could limit British freedom of choice between the various deterrent improvement options that were open at that time: Super Antelope, Stag, Hybrid and Poseidon.[4] The FCO saw no need to change the conclusion that it had reached in 1967 in response to Wilson's review of nuclear weapons policy: it would be essential to retain the UK's deterrent for the time being.

---

[1] Kristan Stoddart, 'The Wilson Government and British Responses to Anti-Ballistic Missiles, 1964–1970', paper prepared for British Rocket Oral History Programme, Charterhouse Conference, 2–4 April 2005.

[2] TNA FCO 46/291, FCO telegram No. 1022 to Washington, 5 May 1969, Guidance for Use in Initial North Atlantic Council Discussion on SALT; FCO 46/292, FCO telegram 177 to UKDel NATO, 15 May 1969.

[3] TNA FCO 46/293, MISC 237 (69) 8, 30 May 1969, Cabinet International Aspects of Nuclear Defence Policy, Note by the FCO Strategic Arms Limitation Talks, 30 May 1969.

[4] TNA FCO 46/292, R.A. Sykes, Defence Policy Dept to Mr Waterfield, Mr Hope-Jones et al., SALT, 12 May 1969, Nuclear Weapons Policy and SALT. For more detailed discussion on Super Antelope etc. see John R. Walker, *British Nuclear Weapons and the Test Ban 1954–1973* (Farnham, 2010), chapter 11.

1969 to 1970: UK First Steps and US Assurances

In the first exchanges with the US in January 1969 – US Ambassador (Bruce) and the Secretary of State (Stewart) – following the announcement that there would be strategic arms limitation talks with the Russians, Ron Spiers (US Embassy) told the FCO that the negotiations would cover the three US Polaris submarines and the UK boats that were committed to NATO.[5] This pretty quickly provoked MOD complaints. It seemed odd, to put it mildly, that the US should propose to discuss the UK deterrent at a negotiating table at which the UK was unrepresented; a move moreover that would compromise the whole of the British strategic deterrent.[6] This proposal raised a range of important questions for British interests and in particular the understanding that the US had given the UK when it had first mooted strategic arms limitation issues in 1964 that, come what may, the US would honour the 1963 Polaris Sales Agreement. MOD urged the FCO to seek urgent clarification from Spiers. This was done and Spiers emphasised that there was no intention that the UK Polaris force would be raised in negotiations with the Russians. If the Russians raised the UK deterrent, then the US would of course consult immediately; the honouring of the 1963 Agreement was not in doubt.[7] In light of these exchanges MOD and Chiefs of Staff anxieties were allayed by prompt FCO intervention.[8]

The Cabinet Office coordinated UK policy towards SALT in a new interdepartmental officials sub-committee (FCO, MOD and Cabinet Office) entitled 'International Aspects of Nuclear Defence Policy MISC 237'. This held its first meeting on 16 May 1969 at which a paper, initially prepared as an internal MOD discussion document by the Defence Policy Staff, was considered.[9] This identified three particular UK interests in the SALT/ABM process. First, it was essential that nothing in any agreement should inhibit UK freedom of policy

---

5 TNA FCO 66/109, FCO telegram 36 to UKDel NATO, 22 January 1969.

6 TNA FCO 66/109, Dec 31, E. Broadbent, AUS (Pol), MOD to J.P. Waterfield, Western Organisations Department, FCO, 23 January 1969.

7 TNA FCO 66/109, J.P. Waterfield, WOD, FCO to Ronald Spiers, US Embassy, London, 24 January 1969.

8 TNA FCO 66/109, J.P. Waterfield, WOD to Sir E. Peck, Strategic Arms Limitation Talks, 27 January 1969.

9 TNA CAB 164/394, R.W.J. Hooper to Sir Burke Trend, Strategic Arms Limitation Talks (SALT), 12 May 1969; Peter Nailor, Head DS 22, MOD to J.C.W. Bushell, Cabinet Office, 17 May 1969 (dated by Cabinet Office registry as 14 May); R.W.J. Hooper, Cabinet Office to Sir Edward Peck, FCO, Strategic Arms Limitation Talks, 14 May 1969. This committee was later designated MISC 237; TNA FCO 46/292, MISC 237 (69) 5th Meeting, Cabinet, International Aspects of Nuclear Defence Policy, Minutes of a Meeting, Strategic Arms Limitation Talks: Programme of Work, 6 May 1969.

or operation. Quite apart from the fact that it would be unacceptable that UK weapons systems should be affected by a negotiation to which the UK was not a party, the comparatively modest size and limited variety of the UK's capability would make any restrictions weigh proportionately much more heavily on the UK than on the US.[10] Second, the relationship between Soviet ABM defences and UK offensive capability was one of much technical complexity. In general terms, however, there was no doubt that any agreement constraining Soviet ABM defences would, in the long run if not the short, tend to the UK's advantage in security or economic terms, if not both. The sooner such a constraint came into force and the more extensive its ambit the better it would be likely to be for the UK. Third, it was assumed that it would be unlikely that SALT would call into question the 1958 Mutual Defence Agreement and the 1963 Polaris Sales Agreement, which were of great benefit to the UK. It was imperative that neither of these should be inhibited. London already possessed explicit assurance from Washington that this would be the case on the Polaris Sales Agreement. It was also very desirable that there should be no future constraint, other than those in the NPT, on transfer of the control of nuclear weapons and the fissile material for them. A UK need for any delivery systems and nuclear know-how was at that stage wholly conjectural, but as the future could not be foreseen it was critical that options should not be foreclosed or seriously limited.

A parallel and equally initial FCO paper was also submitted to the Cabinet Office. This noted that British military interests in SALT/ABM were to preserve the credibility and effectiveness of British nuclear forces at minimum cost.[11] Soviet ABM deployments could in time destroy the deterrent's credibility, but FCO officials worried that if there were to be constraints on ABMs, then the Russians might insist on discussing UK and French nuclear forces.[12] So Whitehall was united in its view on what SALT must not do. There was some discussion of the line that should be taken about UK forces, but officials agreed that it would be premature to attempt to write any paper on this topic at that time – summer 1969 – but that it was certainly a subject that the Committee would need to consider in due course.[13] At the Cabinet MISC

10 TNA CAB 164/394, US/Soviet Talks on Limiting Strategic Nuclear Armaments: European and UK Security Interests D of DP (D) 45/Note 1, Defence Policy Staff (D Division), 30 April 1969.

11 TNA CAB 164/394, Edward Peck, FCO to Sir Robin Hooper, Cabinet Office, 15 May 1969; TNA DEFE 24/770, folio E 140 B. Strategic and Political brief no. 40: SALT.

12 TNA FCO 46/293, C.H.D. Everett, Planning Staff to Mr R. Sykes, Defence Policy European Interests in Soviet/US Strategic Arms Limitation Talks – paper attached to minute, 10 June 1969.

13 TNA FCO 46/292, R.A. Sykes to Sir E. Peck, 22 May 1969.

237 meeting on 17 June, the effect of SALT on Polaris improvement plans and the need for a paper on this was discussed again. However, officials still felt that it would be too soon to put pen to paper. Certain studies undertaken following the inter-departmental paper at the end of 1967 on the UK Polaris force and other nuclear defence matters were still in hand, and, since it was unlikely that the initial US negotiating position would include the allied forces, it would be better not to complicate things at that juncture. The Committee agreed to keep this issue in mind.[14] Overall a considered view of UK policy towards SALT was not possible, a fact recognised at the time, until the UK knew more about US and Soviet attitudes.

MISC 237 produced its first detailed report on 8 July 1969.[15] This was largely a position paper as the UK still did not know much about the US negotiating position, which was still being elaborated. Despite this, officials thought that any likely agreement would serve HMG's interests in a number of ways: it would reduce the risk of destabilisation in the strategic relationship, create a better climate for future arms control, reduce super power tensions, help prevent the further widening of the gap between the US and Europe and, if applied to ABMs, it might well make it easier to maintain the viability of Polaris. There were risks too: it might place restrictions on the UK deterrent. Sir Burke Trend (Cabinet Secretary) submitted this report to the Prime Minister on 9 July and it was also submitted in parallel to the Foreign and Defence Secretaries.[16] There was no undue concern at this stage on the main issues that were to become the main UK anxieties on SALT: ABM constraints and no-transfer clauses, but that is hardly surprising as the negotiations proper between the US and USSR had not yet begun in mid summer 1969.

There was also a more general and fundamental factor in addition to the particular point about constraints on ABMs, namely that the longer the arms race went unchecked, the greater must become the possibility that it would eventually reach into areas where the smaller powers, such as the UK, could not remain in the field. Zuckerman (Chief Scientific Advisor, Cabinet Office) had told the Prime Minister in June 1969 that UK resources were strained and it could not afford a penny more than was already being spent. An acceleration of the arms race, if SALT failed, would simply push the UK further down the league of world powers than the position it currently occupied. Even if the UK were to try to increase or improve its nuclear arsenal at the expense of other

14 TNA FCO 46/293, MISC 237 (69) 7th Meeting, 17 June 1969.

15 TNA CAB 164/394, R.W.J. Hooper to Sir Burke Trend, 8 July 1969.

16 TNA CAB 164/394, Burke Trend to Prime Minister, 9 July 1969.

commitments, it would not make the slightest difference to the UK's strategic position relative to the superpowers.[17]

The UK's main concern at the outset, and one which Harold Wilson made clear to President Nixon at their brief Mildenhall meeting on 3 August 1969, was that every effort should be made to use SALT as a mean of preventing the arms race from entering a qualitatively new and more dangerous stage, which development of Multiple Targetable Re-entry Vehicles (MIRVs) and ABMs could represent.[18] UK officials and Ministers, Denis Healey in particular, believed that SALT had to be used as a means of preventing vertical proliferation from entering a new irreversible course and were anxious that Wilson raise this with Nixon.[19] Risks posed by continuation of the arms race were also a core British preoccupation at this stage. The UK, however, was also anxious to ensure that any early interim moratorium on proof testing flights of MIRVs as part of the SALT process excluded UK Polaris improvement tests.[20] Nixon agreed with Wilson that there should be bilateral UK–US consultations about SALT, both before and during the substantive talks.[21] As the US began to work out its position, the UK shared its questions with the Americans on ABM issues to understand better US views on Soviet intent and capabilities.[22] Constraints on ABMs were to loom large in British preoccupations. The question of specific US assurances on continued supply of Polaris in the context of any agreement was, as we have seen, first raised between the Foreign Office and the US at the request of the MOD before the SALT talks began, and as Healey (Defence Secretary) told Wilson on 5 November 1969, the Americans had given the UK specific assurances regarding the honouring of the agreement for the supply and support of Polaris and that the honouring of the Polaris Sales Agreement was not in doubt.[23] For much of the rest of 1969 UK efforts were targeted on ensuring

17 TNA PREM 13/2569, Solly Zuckerman to Prime Minister, 25 June 1969.

18 TNA PREM 13/2569, E. Youde to Prime Minister, 25 July 1969; see also A.R.M. Jaffray, MOD to J.A.N. Graham, FCO, 23 July 1969 and message from Mr E. Youde to Hal Sonnenfeldt, 30 July 1969.

19 TNA FCO 46/296, FCO telegram 1614 to Washington, 18 July 1969; R.A. Sykes to Mr Burroughs, Private Secretary, 23 July 1969; J.A.N. Graham, FCO to E. Youde, 10 Downing Street, Strategic Arms Limitation Talks, 24 July 1969.

20 TNA PREM 13/2569, A.R.M. Jaffray, MOD to E. Youde, 10 Downing Street, 28 July 1969.

21 TNA CAB 164/394, The Prime Minister's Account of His Talks with President Nixon at Mildenhall on Sunday 3 August 1969.

22 TNA CAB 164/394, J. Freeman, Washington to Sir E. Peck, FCO, 10 September 1969.

23 TNA CAB 164/936, V.H.B. Macklen, MOD to P.J. Hudson, Cabinet Office, SALT – Background Brief, 30 July 1970; TNA DEFE 13/872, MO 11/9, Denis Healey to Prime

that effective consultations on SALT with the Americans started. ABMs and MIRVs became the two immediate areas of British interest at this point.[24] MOD outlined for the FCO its proposed UK aims and initial questions on MIRVs and ABMs in mid September 1969 for the proposed consultations with the Americans in 1969.[25] In essence, the UK's immediate objective was to secure the fullest possible understanding of US thinking on these issues. Consultations would enable the UK to take informed advantage of any opportunities that might arise to influence aspects of the US position that might affect Soviet ABM levels, since these would have implications for Polaris. This soon became one of the interdepartmentally agreed objectives.[26] As for a first meeting with the chief US negotiator, Gerard Smith, this was eventually arranged for 22 October.[27] The implications that SALT might have for the UK deterrent, however, were not raised as the main issue at this meeting.[28]

Nevertheless, discussion of Soviet ABM capabilities and how they might be constrained were clearly of direct relevance to Polaris effectiveness.[29] Vic Macklen (DCA (PN) MOD) described UK thinking on what might be done to overcome defences and since the UK had a small missile force, it made more sense to think in terms of decoys rather than MIRVs because it would not be possible to exhaust Soviet defences. Instead, the better approach for the UK would be to increase the numbers of targets that the defence had to shoot at, hence the advantage to the UK if these could be limited. US impressions of the UK questions were interesting as the Americans now thought that the UK was drawing back in its earlier enthusiasm for SALT to a position of considerably greater caution. Washington supposed that one reason was that London had

---

Minister, SALT, 5 November 1969. In early 1970 the US provided the UK with another assurance that UK strategic forces would not be open to negotiation with the Russians. However, in the US and Soviet agreed work programme for the SALT negotiations there was an item entitled 'third country nuclear forces'.

24 TNA FCO 46/298, D.A. Campbell, Defence Policy Department to Mr Johnstone, Mr Burroughs, Sir E. Peck, 5 September 1969; Edward Peck to The Right Hon J. Freeman, Washington, 8 September, 1969; R.E.L. Johnstone, Defence Policy Dept to Private Secretary, 10 September 1969; J. Freeman, Washington to Sir E. Peck, FCO, 10 September 1969.

25 TNA FCO 46/298, MISC 237 (69) 33, SALT: Anglo-United States Consultation, Note by the MOD, 19 September 1969.

26 TNA FCO 46/299, MISC 237 (69) 35 (Revise), SALT: Anglo-United States Consultation; Background Brief Note by the Secretaries, Cabinet Office, 10 October, 1969.

27 TNA FCO 46/299, Washington telegram 2766 to FCO, 14 October 1969.

28 TNA FCO 46/299, Washington telegram 2866 to FCO, 22 October 1969.

29 TNA FCO 46/299, C.M. Rose, Washington to V. Macklen, MOD SALT – Consultations with the Americans, 24 October 1969; Record of a Meeting Held in Mr Gerard Smith's Office in the ACDA on 22 October 1969.

woken up to the subject's inherent complexities. Alternatively, the Americans thought that British reticence could be explained by concerns about the future prospects for Polaris. Washington wondered which position actually reflected UK thinking; Spurgeon Keeny (ACDA) suggested to the Embassy in November 1969 that there was clearly a link between SALT and future plans for Polaris and it therefore seemed natural that this issue should be discussed bilaterally.[30] Clive Rose (Counsellor in the Embassy) replied that the implications for Polaris were one reason why the UK wanted to keep in touch with the US, but it was not the only one. Furthermore, it would be inappropriate to link the technical discussion on Polaris modernisation and SALT. Keeny agreed that for matters such as implementation of the 1958 Agreement that would be fine, but the political aspects of Polaris modernisation could hardly be excluded from future consultations on SALT. London did not agree and endorsed Rose's response; the two other elements that shaped UK thinking, and which were much higher in the pecking order at that time, were the UK's wider interests in Western security and its wish to help the US towards a successful outcome.[31] London thought too that Keeny was unaware of the US assurance offered back in January 1969 that the US would not raise matters affecting UK Polaris. It was a result of these assurances that the direct implications of SALT for Polaris had not been a determining factor in shaping the initial British approach – although officials were fully aware that the indirect implications might impact on US willingness to help the UK with possible improvements. It simply was not true to say that Polaris considerations had resulted in a much more cautious approach than earlier in 1969. In actual fact from the narrow Polaris perspective, it could be argued that this would make British interests all the greater in a far reaching agreement, including limitations on ABMs. For the British then, the primary objective was to help the US achieve a successful outcome in their negotiations with the Russians. ACDA also accepted that it would be inappropriate for discussion to take place in the SALT UK–US bilateral talks on technical matters concerning Polaris improvement issues.[32] British arms control oriented objectives were thus a defining characteristic in the opening phase, perhaps reflecting the colour of the Labour government at that time. As we shall see, as the political colour changed with the arrival of the Conservatives and with Polaris modernisation decisions becoming more acute and US assistance more critical, UK interests in SALT became much more insular.

30 TNA FCO 46/300, Washington telegram 2989 to FCO, 1 November 1969.

31 TNA FCO 46/300, FCO telegram 2370 to Washington, 5 November 1969; Record of a Meeting Held in Mr Gerard Smith's Office in the ACDA on 22 October 1969.

32 TNA FCO 46/300, C.M. Rose, Washington to R.A. Sykes, FCO, 7 November 1969.

SALT Underway

Healey reminded Wilson, once the start date for the SALT talks had been announced, that the UK thought that the most urgent objective would be to see early agreement on a moratorium on the testing or deployment of MIRVs.[33] He also reported that the US had recognised that, as a nuclear power in its own right, the UK had a special interest in the implications of SALT for Polaris. Accordingly they had given specific assurances that the US would not raise matters affecting British Polaris with the Russians and that honouring the 1963 Polaris Sales Agreement for the supply and support of Polaris was not in doubt. Agreed limitations on ABMs might, in fact help to prolong the effectiveness of Polaris against Soviet defences. This latter issue came up in a discussion on SALT involving the Prime Minister, Foreign and Defence Secretaries and Cabinet Secretary on 7 November.[34] Would the Helsinki talks affect consideration of the question of the hardening of the UK's Polaris warheads? Healey commented that if the development and deployment of ABMs was arrested, then this would reduce the case for hardening, but it was agreed to leave open the further consideration of this issue until it was apparent how the SALT talks would progress.

Despite US assurances that the Polaris Sales Agreement would not be called into question, reports from the early rounds at Helsinki indicated that the Russians did consider that the transfers of delivery systems, as well as provision of technical assistance, were relevant to an eventual agreement's scope. Gerard Smith, chief US negotiator, accepted that at first glance this was indeed a 'loophole' that required a good deal of thought. The Russians main concern was that such activities could provide the US with a strategic advantage after conclusion of a bilateral agreement.[35] The US delegation had told the Russians that such a concept was beyond SALT's scope because it was a bilateral process. At that stage it was not clear to the Americans, or to the British for that matter, how deep Soviet concerns with third country systems actually were. Such concerns could be intended to drive wedges between the allies: wedge-driving was another anxiety in British thinking. Soviet references to 'provocative strikes' indicated to the Americans at least, that China, rather than the UK or France, was the problem. However, the Americans recognised that the relative strengths of the UK and French nuclear forces were of some interest to the Soviet Union. Soviet concerns seemed to focus on the possibility of American evasion through

33 TNA PREM 13/3131, Healey to Prime Minister, 5 November 1969.

34 TNA PREM 13/3131, E. Youde to A.R.M. Jaffray, MOD, 7 November 1969.

35 Gerard Smith, *Double Talk: The Story of SALT 1* (London, 1985), pp. 97–8.

substantial transfers of delivery systems to its allies. So no-transfer would not go away, even though it rarely became a predominant issue in US–Soviet deliberations.

A report to Ministers on SALT progress in January 1970 noted that the US had reported that at the latest Helsinki round the Russians regarded transfers to third parties of strategic armaments and technical assistance in developing such armaments as relevant.[36] London quickly realised that it would have to watch this issue, but for the present the clarification received from the Americans revealed no special cause for concern. Thus there was no need to brief the Prime Minister to raise it with President Nixon when he went to Washington at the end of January 1970.[37]

As the UK had a vital and paramount interest in the success of SALT, it would not be shy in reminding the Americans as we shall see. Clive Rose told Gerard Smith in mid January 1970 that if he was asked at the Joint Congressional Committee on Atomic Energy, nothing should be said to this committee in response to questions about the Russian no-transfer proposal that would prejudice the prospects for the 1958 Mutual Defence Agreement amendment, which was to be sent to Congress for renewal as soon as it reassembled. Smith did not think that the UK had to worry about its prospects, but he would certainly bear in mind the point when he answered any questions.[38] Smith had told the Russians, when they mentioned no-transfer during the first SALT round in Helsinki that this issue, 'was not likely to be conducive to progress'. He told all this to the Foreign Secretary on 27 January, but added that while he did not think that it would be a serious obstacle, it could cause considerable difficulties.[39] The US thought that if the transfer or non-circumvention of the agreement became a real problem, then it was open to a party to withdraw from the treaty. The US did not think that it was appropriate to account for this sort of eventuality by detailed textual provisions inside the agreement.

Now that the Soviet hand had been revealed, MOD officials were anxious that the issue of no-transfer should be discussed at the second UK–US SALT bilateral, scheduled for early March 1970.[40] MOD thought it important to look

---

36 TNA CAB 130/440, MISC 231(70) 1, 8 January 1970; see also TNA FCO 46/591, Discussions between the Soviet Union and United States on Limitations of Strategic Armaments, 1970.

37 TNA FCO 46/591, PMV (W) (70) A3, Visit of PM to Washington and Ottawa; 26–28 January, Strategic Arms Limitation Talks, Brief by the FCO 1970, 23 January 1970.

38 TNA CAB 168/277, Washington telegram No. 135 to FCO, 15 January 1970.

39 TNA FCO 46/591, Extract from Record of Secretary of State's Meeting with Mr Rogers on 27 January 1970 in Washington during PM's Visit.

40 TNA FCO 46/591, Michael Quinlan to R. Sykes, FCO, 23 February 1970.

at the possible implications for the UK. The FCO consequently proposed that this issue be placed on the agenda.[41] Officials noted that their purpose in raising this was not to establish maximum terms which would be acceptable to the UK, but instead to find out what the US line would be if the Russians pressed their no-transfer proposals and to satisfy themselves, as far as possible, that the US line was not inconsistent with UK interests.[42] Senior FCO officials certainly worried that the no-transfer problem would prove to be the most troublesome issue for the UK.[43] At the second bilateral the UK felt that the US had given little thought to the implications of the Russian no-transfer proposal, but American officials repeated the view that it was a non-starter. Moreover, the Americans made clear that they recognised that this issue was likely to come up again at some stage, but they themselves did not intend to raise it; it would not need to be tackled until the negotiations' end-game.[44] The Americans also asked about UK views on the viability of Polaris when discussing Russian ABM levels. However, British officials noted that this depended on a number of related factors and it was impossible in spring 1970 to provide a simple answer.

For the rest of SALT I the Nixon Administration considered the 'withdrawal' clause to be the best alternative to an express provision on no-transfer. In April 1970 the US formulation for third country systems was much tighter, no doubt much to the relief of the UK because it was no doubt couched in general terms. This read that the US, 'will continue to stress that the viability of any agreement would depend on neither side seeking to circumvent the provisions and effectiveness of the agreement through transfers to third countries'. The previous US proposal had simply suggested that viability of any agreement would depend on neither side 'seeking to upset the strategic balance by transfers to third countries'.[45] Nevertheless, the Russians continued to press on this point, and in April 1970 they called for a prohibition on the transfer of offensive and defensive systems plus related technology to any country. These would take precedence over existing international commitments. Given the on-going US assistance to the Polaris Improvement Programme, this suggestion was hardly likely to appeal to the MOD. In response to the Soviet proposal, the US maintained that the withdrawal clause was still the best way to deal

---

41 TNA FCO 46/591, FCO telegram 453 to Washington, 24 February 1970; FCO telegram 468 to Washington, 25 February 1970.

42 TNA FCO 46/593, MISC 237 (70)7, Policy Briefs for UK/US Bilateral Discussions, Washington, 5 March, Note by the Secretary, 2 March 1970.

43 TNA FCO 46/593, R.M. Tesh, Defence Dept to Mr Bendall and Mr Williams, 2 March 1970, Bendall Manuscript Note, 3 March 1970.

44 TNA FCO 46/593, Washington telegram 673 to FCO, 6 March 1970.

45 TNA FCO 46/593, R.M. Tesh, Defence Dept to Mr Bendall, 24 April 1970.

with any British concerns. However, the Americans told London that it must expect Washington to consider any future UK request for a new delivery system in the context of SALT, but it was still too early to predict exactly what that context might be. And indeed this appears to have been a factor in the parallel and highly compartmented consideration, especially on the US side, of the UK's Polaris modernisation options, when this factor played a part in inhibiting the sale of Poseidon to the UK.[46] Indeed John Baylis and Kristan Stoddart argue that any major sales agreement concerning a MIRVed system would have no doubt harmed the improving bilateral relationship between the superpowers.[47]

The Administration kept reminding UK officials of the various private assurances given before SALT began that the continuation of the Polaris Sales Agreement was not in doubt. But this was now qualified. Washington, in deciding whether to provide what might be interpreted as an addition to the UK deterrent in numerical terms, would have to look at the implications for any SALT agreement that imposed quantitative limits on US and Soviet strategic delivery systems. Article 9 of the Polaris Sales Agreement appears to have been the key as this covered the continuing nature of the agreement since it dealt with modifications and set no terms to these. London had to concede that it was for the Americans to decide how to interpret this Article. Since there could be no absolute guarantee, neither could there be absolute peace of mind either.

The Conservatives Take Over: June 1970

GEN 70 was the name given to the new Cabinet official committee on SALT in 1970, replacing MISC 237; its remit however was essentially the same. It prepared a draft report for the new Conservative government in June 1970 on the latest state of play in the negotiations, then in their second round.[48] This noted at the end that the UK was keeping a careful watch that its weapons systems were not constrained by an agreement to which it was not a party. The Russians were insisting, for instance, that there should be no transfers of strategic nuclear weapons or technology to third countries and this would clearly have implications for the Anglo-American military nuclear relationship. Thus far the US had rejected this as outside the scope of the talks and seemed ready to

46 John Baylis and Kristan Stoddart, 'Britain and the Chevaline Project: The Hidden Nuclear Programme, 1967–1982', paper prepared for Mountbatten Centre Fifth Seminar on the History of the UK Nuclear Weapon Programme, Thursday 9 October 2003, p. 13.

47 Baylis and Stoddart, 'Britain and the Chevaline', p. 13.

48 TNA CAB 164/936, GEN (70), Cabinet, International Aspects of Nuclear Defence Policy, Strategic Arms Limitation Talks (SALT), Report to Ministers, 29 June 1970.

stand firm. London would encourage them to do so. Moreover, the UK also had a specific interest in keeping Soviet ABM deployments to the minimum. The lower this level the more readily would the UK be able to maintain Polaris' effectiveness at minimum cost.[49]

Soviet ABM deployments were by now assuming much greater significance in UK thinking.[50] So when the US proposed a revised package on ABM constraints to the Russians in Vienna in the summer of 1970, which would limit deployment to 100 launchers, London felt that the UK should support zero deployment because it would give greatest relief to the defence budget and would best ensure the UK deterrent's future viability.[51] The greater the limitations on Soviet ABM deployment, the longer the effective life of the UK's own nuclear forces in their then present form.[52] ACDA was certainly aware that any significant increase in ABMs could have serious implications for the UK Polaris force, and the UK had made this very clear at the formal consultative bilateral talks in Washington in July when the sorts of cooperation required for Polaris modernisation were discussed.[53] Indeed the UK also worried that an ABM no-transfer agreement might prevent the US from further cooperating with the UK in the realm of penetrating ABMs.[54] And this was the line taken by the UK at the NATO North Atlantic Council (NAC) consultations on SALT on 15 July 1970.[55] It was not just a question of ABM launchers: Soviet radar capabilities were an issue too.

A ban on US transfers would have very serious consequences for UK–US bilateral exchanges since the maintenance and possible improvement of the

---

[49] TNA FCO 46/593, R.M. Tesh, Defence Dept to Planning Staff, Preparation of Papers for an Incoming Government, 3 June 1970.

[50] TNA PREM 15/289, Burke Trend to Prime Minister, Strategic Arms Limitation Talks (SALT), 14 July 1970.

[51] TNA PREM 15/289, Burke Trend to Prime Minister, Strategic Arms Limitation Talks (SALT), 14 July 1970 and R.T. Armstrong to Burke Trend, Cabinet Office, 15 July 1970; TNA DEFE 13/1032, W.F. Mumford, Head of DS 2 to PS to Secretary of State, 14 July 1970.

[52] TNA PREM 15/289, Presentation by Ambassador Parsons on 29 October at Number 10, Defence Department, 27 October 1970.

[53] TNA PREM 15/289, Washington telegram 2525 to FCO, 23 July 1970; TNA DEFE 24/775, Speaking Notes, SALT; UK/US Bilateral, 23 July 1970; see TNA DEFE 24/770, UK Interest in SALT, 9 July 1970, F.H. Panton ACSA (N) to DS 22 (Mr Mumford); TNA DEFE 24/776, SALT Record of Anglo/US Bilateral Discussions Held in the State Department, Washington on Friday, 23 July 1971.

[54] TNA DEFE 24/775, (Bilateral Talks with US), Agenda for Discussion on 23 July 1971.

[55] TNA DEFE 13/1032, FCO telegram 263 to UKDel NATO, 15 July 1970.

deterrent was critically dependent on their continuation.[56] The Assistant Chief of the Defence Staff (ACDS) (Pol) told his boss, the Chief of the Defence Staff, in mid July 1970 that any US–Soviet agreement would inevitably have some effect on US freedom on future transfers.[57] Neil Cameron felt that there was bound to be an informal understanding between the two that neither side would use assistance to allies as a way of circumventing bilateral agreements. Such an understanding would tend to make the US look particularly cautiously at any subsequent assistance the UK might, conjecturally, want to request one day, though it would not strictly prevent the US, if they were otherwise inclined, from helping the UK to construct and maintain a successor force of equivalent size and relative capability. Across the street FCO officials were also anxious, perhaps even more so. Whatever the precise terms of the Polaris Sales Agreement, the Americans might in future undoubtedly find justification for a refusal to help Britain if the Russians persuaded them that such help would prejudice a SALT agreement. Moreover, if there were ever restrictions on ABMs, then the Russians would not simply be interested in numerical constraints on third country systems; they might be more concerned about efforts to improve Polaris qualitatively through better penetration aids and/or hardening, even if the numbers of UK missiles and warheads remained unchanged.[58] Underlying anxiety refused to go away despite US repeated assurances and reflected clearly the level of UK dependence on American assistance to sustain an effective deterrent. Worries about the future durability of this dependence is a recurring feature of UK thinking throughout this period and is hardly surprising given the extent of British dependence on the US for nuclear assistance.

The US gave the UK a further update on the state of play in SALT towards the end of June 1970.[59] The Russians had made no effort in Vienna to elaborate their no-transfer proposal and the US had the impression that this would not become a major issue. Some way, however, might have to be found of covering this issue in an agreement. Washington accepted that it was on notice that if, for example, the US attempted to deploy new Medium Range Ballistic Missiles (MRBMs) in Europe, the Russians would raise strong objections. Beyond this, however, the US did not think that the Russians proposed to interfere with existing allied nuclear sharing or basing arrangements.

---

56 TNA DEFE 24/770, F.H. Panton ACSA (N) to DS 22 (Mr Mumford), 9 July 1970.

57 TNA DEFE 24/770, Air Vice Marshall Neil Cameron, ACDS (POL) to CDS, SALT, 14 July 1970.

58 TNA FCO 46/597, A.B. Urwick, Defence Department to Mr Bendall, SALT and the British Polaris Force, 5 August 1970.

59 TNA PREM 15/289, Washington telegram 1952 to FCO, 26 June 1970.

A further round of UK–US consultations on SALT took place in Washington on 23 July 1970.[60] Phil Farley (ACDA) told UK officials that the non-circumvention clause that the US was proposing for the ABM agreement would prohibit transfers to third countries of systems and components limited in the agreement. The Russians wanted to include technology, but this was unacceptable to the Americans. Farley emphasised that while the US was not proposing such a clause for the offensive forces agreement, there would inevitably be some inhibition on political grounds on the US Administration's ability to assist third countries in areas covered by either agreement. British officials took the opportunity to remind the US again of UK concerns on possible or potential inhibitions on US ability to provide Britain with information and technical assistance for its nuclear weapons programme. Fortunately, London assessed that there were unlikely to be any significant early developments that would affect the UK nuclear armoury.[61] London, however, viewed the situation in the US–Soviet talks in mid summer 1970 with growing anxiety. It appeared that an agreement would leave both ABM launchers and associated radars deployed in such numbers and capabilities, actual or potential, that the credibility of Polaris would be seriously affected unless steps were taken to improve it.[62] MOD knew that the Russians had a large missile defence research, development and flight test programme aimed at improving radars and missile interception outside the atmosphere. It was still too early to say (end August 1970) what impact this R&D effort would have on future Soviet defences.[63] Nevertheless, officials (FCO, MOD and Cabinet Office) felt that SALT had to be kept to the fore as Ministers were scheduled to consider Polaris modernisation options later in the year – certainly the Foreign Secretary was being kept informed by his officials even to the extent of their trying to arrange sending highly classified briefs to him on holiday in Scotland in August.[64]

Meanwhile, in SALT itself by late summer 1970 it appeared as if the Russians were moving closer to Smith's formulation on third country forces. At the same time the Russians dropped their earlier proposals for a specific no-transfer provision. The Soviet draft proposal contained no clause on this issue and referred only to Intercontinental Ballistic Missiles (ICBMs). Both sides were now, or at

---

60 TNA PREM 15/289, Washington telegram 2525 to FCO, 23 July 1970.

61 TNA PREM 15/289, Alec Douglas-Home to Prime Minister, 4 August 1971.

62 TNA DEFE 24/770, F.H. Panton ACSA (N) to DS 22 (Mr Mumford), 9 July 1970.

63 TNA DEFE 24/770, Re-draft of Sub-paragraph 11(e) of the Draft Report to Ministers on SALT, Attachment to FHP/751/70 dated 25th August 1970. Omitted by mistake from final report; F.H. Panton ACSA (N) to Head DS22, 25 August 1970.

64 TNA FCO 46/597, A.B. Urwick, Defence Department to Mr Bendall, 5 August 1970; R.M. Tesh, Defence Department to Private Secretary, 21 August 1970.

least as reported to the British, talking about an article in the main agreement to the effect that its object and purpose should not be vitiated by relations with third countries. A progress report for Ministers noted in August 1970 that the Russians had proposed a limitation of the operational activities or patrol areas of strategic nuclear forces and this could conceivably have interfered with the operational deployment of Polaris because of the difficulty of distinguishing UK submarines from American ones. However, the Russians did not pursue this. Another possible constraint might have been a MIRV flight ban and this could have complicated British Polaris test flights from the US test range off Cape Canaveral.[65]

The Prime Minister, Foreign Secretary and Defence Secretary considered during a meeting on 30 September 1970 whether the UK should directly participate in SALT. After discussion, Ministers agreed that the UK should not seek for the present to participate directly, though further consideration would be given at the appropriate moment in future.[66] Several dangers in attempting to participate were foreseen, quite apart from the question of whether the participants would ever be likely to welcome any such initiative. First and foremost the UK arsenal was very small in comparison to that of the superpowers, but readiness to take part in the talks would imply that the UK was prepared in principle to accept some limitations on its forces. The effect of such limitations could not be foreseen, but there was a clear possibility that they would seriously undermine the credibility of the deterrent: force levels were at the minimum to meet the Joint Intelligence Committee's (JIC) deterrence criteria that Moscow must be held at risk. UK interests were best served by remaining outside and thus retaining maximum freedom of action.[67] This assumed of course that the US did not agree to any constraints that would impinge on UK freedom.

Although the Americans repeated their assurances to London that there was no question of discussing third country forces in SALT, they said that there would have to be an escape clause that would, amongst other things, cover the no-transfer question. A key concern for British officials at this particular moment was the renewal of the 1958 Mutual Defence Agreement, which had expired in 1969 and which was extended to 1974, so anything that might compromise this was viewed with alarm. In the mid 1960s it seemed that the 1958 Agreement

---

65 TNA DEFE 13/1032, Report on Progress to Date, Cabinet Office, 20 August 1970, paragraph 25; see also TNA DEFE 24/770, W.F. Mumford Head of DS 22 to PS to Secretary of State, 24 August 1970.

66 TNA CAB 164/936, Draft Brief, Strategic Arms Limitation by Ambassador Parsons on 29 October at No. 10.

67 TNA CAB 164/936, Presentation by Ambassador Parsons on 29 October at Number 10, Defence Department (FCO), 27 October 1970.

would expire, but with Polaris modernisation being pursued, this agreement assumed a new and much more important role in British nuclear weapons policy. As the renewal was being discussed in Congress, the Administration planned to say that the no-transfer issue was not relevant to SALT if pressed by inquisitive Congressmen. However, Smith tried to persuade UK officials that he thought that the Soviet position on no-transfer was not entirely unreasonable; for example, if the US proposed to transfer a 1,000 missiles to the UK, the Russians would be entitled to object. This was probably not seen as a very reassuring comment. Smith also believed that the Russians were more concerned about possible transfers to Germany than existing Anglo-American arrangements, and in any case were more animated about major strategic delivery systems. Despite this the US were still unclear on how the Soviet Union would actually want to interpret 'strategic' in the context of a ban on transfers to third parties, they did not probe further through fear of giving the Russians the impression that they were willing to consider the issue. With such a background it is hardly surprising that Heath told Nixon in October 1970 that the question of zero ABMs was very important.[68] However, on the no-transfer front, Whitehall was much more relaxed by this time. The Foreign Secretary told the Prime Minister in August that in light of further consultations with the US, he did not expect any significant early developments which would affect the UK's existing nuclear armoury.[69]

UK Attitudes and Possibilities: 1971

By early 1971 the UK's main concern was that the Russians would use the no-transfer clause in the emerging Anti Ballistic Missile Treaty (ABMT) as a precedent for offensive forces. Any Soviet attempt – so the US insisted – would be firmly rejected; any no-transfer clause would only apply to the ABMT and not to offensive forces then, or in the future. Moreover, the Americans did not envisage that the temporary offensive forces freeze (what would become the Interim Agreement SALT I) would include a formal or informal provision on third country forces. Nevertheless, the Russians were still seeking prohibitions on transfers of 'related technology'; this was close to the bone given the critical exchanges in hand or envisaged on the Polaris Improvement Programme. In discussions with the US in February 1971 one of the technical issues under

---

68 TNA DEFE 24/771, A.W. Stephens, APS Secretary of State to Head of DS 22, Visit of President Nixon, Attaching Extracts from Record of Discussion at Chequers on 3rd October, 6 October 1970.

69 TNA DEFE 24/775, Foreign Secretary to Prime Minister, 4 August 1971.

consideration was the differing views on the role of the Soviet HENHOUSE (the NATO codename for the structures) early warning radars. British experts were not convinced that these were a fundamental part of Soviet ABM plans, although they could certainly provide assistance to the battle management radars once incoming missiles were detected.[70] HENHOUSE was to loom rather larger later on in UK anxieties.[71] Officials felt that it was most important that the UK did not give the US an impression that British interest in limiting Soviet ABM systems had diminished.[72]

At a NAC briefing meeting in Brussels on SALT on 24 February 1971 Ray Garthoff (Secretary to the US delegation) said something that caused the UK delegation to prick up its ears when he referred to an American proposal to prohibit 'significant transfers' to third parties of central strategic systems. The delegation thought that this would clearly need to be a subject for the forthcoming UK–US bilateral consultations on SALT scheduled for 2 March.[73] And as Frank Panton (ACSA (N)) pointed out separately, this idea taken alongside the 'generalised provision' binding each side to refrain from making substantial changes in their non-limited systems (namely in US Forward Based Systems [FBS]), would make it very difficult for the US to assist and cooperate in any way with its allies in nuclear systems.[74] There was always a niggling doubt that language on FBS, although not directly affecting the deterrent, might set an unwelcome precedent or increase the possibilities for Soviet pressure on UK–US nuclear exchanges, especially if coupled with no-transfer language.[75] Peter Hudson (Cabinet Office Secretary of GEN 1) informed his colleagues Philip Adams and Burke Trend that at the forthcoming bilateral with the Americans the UK would want to know precisely what the US offer contained and how it impinged on their assurances given in 1969 in advance of the first SALT round that the future of the Polaris agreements was not in doubt.[76] In Adams' view this could be a sinister development requiring careful watching. Hudson thought

[70] TNA DEFE 24/772, United Kingdom/United States Bilaterals, Note by the Ministry of Defence and the Foreign and Commonwealth Office, 11 February 1971.

[71] Personal Correspondence.

[72] TNA DEFE 24/772, W.F. Mumford, Head of DS 22 to DUS (P), 15 February 1971.

[73] TNA CAB 164/936, W.F. Mumford, Head of DS 22, MOD to DUS (P), SALT: NATO Discussion February 24, 25 February 1971.

[74] TNA CAB 164/936, F.H. Panton, ACSA (N) to DCA (PN), SALT; NATO Discussion February 24, 25 February 1971.

[75] TNA DEFE 24/774, F.H. Panton, ACSA (N) to DS 22, 25 May 1971; TNA DEFE 24/774, Strategic Arms Limitation Talks (SALT), W.F. Mumford to P.J. Hudson, Cabinet Office, SALT – FBS, Covering Letter for Paper Entitled 'SALT: Forward Based Systems (FBS)', pp. 13–14, paragraph 3, 11 June 1971.

[76] TNA CAB 164/936, P.J. Hudson to Sir Burke Trend, 1 March 1971.

too that the upshot of UK inquires might well affect Ministerial views on Anglo-American nuclear collaboration. It would be possible by the end of the week to better judge whether there was any relationship between the encouraging noises made by Nixon on the subject of Anglo-French cooperation in December 1970 and the fact that the US and USSR had evidently been discussing an agreement that might curtail further Anglo-American collaboration. At the Washington consultations the US claimed, contrary to UK recollections, that there was nothing new in their statement to the NAC and that they had advanced a formula that recognised that substantial transfers to third parties could undermine an agreement on central systems.[77] For the Americans the key word was 'substantial', which the Russians had tried to delete. The US had, however, stuck to its guns. Nevertheless, in a situation after SALT each side would have to assess 'what the traffic would bear' in the way of new assistance to third parties, taking into account international and internal political factors and a necessity to study each specific move given external and domestic critics.[78] Despite this, it seemed to the UK that there was evidence that the US and Soviet positions on this matter were becoming closer. While there could be no definite provision that existing agreements would stay in force, there would also be no specific provision to the contrary.[79] The Americans told the UK in April 1971 that a no-transfer clause would at least be a further complication in the UK/US nuclear relationship. Taking such a clause together with any constraints on FBS, the Russians would be strongly equipped to argue that the Americans should cease to help the UK in the nuclear field.[80]

ABM Treaty Becomes the Main Focus: May 1971

Following the US–Soviet announcement of May 1971 that they would concentrate on concluding an ABM agreement in the coming year, the likelihood that there would be a positive outcome increased. This came out of the blue; it was a result of the back channel between Henry Kissinger (National Security Advisor) and Anatoly Dobrynin (Soviet Ambassador in Washington) by-passing the delegations in Vienna where there was complete stalemate. MOD was reasonably confident that the Polaris Sales Agreement and 1958 Mutual Defence Agreement would not be impeded in anyway, but the position on

77 TNA DEFE 13/1032, Anglo/US Bilateral Talks, Washington, 2 March 1971.

78 TNA CAB 164/936, Draft SALT: Anglo/US Bilateral Talks, Washington, 1 March 1971; TNA FCO 46/697, Anglo/US Bilateral Talks, Washington, 2 March 1971.

79 TNA PREM 15/289, Washington telegram 748 to FCO, 2 March 1970.

80 TNA CAB 164/937, Clive Rose to Private Secretary, 22 April 1971.

improvements to, or replacements for, present systems was much less clear.[81] Under the Polaris Improvement Programme the UK required US information, advice and facilities throughout the definition phase of the Programme (up until the end of 1972) and throughout the development and production phases early in 1977–78, if approved. Facilities required included space in US underground tests and use of US flight test ranges. The Polaris Improvement Programme's objective was to enable the present force to penetrate Soviet defences up to a level of hundreds of ABMs – probably more than a credible ABM Treaty might contain. However, the concern was ever present that it was not inconceivable that the US might feel inhibited from continuing to help Britain improve its deterrent to penetrate Soviet defences constrained to levels that might be designed to give it some protection against medium nuclear powers. It was by no means clear whether in the absence of an agreement the existing agreements with the US would enable the UK to buy the Poseidon SLBM – the US successor to Polaris. An amendment might be needed to the Polaris Sales Agreement and the 1958 Mutual Defence Agreement. As far as the Polaris Improvement Programme was concerned, the Department of Defense and the White House, including the President, knew of the areas where the UK required continued cooperation and that cooperation was already underway. On the surface then it would seem that there was nothing to worry about, but the nagging doubts did not go away as subsequent British attitudes and actions reveal.

As of May 1971 the US had put three options to the Russians for ABM limitations: complete elimination, Moscow and Washington only and asymmetrical limitations. However, the Russians rejected the first and last, which London thought meant that the second would be pursued. This would clearly be better for the UK than no limitations.[82] The Polaris Improvement Programme took into account all the possible permutations and combinations of Russian ABM development and deployment with or without SALT. However, the MOD did not want to discuss this or that postulate in a purely SALT forum. Nonetheless, MOD officials – primarily Macklen – thought that they should discuss ABM limitations with ACDA to find out exactly what US and Russian views were so that they could take them into account in the highly classified Polaris Improvement Programme context. For the MOD, only the complete dismantling of the Soviet ABM system would remove the threat to the UK deterrent unless essential improvements were made. Such an outcome was exceptionally unlikely. MOD Officials believed that there was no prospect that the Russians would concede any prohibition on all round radar coverage

81 TNA DEFE 24/774, F.H. Panton, ACSA (N) to DS 22, 25 May 1971.

82 TNA CAB 164/937, N.J. Barrington, FCO to P.J.S. Moon, 21 May 1971.

as the US ABM system would have this facility; neither did it matter, for the two existing sites effectively covered the patrol areas accessible to the UK. As long as the Russians had a few tens of ABMs, the UK would have no serious problems; these would not be made worse by 64, 100 or 128, but a number above 500 might well be different. MOD officials were most concerned that the US, by pressing limits on ABMs to two Safeguard sites (or four), would give the Russians freedom to increase ABM numbers well above 128 and to introduce an atmospheric defence system such as Sprint. The latter development would have serious implications for the UK as the Polaris Improvement Programme was predicated on defeating exo-atmospheric threats only. Overcoming terminal defence was not part of the equation. Even if the Russians reduced their present deployment by a factor of three, the UK would still need to improve Polaris.[83] Indeed it did not make any difference whether the Russians were limited to 64 or 100 ABMs: neither figure would remove the need to improve Polaris.[84] Considerable modification was required to restore the credibility of Polaris against even the then deployed Soviet ABM system.[85] No doubt for this very reason MOD worried that a more precise formulation on no-transfer might put the UK in difficulty.[86] Neuroses would not have been eased by Kissinger's comment to senior Whitehall officials in London on 25 June, that although the Americans did not contemplate discussion of no-transfer in the negotiations for an initial agreement on offensive systems, 'they had not formed a definitive position on it'.[87]

At the regular six-monthly UK–US bilateral meeting on SALT in Washington on 23 July 1971 the UK team came away with valuable clarifications of US positions and thinking. Farley (ACDA) said that the Russians had consistently pressed for constraints on the transfer of related technology, which would enable a wide range of assistance to be challenged even in the non-military field.[88] He added that, although the US was not proposing a no-transfer clause, there would

---

[83] TNA DEFE 13/1032, V.H.B. Macklen Deputy Chief Adviser (Projects and Nuclear) (DCA (PN)) to Head of DS 22, 29 October 1970.

[84] TNA DEFE 24/770, Neil Cameron ACDS (POL) to CDS, 14 July 1970.

[85] TNA DEFE 24/770, V.H.B. Macklen DCA (PN) to Head of DS 22, SALT: Background Brief to Ministers, 28 July 1970.

[86] TNA FCO 46/700, D.C. Humphreys, Director of Defence Policy, MOD to A.E.D. Chamier, Defence Department, FCO, 21 May 1971.

[87] TNA FCO 46/701, KV (71), 3rd Meeting, Visit of Dr Henry Kissinger to London 24–26 June, Note of a Meeting Held on 25 June 1971.

[88] TNA DEFE 13/1032, Record of Anglo/US Bilateral Discussions Held in the State Department, Washington on Friday, 23 July 1971; TNA FCO 46/702, Washington telegram 2525 to FCO, 23 July 1971.

inevitably be some inhibition on political grounds on an Administration's ability to assist third countries in areas covered by the ABM or interim offensive forces agreement. This was the root cause of continuing UK concerns about SALT; that the US would fall victim to a self-denying ordnance. Macklen said that the UK was vitally concerned with maintaining the capability of its small deterrent force. To do this some US assistance was necessary, both in the technology of offensive systems and information on Soviet ABM defences. Farley replied that whatever the letter of a SALT agreement, any requests for US assistance on strategic weapons systems would inevitably be subjected to close political scrutiny and assessment as to whether they could be held to be contrary to the spirit of agreements reached in SALT. Even if SALT did not extend beyond the first phase agreements, the same political question would arise whether it would be consistent with an ABM agreement to help the UK improve its offensive capability. However, UK special interest had been noted and brought to the attention of 'higher authority'.[89] Presumably this meant Kissinger and the President. One unnamed but senior US official volunteered to the UK after the talks that the nature of the compromise on ABMs was evidently of much greater importance to the UK than the US and expressed the personal view that it would do no harm for the UK to point this out. On this basis the Embassy suggested that the UK might indeed like to remind the Americans formally of that interest.[90]

The US ABM proposal at the Helsinki round (this had ended in September 1971) was that there should now be an asymmetrical ABM agreement; the US would be limited to two Safeguard sites and that the Russians should have one National Command Authority (NCA) site. At the same time the UK's plans for improving Polaris took account of predicted possible Soviet ABM levels whether or not an ABM agreement was concluded, but the higher the level of defences, the greater the operational strain of putting sufficient Resolution class boats to sea as necessary to pose a credible deterrent.[91] Britain's interest, given

[89] TNA DEFE 24/776, G.E. Millard, British Embassy, Washington to Sir Stewart Crawford, FCO, 9 September 1971, re Conversation between Michael Butler and Tom Pickering on ABMs, 7 September 1971.

[90] TNA FCO 46/703, Report to Ministers Draft by the FCO, SALT: Implications for the UK National Deterrent of an Agreement to Limit Anti-Ballistic Missile Defences (ABMs) J. Laughton, Defence Department to Mr Rose, Meeting of GEN (1), 16 September 1971.

[91] TNA DEFE 24/777, F.H. Panton, ACSA (N) to DS 22, SALT: Tesh's Letter to Nairne of 23 September, 23 September 1971; Strategic Arms Limitation Talks (SALT) – ABM Limitation, J.P. Groom, for Secretary of COSC to Chiefs of Staff, Annex A, Progress in the Strategic Arms Limitation Talks: ABM Limitations, paragraph 9.

that a complete ban was not now likely, was in keeping ABM deployments as low as possible, and a level of 100 would be much preferable to 200 or more. London felt that this was now the psychological time to underline with the US Secretary of State, William Rogers, when the Foreign Secretary met him in New York on 1 October 1971, that the UK national interest was in as low a level as possible of ABMs.[92] The possible risk of causing irritation by a UK intervention was somewhat reduced by evidence that the Americans themselves were reasonably open minded on the question and in any case was largely offset by the genuine importance of the point for the UK, which the Americans recognised. The FCO felt that the most natural way for the UK to get its view across would be for the Foreign Secretary to mention it when he saw Rogers on 1 October 1971.[93] However, the advice from Washington was that it would be better for Home to speak to the President himself because Rogers was not closely involved in the key decision making on SALT.[94] As for the advice that should be given, the Foreign Secretary might say that, whilst the President deliberated on how to tackle the ABM problem, the UK would be very grateful if its own strong interest could be borne in mind. For the UK, with its small force, the point was of genuine importance; it did not really matter to the US whether there were 100 or 200 interceptors. By far the best solution for the UK would be one that limited the Russians to 100 missiles of within 100 kilometre radius of Moscow.[95] London agreed with this advice, but preferred that no specific number should be mentioned – merely that the UK would wish them to be fixed at the lowest possible level. Any specific numbers might cause irritation as it might appear that the UK was intervening too closely in the detail of the President's bilateral negotiations and on the strength of information given to the UK at official level.[96] In the end the FCO decided to follow the Embassy's advice that it would be better if Home put this to Nixon rather than Rogers and to do so in stronger terms.[97] The Foreign Secretary would make clear that the UK had a special interest in the level of ABMs that might emerge from SALT and that

---

[92] TNA DEFE 27/777, Washington telegram 11 to UKMIS, New York, 28 September 1971.

[93] TNA PREM 15/289, FCO telegram 2499 to Washington, 24 September 1971.

[94] TNA PREM 15/289, Washington telegram 11 to UKMIS, New York, 28 September 1971.

[95] TNA PREM 15/289, Washington telegram 12 to UKMIS, New York, 28 September 1971.

[96] TNA PREM 15/289, FCO telegram 702 to UKMIS, New York, 29 September 1971.

[97] TNA PREM 15/289, FCO telegram 702 to UKMIS New York, 29 September 1971.

the level should be as low as possible. This was of vital concern to the UK for the maintenance of its deterrent. The UK therefore hoped that the President would take UK interests into account when he decided what proposals should be put to the Russians. Home raised all this with Nixon on 30 September; the main message that the UK wished to convey clearly was its interest in seeing any limit on ABMs in any agreement being as low as possible.[98] No Aide Memoire to this effect was handed over.[99] Nixon said that the number of ABMs was likely to be 100, which no doubt provided some comfort to fretting officials back in London. At round about the same time an MOD assessment noted that nothing that had emerged so far from the negotiations to suggest that the Russians would stop, though within certain limits, extending and improving their ABM defences. An ABM treaty might modify this assessment, but the range of alternatives under consideration in the negotiations would not alter significantly the summary of the threat to the effectiveness of Polaris. The UK was thus faced with the prospect that from about 1975, the effectiveness of the present force was likely to be degraded unless it was effectively improved, or the deterrent criteria were changed, i.e. the destruction of the five largest Soviet cities including Moscow.[100] A no transfer clause that included offensive missiles, depending how it was drafted, could bite harder in the case of a new missile such as Poseidon.

Subsequently, speaking notes upon which Home's message was based were also handed to both State Department (Spiers) and the National Security Council (NSC) (Hal Sonnenfeldt).[101] The Americans got the point: Gerard Smith told Lord Cromer (Ambassador) in November that this was understood.[102] And as an interesting aside, Sonnenfeldt commented to Michael Butler that it could be argued that the UK interest in keeping the Russians down to 100 ABMs could by derivation be a US interest too. Such constraints might reduce the pressures on the UK to seek US assistance to penetrate Soviet defences and might be helpful in the future if it became difficult to help the UK – not quite the reaction that

[98] TNA FCO 46/703, R.M. Tesh. Defence Department to Mr Graham, 10 November 1971. Tesh was asking if Graham had a record of this meeting as Defence Department had none and neither had State Department.

[99] TNA FCO 46/822, C.M. Rose to M.D. Butler, Washington, 2 February 1972.

[100] TNA DEFE 11/792, Chief of the Defence Staff to Secretary of State, 22 December 1971.

[101] TNA DEFE 24/778, Brief for CSA's Visit to Washington, W.F. Mumford to DCA (PN) and AUS (D. Staff), 30 November 1971.

[102] TNA DEFE 24/778, Record of a Conversation between Lord Cromer and Mr Gerard Smith on 8 November 1971.

London would have hoped for.[103] Back at the end of October Aaron (ACDA) had mentioned to Butler that the US had not wanted to use a general formula on no-transfer in the separate agreement on ABMs, such as the one that they had previously had in mind for a comprehensive SALT agreement.[104] However, London was not convinced that this was the whole story.

Butler did not rise to the bait, but Clive Rose (now Assistant Under Secretary Defence, FCO) back in London thought it serious enough to suggest to MOD that there might well have been more discussions and possible agreements between the US and USSR than the British were aware of. It was troubling too that a US official had asked why the UK placed so much importance on Moscow as a key strategic target. No-transfer was an issue that the UK had accepted would appear in a definitive agreement on offensive forces, but it was essential that a future pitch should not be queered by unfortunate precedents set in the present. Therefore it might now be necessary to brief the Prime Minister to raise this matter with Nixon when the two met in Bermuda in December 1971. Nixon, however, was already fully aware of the seriousness for the UK of anything that might restrict the availability of facilities required from the US for the completion of the Polaris Modernisation Programme. This included underground nuclear effects testing, information on hardening and re-entry characteristics of re-entry body materials, critical reviews of UK designs, advice on flight test programmes to prove these designs and information about the impact of UK re-entry systems proposals on the remainder of the weapon system.[105] All of which was being pursued under the aegis of the 1958 Mutual Defence Agreement. Rose remarked that on the Polaris Improvement Programme, the UK would need a free run, if possible, until 1974 – although he thought it might be possible to adapt to a shorter period.[106]

However, London was still anxious about this. Reports that the Russians might now still insist on a no-transfer clause in the interim agreement on offensive systems only magnified these anxieties. Indeed it had become evident to UK officials that there were practical and political pressures that might influence the US to fall back to a 'two for two' solution (two ABM sites per country), which would give the Russians 200 launchers. Depending on where these were located, this new number could cause the UK serious difficulties, even with a completed

103 TNA DEFE 24/778, M.D. Butler, Washington to R.M. Tesh, Defence Department, FCO, 18 November 1971.

104 TNA FCO 46/704, M.D. Butler, Washington to R.M. Tesh, Defence Department, 27 October 1971.

105 TNA DEFE 24/778, Brief for CSA's Visit to Washington, W.F. Mumford to DCA (PN) and AUS (D. Staff), 30 November 1971.

106 TNA FCO 46/703, C.M. Rose to Sir S. Crawford, 30 November 1971.

Polaris Improvement Programme.[107] There were clearly still doubts that the US did not fully grasp the critical nature of UK interests in the no-transfer issue. Although the UK had made its views known, London still was not sure that its message had been understood. It was thus essential that the US Administration should be aware at the highest level of the possible implications of a no-transfer clause not only for the Polaris Sales Agreement, but also in connection with US support needed for the Polaris Improvement Programme. Some of the key US officials involved in the SALT process were unaware of UK essential defence requirements; namely the separate and tightly held exchanges on Polaris modernisation.[108] Officials therefore decided to recommend that the Prime Minister should raise UK interests with Nixon. Lord Cromer in Washington agreed, suggesting that Heath might ask the President to indicate his views on no-transfer and ABMs, which would help UK defence planning – assuming that Nixon was disposed to respond favourably of course.[109] In parallel, Sir Herman Bondi (the new MOD Chief Scientific Adviser) would take advantage of his visit to Washington to make the case to Kissinger direct, assuming that he was available, just to make sure that the White House was fully up to speed.[110] In the event he only saw Sonnenfeldt.[111] Sonnenfeldt (NSC) said that the UK had nothing to be concerned about as the Russians had not raised the no-transfer question in relation to the freeze on offensive systems.[112] But British interests were related to future decisions that might have to be taken on the UK deterrent and the need to have all the relevant information available in order to advise Ministers on the choices.[113] Ministers would have to take some major decisions in 1972 on the UK's nuclear forces future. Sonnenfeldt told Rose that the

107 TNA FCO 46/704, C.M. Rose to Sir S. Crawford, 30 November 1971.

108 In February 1972 Butler could report from Washington that some parts of the bureaucracy still had not acquired the habit of automatically including in the papers that they wrote on this subject a reference to UK interests, which suggested that London was right to worry. However, the NSC claimed that they made good this deficiency as papers progressed through the bureaucracy to the President and other principals. TNA FCO 46/822, M.D. Butler, Washington to C.M. Rose, London, 1 February 1972.

109 TNA FCO 46/704, Washington telegram 4008, Bermuda Meeting – Nuclear Policy and SALT, 30 November 1971.

110 TNA DEFE 24/779, Foreign Office telegram 3071 to Washington, 2 December 1971.

111 TNA DEFE 24/779, Record of a Discussion between CSA and Mr Sonnenfeldt in Washington, 7 December 1971.

112 TNA DEFE 24/779, Washington telegram to Washington 4059, 3 December 1971.

113 TNA DEFE 24/779, C.M. Rose, SALT (Record of Conversation with Sonnenfeldt in Bermuda, 21 December 1971), 23 December 1971.

UK would have to reckon with a no-transfer clause in a second phase strategic arms limitation agreement and this might be relevant to negotiations for the extension of the weapons components article in the 1958 Defence Agreement, which was set to expire on 31 December 1974. Rose was worried that the US was now feeling it necessary to restrict their support even before a definitive agreement had been reached.[114] He was not the only one: Dr Bob Press, Cabinet Office, wondered in light of the clear warning that no-transfer would come up in a second phase of SALT and its possible impact on the 1958 Mutual Defence Agreement whether the US had already gone further in its thinking about its termination than London was hitherto aware.[115]

Bermuda Summit: British Pleas

Heath met Nixon at Bermuda on 20 to 21 December 1971. FCO briefing papers for this meeting set out core UK concerns that they wanted the Prime Minister to convey clearly to the President.[116] Heath had to emphasise the importance that Britain attached to maintaining present levels of co-operation in nuclear matters, specifically continued US support for the Polaris Improvement Programme and continued official-level bilateral consultations on SALT. Nixon also needed to know that it was Britain's intention to contribute to the defence of Europe by maintaining a credible British deterrent for the foreseeable future. To achieve this, the UK deterrent had to retain an ability to devastate Moscow – as well as other cities and such areas of Western Russia as an improved ABM system might in future be able to cover. A first phase SALT agreement might put this in jeopardy. London recognised that there was no hope of a zero level ABM agreement, which would have suited the UK best, but numbers should be as low as possible. If permitted levels were to exceed 100 missiles then this would involve the UK in even greater expenditure to ensure Polaris' continued effectiveness. London assumed that the US would be able to maintain its rejection of any no-transfer undertaking on offensive systems, and wanted, moreover an undertaking from the Americans to be consulted well in advance about the terms of any eventual no-transfer agreement on offensive systems. Maintenance of a credible deterrent now depended critically on US support. If it did not the

[114] TNA FCO 46/704, C.M. Rose to PUS, 23 December 1971.

[115] TNA FCO 46/704, R. Press, Cabinet Office to R.M. Tesh, Defence Department, 29 December 1971.

[116] TNA PREM 15/289, Talks between the Prime Minister and the President of the United States of America at Bermuda, 20–21 December 1971, Nuclear Matter Brief by the FCO PMV (BER) (71) 9, 18 December 1971.

UK could have afforded to have been much more relaxed about the direction and content of US-Soviet negotiations, which of course it was anything but. Heath was briefed to indicate to Nixon that the UK would welcome seeing, at the earliest possible stage, the text of any no-transfer agreement, since the wording would be likely to influence the wording of a corresponding clause on offensive systems.[117] Finally, Heath could inform Nixon that Britain hoped to be able to reach key decisions during 1972 on Polaris improvements. It seems that Heath made all the points requested, including the one on consultation on no-transfer, in the Aide Memoire prepared for the occasion, although Heath did not hand this over. There appears to have been no record of whether Nixon gave the assurance requested, and if so, in what terms.[118] Rose later told the Cabinet Office that he still had not seen a record of what had actually been said at Bermuda. In fact it appears that the issue was dealt with only briefly at the meeting and it was not certain whether Nixon fully understood the problem and in any case no specific assurances were given.[119]

Rose summarised the position in SALT for Sir Denis Greenhill, the Permanent Under-Secretary at the FCO, in relation to British interests in late December 1971. If, as was now possible, the US dropped its insistence on including submarines in the interim freeze, then the UK would probably not have to face a no-transfer clause on offensive missiles in SALT phase one. Conversely, a no-transfer clause in the ABM agreement was likely to be quoted by the Russians as a precedent for the second phase negotiation on offensive missiles however much the US would wish to deny this. Although, the US was thinking in terms of five years for the likely duration of the second phase negotiation, Rose feared that Americans might be inhibited before then in their support for the Polaris Improvement Programme, depending on how the negotiations went. As a result London would have to keep in very close touch with Washington over possible implications for the 1958 Mutual Defence Agreement, which was due to expire in 1974. In such circumstances it would be better to arrange a further round of formal bilateral consultations in Washington in January.[120] Despite Rose's concerns, Lord Cromer advised that it would be a mistake to ask the Americans to host such a meeting at that point in time.[121]

---

117 TNA DEFE 24/780, J.D. Laughton, Defence Department to B.M. Webster, Cabinet Office, 11 February 1972.

118 TNA DEFE 24/780, C.M. Rose, FCO to A.P. Hockaday, Cabinet Office, 29 February 1972.

119 TNA FCO 46/823, C.M. Rose to Mr Laughton, 6 March 1972.

120 TNA DEFE 24/779, C.M. Rose to PUS, 23 December 1971.

121 TNA DEFE 24/779, M.D. Butler, Washington to C.M. Rose, FCO, 23 December 1971.

The Run-up to Moscow: January to April 1972

At the turn of the year, as the US still mulled over elements of a possible ABM agreement, the Embassy did its best to keep the chances of a one-for-one ABM system alive by saying to all concerned that to go for a two-for-two instead would be to ignore British interests without gaining any strategic advantages for the US.[122] GEN 1 agreed at its sub-group meeting on 11 January that the FCO should now prepare and circulate a draft minute from the Foreign Secretary to the Prime Minister that would seek approval for a further bilateral meeting with the US at official level to discuss: the future of SALT, ABM deployment, radars, Hardsite (a new US ABM system for defence of fixed land-based missiles such as Minuteman), exotic ABM systems, no-transfer and the US–USSR Prevention of Accidental War Agreement.[123] London no longer saw the bilateral talks, however, as the best forum for representations on urgent issues of major national importance. Instead they provided a useful means of following up high-level representations, when the UK could take the Americans through British thinking in detail.[124] Moreover, such talks held at regular intervals maintained the principle that HMG had a special claim to such consultations separate from those enjoyed by other American allies. And as before, UK interests would not be served by any move which produced arguments for greater deployments of missiles than were then being discussed in SALT, or in more than one location as far as the USSR was concerned. UK requirements remained in keeping launcher/missile numbers as low as possible. Hence the slight anxiety over US interest in Hardsite.[125] Butler and Cromer accepted that the proposed agenda was substantial enough to justify a visit by senior officials.[126] Farley (ACDA) told Butler at the same time that there had been no developments on no-transfer on offensive systems and that this subject was not discussed at the US-USSR talks that had just ended in Vienna. The Americans accepted the UK proposals for a meeting and tentatively fixed 23 or 25 February 1972.

Meanwhile, the Russians and Americans provisionally agreed on a no-transfer clause for the ABM Treaty in the February round in Vienna. This

---

122 TNA DEFE 24/779, M.D. Butler, Washington to R.M. Tesh, Defence Department, 7 January 1972.

123 TNA DEFE 24/779, B.M. Webster, Cabinet Office to J.D. Laughton, Defence Department, 18 January 1972.

124 TNA DEFE 24/779, C.M. Rose, FCO to M.D. Butler, Washington, 19 January 1972.

125 TNA DEFE 24/779, F.H. Panton, ACSA (N) to DS 22, 19 January 1972.

126 TNA DEFE 24/779, M.D. Butler, Washington to C.M. Rose, FCO, 28 January 1972.

provided that each side would undertake not to transfer to other states, and not to deploy outside its national territory, ABM systems or their components limited by the Treaty. An explicit ban on technical assistance was not included. However, the negotiating record would make it clear that this prohibition was to be interpreted as precluding a transfer of technical descriptions or blueprints specially worked out for the construction of ABM systems and their components. London needed time to assess this provision's implications for UK interests and the extent to which it might set a precedent for no-transfer provisions in later agreements on offensive systems.[127] This would therefore need to be a priority topic for exploration at the forthcoming SALT bilateral in Washington. In his regular report to the NAC on 2 February 1972, Gerard Smith reported on the latest state of play.[128] On the no-transfer issue the Russians had not raised this in the context of offensive systems, but had proposed that no ABM systems should be transferred to third countries, that there should be no deployment of ABM systems except in either side's national territory, and that there should be no transfer to third countries of technical assistance. The US had resisted the last proposal and also an 'encouragement and inducement' ban. In the end, agreement was reached *ad referendum* on the first two points. The third would be dropped on the basis that the negotiating record would show that 'technical description and blueprints' would also be excluded from transfer. Whatever construction was to be put on this phrase was crucial in MOD eyes.[129] (And it became clear that the US saw no reason why this clause should prevent intelligence exchanges on ABM systems or general technical discussions of such systems. These were critical for threat assessment, which contributed significantly to Polaris modernisation decisions.[130]) In response to a British question, Smith said that there was no indication that the Russians linked the question of how wide the coverage of the interim freeze should be with the question of no-transfer.[131] Despite this, the underlying question continued to bother London, and Rose informed Butler that the UK team would want to press the US, first, for more detail about what the Americans had thus far agreed with the Russians and, second, about the likelihood of the latter demanding a no-transfer clause in an agreement on offensive systems if the US maintained its insistence on

127 TNA FCO 46/823, Visit of President Pompidou, SALT, Annex B, SALT: UK National Interests, B.M. Webster, Cabinet Office to J.D. Laughton, Defence Department, 17 February 1972.

128 TNA DEFE 24/780, UKDel NATO telegram 70 to FCO, 3 February 1972.

129 TNA DEFE 24/780, W.F. Mumford, Head of DS 22, MOD to J.D. Laughton, Defence Department, FCO, 14 February 1972.

130 TNA DEFE 24/780, V.H.B. Macklen, DCA (PN) to DUS (P), 28 February 1972.

131 TNA DEFE 24/779, UKDel NATO telegram 71 to FCO, 3 February 1972.

inclusion of SLBMs in the interim agreement. There was also the precedent issue: London was never convinced by the US argument that the terms agreed in an ABMT would not be used as a precedent for a no-transfer clause in a definitive agreement on offensive systems.[132] In addition, the UK hoped that the US would at least let the British team have sight of the relevant SALT records on no-transfer; London was a little concerned that the US had even gone as far as agreements on this issue, even if it had done so only *ad referendum*, without giving the UK a chance to comment, or even giving prior warning. No-transfer was, in the MOD view, likely to be the most important part of the bilateral meeting. The US acceptance of a no-transfer text in an ABM Treaty could be a most awkward precedent in future years.[133] That the US had not met the Prime Minister's request at Bermuda for prior consultations over texts only amplified the anxiety that was never far from the surface in British thinking.

Consultations continued with the Embassy in Washington keeping it busy in picking up details from US officials on the latest state play and on US thinking on outcomes and options. Phil Odeen (National Security Council [NSC]) told Butler that the White House was paying great attention to UK interests, which had been made at every level, including the highest. Ashton Carter (NSC) said that the US had recently completed some studies of British penetration capability and had come to the conclusion that the British ought to be reasonably happy with their capability, even without improving Polaris, if there were only 100 missiles around Moscow. This assumed two submarines with 96 re-entry vehicles on station. Carter wanted to know how the UK would react to 150 or 200 ABMs. Presumably this would not cause anxiety if Super Antelope (the new name for the Polaris Improvement Programme) progressed. Butler thought the UK would much prefer 100. After all it was not inconceivable that a CTBT might interfere with the Polaris Improvement Programme.[134] This did not go down too well in Whitehall. Officials were surprised to learn that the US had made such calculations without checking the facts with London, or their conclusions for that matter.[135] The assumption that two boats would be on station at all times was unwarranted; nor was it correct to say that 200 missiles would not cause anxiety if Super Antelope went ahead. Butler advised the FCO

132 TNA DEFE 24/779, C.M. Rose, FCO to M.D. Butler, Washington, 7 February 1972.

133 TNA DEFE 24/780, P.D. Nairne, DUS (P) to PS/PUS, 9 February 1972; C.M. Rose, FCO to M.D. Butler, Washington, 14 February 1972.

134 TNA DEFE 24/780, M.D. Butler, Washington to R.M. Tesh, Defence Department, FCO, 7 February 1972.

135 TNA DEFE 24/780, C.M. Rose, FCO to M.D. Butler, Washington, 14 February 1972.

that it would be best if the UK could let the US have a full statement of what its requirements would be in order to ensure that any future no-transfer provisions excluded such things. However, he recognised that might not be possible until decisions about Polaris' future had been taken and these were necessarily sometime away in the future.[136] Laughton in the FCO's Defence Department observed that Butler's remarks were a further powerful argument for dithering no longer over the Poseidon option for the UK deterrent. In his view it would be better to work out precisely what was needed for Super Antelope so that it could be secured at least at short notice. Total distraction by higher ambitions could well jeopardise the attainable minimum.[137]

In any event the formal US–UK bilateral meeting on SALT issues took place on 23 February and enabled the UK to put formally on record its interest in several points that were likely to arise during the preparation of the US negotiating positions for the Helsinki round, due to open on 28 March 1972.[138] At NATO consultations prior to this session, the UK Ambassador – Sir Edward Peck – emphasised to Gerard Smith the importance of keeping to a minimum the risk that no-transfer provisions in the ABMT did not become a damaging precedent in follow-on negotiations.[139] On the no-transfer issue, Britain proposed that the US should consult before it entered into serious negotiations with the Russians on no-transfer in an offensive missile limitation agreement, but the US only noted British wishes without giving any specific commitment. The UK stressed that the Russians should not be allowed to use ABMT language as a precedent for a future definitive agreement on offensive missiles.[140] US officials were adamant that this was already sufficiently clear from the negotiating record, but they subsequently did make a formal unilateral statement to the effect that they did not regard the provisions of the ABMT as constituting precedents for a definitive agreement on offensive systems. The Russians did not comment.[141] However, the Americans were interested in an informal UK suggestion that a distinction might be drawn in future between assistance to improve third country forces' quality instead of assistance leading

136 TNA DEFE 24/780, M.D. Butler, Washington to C.M. Rose, 14 February 1972.

137 TNA FCO 46/823, J.D. Laughton, Defence Department to Mr Rose, 16 February 1972.

138 TNA DEFE 24/780, Washington Saving telegram 20 of 24 February to FCO; TNA FCO 46/823, Record of UK/US Bilateral Talks on SALT Held at ACDA in Washington on Wednesday 23 February 1972.

139 TNA FCO 46/824, UKDel NATO telegram 178 to FCO, 27 March 1972.

140 TNA DEFE 24/780, W.F. Mumford, Head of DS 22 to DUS (P), 25 February 1972.

141 TNA DEFE 24/782, Washington telegram 1857 to FCO, 6 June 1972.

to increased numbers. So despite Heath's pleas at Bermuda it was evident to UK officials following the bilateral that, thus far, those US officials handling SALT had received no directive from the White House requiring that the UK should be consulted about no-transfer.[142] If further high-level UK representations were needed on no-transfer, they would have to be made before the next Helsinki round started.

Vic Macklen explained to the Americans the basis of UK ABM concerns in relation to Polaris. Taking a pessimistic assessment of the Soviet system, assuming an interception range into Russia of 300 to 400 miles and assuming the system to be easily saturated by incoming warheads, the UK calculated that for a two-boat force with its current loadings, it only required 33 Soviet ABMs to provide a 80 per cent chance of destroying all incoming UK warheads; and that the expenditure of 56 ABM launchers would raise this to 96 per cent. If the Polaris Improvement Programme aims were achieved, a UK threat from two boats would meet deterrent criteria if those parts of Western Russia within range were not defended by more than about 400 ABMs. However, with only one boat, more than 130 ABMs around Moscow would reduce the effectiveness below UK deterrence criteria set by the Joint Intelligence Committee. Even if a more restricted view of range capability was taken, Britain could still not attack Moscow with unimproved Polaris. For this reason the UK was therefore glad that US–USSR negotiations were turning toward a somewhat lower aggregate ABM level that might limit the Moscow complex to less than 100 launchers.

UK officials welcomed the *ad referendum* agreement on ABM no-transfer and sought clarification of the phrase 'technical descriptions and blueprints'. The UK assumed that this would not inhibit US/UK exchanges of intelligence assessments about Soviet ABMs which depended to some extent on estimates based on US ABM technology. The US told Macklen that it had inserted into the SALT negotiating record language that made it clear that any restriction on ABM transfers would not prevent the exchange of general technical information, such as might be needed for threat analysis.[143] This latter activity was particularly important for UK deterrent modernisation work. From the way SALT had developed thus far, British officials noted, limitations would be generally expressed in numerical terms leaving both sides free to improve qualitatively. Perhaps, the UK thought, this could be used as a precedent for no-transfer relating to offensive systems. Looking to the future the UK hoped that the US would not contemplate, in relation to offensive systems, provisions either in the

---

142 TNA DEFE 24/780, C.M. Rose, FCO to A.P. Hockaday, Cabinet Office, 29 February 1972.

143 TNA FCO 46/823, Record of UK/US Bilateral Talks on SALT Held at ACDA in Washington on Wednesday 23 February 1972.

treaty or in the negotiating record that would have the same inhibiting effect as that which had been agreed for ABMs.

Farley assured the British that the Americans understood and accepted UK interests in no-transfer. (It would be instructive to see whether US archives reveal any frustration or exasperation with continuing UK fretting.) He emphasised that the negotiating record would show that timely account had been taken of these; the US would not accept the ABM clause as a precedent for future agreements on offensive systems. Furthermore, the language on blueprints was drafted expressly to exclude activities such as threat analysis. Although the UK requested bilateral consultations before any final decisions were made, the US team only noted UK wishes without giving any specific commitment. Nixon had evidently still not given any express instructions to his officials to accommodate UK concerns. Anxiety levels in London were thus on the rise again.

Officials once back home prepared a draft letter in early March for Heath to send to Nixon on the no-transfer issue, reflecting the neuroses that still existed in Whitehall corridors.[144] Rose thought that the message's main point should be that the Prime Minister should recall his conversation with Nixon in Bermuda the previous December and say that in view of his close interest in this aspect of the SALT negotiations, the UK hoped that the Americans would agree to consult before they committed themselves to any form of words in SALT that might affect future UK plans.[145] The heart of the first draft letter prepared by MOD noted that the UK was particularly anxious to avoid any prejudice to the invaluable cooperation with the US flowing from the 1958 Mutual Defence Agreement and the 1963 Polaris Sales Agreement. Herein lay the UK's recurring concern, namely that arrangements covering no-transfer on ABMs should not to be allowed to constitute a precedent for subsequent agreements on offensive strategic arms. A possible answer might be to apply to third party assistance the same distinction between qualitative improvements and quantitative restraints that had emerged as an important feature in the negotiations; under this approach, either side would be free to help improve the quality of an ally's forces so long as it did not assist that ally to increase its total number of offensive weapons. Obviously Britain was thinking only in terms of missiles, not warheads given that separate consideration was being given to Poseidon as an option for UK modernisation. Consequently, the UK wanted to express its strong hope that it would be possible to be consulted in the formative stages of US thinking and before the Americans reached the stage of putting any texts to the Russians.

---

144 TNA DEFE 24/781, W.F. Mumford, Head DS 22, MOD to B.M. Webster, Cabinet Office, 9 March 1972.

145 TNA FCO 46/823, C.M. Rose to Mr Laughton, 6 March 1972.

The first GEN 1 meeting in 1972 at the end of March decided a further approach to Nixon would not be needed until after the President's trip to Moscow in May.[146] However, the Committee did note that the Prime Minister might wish to write at some later stage in the context of a decision that would be required in the autumn on Polaris Improvement.[147] GEN 1 however remained far from certain that the UK's major interest, especially in avoiding a damaging no-transfer provision in a definitive agreement on offensive systems, would be protected without regular reminders to and consultations with the Americans, if necessary at a high level. A further approach would be needed after the President's visit to Moscow.[148] The level of persistent distrust in US assurances is striking and the chickens from Macmillan's 1958 great prize – resumed close collaboration with the US on nuclear weapons – were coming home to roost.

As the Moscow summit drew near, Sir Burke Trend (Cabinet Secretary) at least was not convinced in mid April 1972 that all was rosy in the garden as far as UK fundamental interests in any outcome of the Interim Agreement on offensive systems and ABMs were concerned. In particular, if a no-transfer conflict provision appeared in a definitive offensive systems agreement, following a precedent established by the ABM agreement, the extension of the 1958 Agreement might be affected; and this could be a significant factor in the decisions on the future of the UK deterrent that Ministers would have to deal with later that year.[149] Trend thought too that the interdepartmental study on SALT was too optimistic. There was a clear warning in the main document that the US might still come under considerable pressure in the final phase of the negotiations to make concessions which might affect UK interests. GEN 1 felt that there could be no guarantee that larger political imperatives might not at some stage compel the US to modify its position on no-transfer and perhaps too on other issues.[150] Trend himself thought that it might therefore be prudent to send a further message to Nixon to underline British concerns before he went to Moscow. Heath asked both Home and Carrington whether they agreed with this recommendation.[151] Carrington thought that it would still be better to wait

146 TNA DEFE 24/781, Lord Bridges, 10 Downing Street to M.O.D.B. Alexander, 19 April 1972; FCO 46/824 R.M. Tesh, Defence Department to Sir Stewart Crawford, 13 April 1972.

147 TNA CAB 130/547, GEN 1 (72), 1st Meeting, Item 1: SALT: Report to Ministers, 29 March 1972.

148 TNA CAB 130/547, GEN 1(72), 2nd Meeting, 11 April 1972.

149 TNA PREM 15/1365, Burke Trend to Prime Minister, 14 April 1972.

150 TNA CAB 130/547, International Aspects of Nuclear Defence Policy 1972–73 GEN 1 (72) 1 (Final), SALT: Report to Ministers, Note by the Secretary, 12 April 1972.

151 TNA PREM 15/1365, Lord Bridges to M.O.D.B. Alexander, 19 April 1972.

and stick with the advice from the officials' progress report on SALT. In short, the UK should wait until after the Moscow summit before approaching Nixon again.[152] Home took the same view.[153] The Foreign Secretary argued that the UK should take the opportunity of US Secretary of State Rogers' visit in May to repeat for the record the UK's special interests. It would be a mistake to exclude SALT from the agenda when the UK was making its views known on a range of issues that might come up during the Moscow summit. Heath happily accepted this advice.[154] By this stage of the game there were two main UK concerns.[155]

The first of these, and the most important, related to the agreement on ABMs; defence levels in Western Russia and particularly around Moscow should be kept as low as possible. This would mean that if the Americans were obliged to accept that, within the figures then under consideration, the Russians would be able to defend ICBM sites and Moscow. It would, however, be important to ensure that the ABM defences should be located sufficiently far from Moscow not to thicken the area defences around the capital. Britain's second major interest was that continuing supplies of US equipment and technical information should not be cut. This meant that there should be no prejudice to US provision of information, components and materials to the UK under the 1958 Mutual Defence Agreement and the 1963 Polaris Sales Agreement.[156] Furthermore, any provision in the ABM agreement regarding no-transfer should not be regarded as setting a precedent for a subsequent definitive agreement on offensive missiles – the same old song. It was also in UK interests that the period within which a definitive agreement on offensive systems had to be concluded if the ABM agreement on offensive systems was not to be open to withdrawal should not be too short. In the event, the US proposed new text to the Russians on ABMs that would limit the number of launchers to 100 around Moscow and 100 at another site east of the Urals, in other words far enough away so as not to provide any overlap coverage; this met essential UK requirements.[157] It is worth noting here that Odeen, in a conversation with Butler in early April, had said that the US would categorically not accept more than 100 launchers around Moscow and that this position had been adopted largely to take account of the strong wishes

[152] TNA PREM 15/1365, R.J. Andrew, MOD to the Lord Bridges, 20 April 1972.

[153] TNA PREM 15/1365, J. Graham to Lord Bridges, 10 Downing Street, 21 April 1972.

[154] TNA DEFE 24/781, Bridges, 10 Downing Street to J.A.N. Graham, 25 April 1972.

[155] TNA PREM 15/1365, Burke Trend to Prime Minister, 28 April 1972.

[156] TNA CAB 130/547, GEN 1 (72) 1 (Final), SALT: Report to Ministers, Note by the Secretary, Annex B, Principal British and European Interests, 12 April 1972.

[157] TNA FCO 46/825, C.M. Rose to Private Secretary, 3 May 1972.

which the UK had expressed on this matter.[158] UK lobbying thus appears to have made its mark, but as ever concerns lingered.

At the Prime Minister's meeting on 4 May 1972 Rogers outlined the nature of the agreement that would be reached in Moscow. As for the executive agreement on offensive systems, he noted that the reasons for the asymmetry in the numbers of launchers allowed for the US and USSR were the existence of Forward Based Systems and allied missile systems.[159] This set alarm bells ringing in the Foreign Office. For Rose the implications of this were disturbing. He sent a personal telegram to Butler in Washington.[160] Rose understood that Rogers had said nothing directly to suggest that this argument had been, or was likely to be, advanced in negotiations with the Russians. But even if it was no more than a justification which the Americans intended to use for domestic purposes for accepting asymmetrical agreement, it seemed to Rose to involve a major concession to the Russians. The suggestion that allied missile systems had in some way been taken into account in calculating the extent of the asymmetry that they could accept seemed inconsistent with assurances to the UK that British nuclear forces would not be involved in SALT. Even if it was not the US intention to mention allied systems to the Russians as a reason for conceding asymmetry, there would be considerable danger in using this argument in Congress. This would put the calculations on the public record and could be exploited by the Russians in the second phase of negotiations with implications for no-transfer. Given all this it was essential to clarify exactly what Rogers meant. It was possible that he had not accurately followed his brief, so Butler was tasked to make enquiries as the UK needed to clarify this as soon as possible. London might have to probe the point at the next NAC briefing, or request more official clarification from Washington.

Butler raised this with Spiers in ACDA; he confirmed that Smith had used this argument in internal debates in order to convince the doubters in the inter-agency process.[161] However, Spiers stressed that the US was very well aware of the need to avoid using this argument in public, though he could not guarantee that it would not appear in newspaper articles inspired by people inside the Administration, he was confident that it would not be used officially. As for the Russians, they had deployed the argument themselves, but the US had not accepted it. Spiers promised to put a note to Rogers and Smith so that they

158 TNA FCO 46/824, Washington telegram 1269 to FCO, 11 April 1972.

159 TNA PREM 15/1365, Record of a Conversation Held in No. 10 Downing Street between the Prime Minister and the US Secretary of State, Mr William P. Rogers at 4.00 PM on Thursday 4 May 1972 (Extract).

160 TNA PREM 15/1365, FCO telegram 1065 to Washington, 8 May 1972.

161 TNA PREM 15/1365, Washington telegram 1558 to FCO, 9 May 1972.

were aware of the problem; Butler asked that this make clear that he had not been making an official demarche. However, it seemed to Butler that ACDA and the State Department were not fully informed about what was taking place. The Embassy had learned separately that NSC officials were considering the possibility that the text on offensive systems might refer to British and French forces as a justification for the asymmetry. Another possibility was that there could be a provision that if the UK and French forces were increased beyond the expected nine boats, the Russians would become entitled to increase their submarine force pro rata.

Given this and other reporting, London thought that it had to seek to avoid anything in the Interim Agreement, in the negotiating record, or in informal American statements, which could enable the Russians to claim that the FBS or UK Polaris had been officially taken into account in arriving at the terms of the interim freeze. Similarly if the Russians made some formal unilateral statement about FBS or allied nuclear forces, then London would naturally hope that the Americans would dissociate themselves from it.[162] UK approaches would have to be made to the State Department since it was Rogers' comments to Heath that had raised the anxiety levels in the first place, notwithstanding the other unwelcome reporting of options from Washington. In these circumstances it was now essential that the UK put its views across at the appropriate level in the White House before the President went to Moscow. Butler was accordingly instructed to make the following points, which were to be made to Spiers, Sonnenfeldt and Farley and he was to leave a copy of his speaking notes for good measure.[163]

First, the introduction of FBS and allied systems into US calculations seemed to imply some concession to the Russians that these should be taken into account in determining the terms of the freeze. Second, the UK recalled the assurances given to London before SALT began that British forces would not be taken into account. London was thus concerned that the Russians should not be given any grounds for claiming the right to discuss limitations on UK nuclear forces in the negotiations. Third, the UK assumed that the US would not accept any formal recognition or mention of UK forces in formal documents or official statements made in the negotiations. Furthermore, London hoped that there would be no possible reference to UK systems as a justification for the asymmetrical agreement in launchers. Such statements would be seized upon by the Russians. Spiers told Butler on receipt of this message that the UK point was now well understood and accepted by all concerned. He thought it certain that

162 TNA PREM 15/1365, FCO telegram 1130 to Washington, 15 May 1972.

163 TNA FCO 46/825, FCO telegram 1130 to Washington, 15 May 1972; C.M. Rose to Permanent Under-Secretary, 15 May 1972.

the Americans would dissociate themselves from any Soviet statement linking allied systems or FBS with the freeze.[164] Spiers was also confident that US official spokesmen would make no use of FBS or allied systems in presenting the freeze.

The Defence Secretary was briefed to raise the no-transfer issue with his US counterpart Melvin Laird on 19 May. London wanted to make clear its concerns lest the constraints on 'technical descriptions and blueprints' should be regarded as setting a precedent when the question of no-transfer of offensive weapon systems came up – as it was bound to do so in the follow-on SALT negotiations.[165]

The Moscow Summit May 1972: The Interim Agreement and the ABM Treaty

SALT I and the ABMT were signed in Moscow in May 1972. And as anticipated by the US, there was no mention of third country forces or non-circumvention clauses in the SALT Interim Agreement. The question of no-transfer of strategic arms had not been raised in connection with the Interim Agreement, and the US had made a firm unilateral statement that the no-transfer clause in the ABMT (Article IX) constituted no precedent.[166] Both sides made unilateral interpretative statements purporting to clarify their views on these issues. On April 18 the US had made a statement on the ABMT's no-transfer Article. Basically this provision, in the US view, set no precedent for the question of strategic offensive arms. On 17 May the Russians made a unilateral statement to the effect that they were entitled to a corresponding increase in the number of their submarines if France or the UK increased theirs. The Americans emphatically rejected this after consultation with UK officials. The US SALT delegation was instructed on 20 May to state, in reply to a Soviet attempt to place on record their view of the link between allied systems and the forward basing of US submarines on the one hand and the SLBM freeze on the other, that the US did not accept the validity of Soviet considerations. The simple and categorical form of this statement had, according to Carter (NSC) been considerably influenced by the UK demarche.[167] Two days later the Russians repeated their statement, but the Americans rejected this on the same day. However, the Russians gave notice that

164 TNA PREM 15/1365, Washington telegram 1675 to FCO, 17 May 1972.

165 TNA DEFE 24/782, Secretary of State's Private Meeting with Mr Laird: SALT Speaking Notes, 15 May 1972.

166 TNA PREM 15/1365, UKDel NATO telegram 298 to FCO, 19 June 1972.

167 TNA PREM 15/1365, Washington telegram 1734 to FCO, 22 May 1972; TNA DEFE 24/782, W.F. Mumford Head of DS 22 to PS Secretary of State, 25 May 1972.

both French and British submarines would need to be considered in the next phase of the negotiations. The UK therefore had reason to be satisfied with the extent to which the US had taken into account UK concerns and fulfilled their undertakings. London publicly welcomed the agreements and Heath wrote to Nixon on 2 June to congratulate him.[168]

In his reporting despatch on the outcome of the SALT process, Lord Cromer wrote from Washington that the Americans had lived up to their undertakings in November 1969 to keep the UK more fully informed than any of the other allies, but the Americans did not actively seek British views even on subjects such as no-transfer in which Britain was known to have a direct interest.[169] However, in his view British views were listened to carefully and on a number of points under negotiation had exerted a certain influence on US thinking. Thanks partly to this the negotiations' outcome was reasonably favourable to the UK on such key issues as ABM limitations, no-transfer and the interpretative statements on these topics. However, FBS and no-transfer were more than likely to feature more prominently in Phase II of SALT, but at least with the Nixon Administration, the UK started from a basis of close consultation.

In short, the UK felt that at the end their interests in no-transfer and in the location of Soviet ABMs were adequately safeguarded.[170] In particular, with the Russians limited to 100 missiles around Moscow, one boat load of Super Antelope missiles against 100 ABMs would fulfil the JIC deterrence criteria; with an exchange ratio of 15 to 1, the Russians would need more than 130 ABMs to neutralise a boat-load of Super Antelope armed missiles to the point where the JIC criteria could not be fulfilled.[171] Jerry Stocker argues that the ABMT, even though the UK was not a party, played an absolutely pivotal role in UK policy towards ballistic missile defence. This is clearly true. It was the determining event in the UK's responses to both US and Soviet ABM deployments.[172] The treaty ensured the limitation of Soviet defences to a scale that Britain could realistically work to overcome, but also meant that Britain had to overcome them as they defended the all important target – Moscow.[173]

---

168 TNA PREM 15/1365, Burke Trend to Prime Minister, 30 June 1972; see also TNA DEFE 24/784, Note by the GEN 1 Group of Officials, 30 June 1972.

169 TNA DEFE 24/784, Lord Cromer to The Rt. Hon. Sir Alec Douglas-Home, Secretary of State for Foreign and Commonwealth Office, 22 June 1972.

170 TNA DEFE 24/782, Note for the Prime Minister's Visit to Copenhagen, 7 June 1972.

171 TNA DEFE 24/782, F.H. Panton ACSA (N) to DUS (P), 7 June 1972.

172 Jeremy Stocker, *Britain and Ballistic Missile Defence 1942–2002* (London, 2004), p. 6.

173 Stocker, *Britain and Ballistic Missile Defence*, p. 241.

Kissinger assured Sir Burke Trend that nothing in SALT would prevent US assistance to the Polaris Improvement Programme. He reminded the UK that the US position remained as expressed by Nixon to Heath at the Bermuda meeting in December 1971. The US wanted the British to retain an independent nuclear capability which was credible and, therefore effective; and that they would not be lightly deterred from this policy. Yet in the following month the President's Press Secretary, Ron Ziegler, in responding to a question on Senator Henry Jackson's amendment to the Interim Agreement, gave the impression that FBS and allied nuclear forces would be dealt with in SALT II. Sonnenfeldt (NSC) informed the UK immediately that nothing had changed since the US interpretative statement in May.[174] There was no need for the UK to take any action to persuade the Administration to set the record straight.[175] The US position was still as placed on record in the response to the Soviet unilateral statement.[176]

SALT II: June 1972 to December 1973

Macklen, by now the key MOD official dealing with the Polaris Improvement Programme, offered his views to Lord Carrington in mid June on the implications for the UK of the SALT Interim Agreement and ABM Treaty.[177] It was important that the ABM Treaty included a separation distance between the two permitted ABM complexes of 1300 km, which meant that in the worst case the UK deterrent would only have to face the 100 launcher no reload capability ABM complex around Moscow. However, the permitted radar coverage was also sufficient to allow the Russians to use the ABM system around Moscow to cover large areas of the Western USSR. The 100 ABM limit would be sufficient to negate either one or two boats' worth of unimproved Polaris missiles. The warheads dispersal pattern was close enough to be rendered ineffective by one successful ABM explosion. Since the UK could expect on reliability grounds either 13 or 26 missiles to threaten the defended area of Western Russia, there were more than sufficient ABM launchers available after allowance for their reliability factor. Therefore, the ABM Treaty did not remove the need for the Polaris Improvement Programme. However, the Treaty did set a limit to the UK task over a longer period. Both Super Antelope and Poseidon (with 12 warheads)

---

[174] TNA DEFE 24/785, Washington telegram 2395 to FCO, 14 August 1972.

[175] TNA DEFE 24/785, FCO telegram 1709 to Washington, 15 August 1972.

[176] TNA DEFE 24/783, D.C. Tebbit, Washington to A.E. Davidson, Defence Department, FCO, 16 August 1972.

[177] TNA DEFE 24/783, V.H.B. Macklen to PS/Secretary of State, 14 June 1972.

would fulfil the JIC criteria for an effective deterrent threat. Poseidon was marginally superior solely on penetration arithmetic grounds. Thus the ABM Treaty enabled the UK to define more closely its improvement requirements against the then current Soviet systems.

On the non-technical side, Macklen was worried about the potential implications arising from the reference in the ABM Treaty's Article X to the requirement on each party not to assume international obligations that would conflict with its object and purpose. (This was also picked up by Clive Rose in the FCO when he spoke to Spiers, ACDA on 20 June.[178]) Although the US had insisted that this would not interfere with existing UK–US nuclear agreements, there were still hints from the Americans that there would be increased domestic difficulties that would have to be overcome if the UK sought continued or additional help. Therefore, Rose felt that the UK would have to watch the future SALT process intently. Officials thought it would be foolish not to recognise that political considerations, both internal and international, might inhibit US assistance to the UK. All the more so if that assistance had the appearance of a straight US deal to improve UK deterrent effectiveness, which might reflect on the delicate balance in the ABM Treaty, particularly if it were true that the National Command Authority defence systems were aimed at preserving the hegemony of the two superpowers. Macklen did not make clear why he thought this, but it is difficult to believe that this was the real motivation in Moscow or in Washington. So even though the UK had secured a good outcome, officials still appear to have been unable to sleep easier in their beds. They had good reason.

The Russians put the UK and US (and France) on notice that British and French submarines would have to be considered in the next phase of the negotiations.[179] Heath wrote to Home to say that he was sure that the Foreign Secretary shared his view that the UK would have to make absolutely clear to the US, and perhaps also the Russians, that HMG would not regard itself bound by provisions reached between the US and USSR; and that if the UK were to enter any obligation, then it would have to participate in negotiations for that purpose. The Prime Minister thought that there would be advantage in making common cause with the French on this subject and the possibility of a joint demarche should be looked at urgently. Home agreed absolutely; if there was any doubt about the UK position, then it was important to make sure that it was clearly understood.[180] However, he did think that the US Administration

[178] TNA DEFE 24/784, Record of a Conversation between Mr C.M. Rose and Mr R. Spiers (Director Bureau of Politico-Military Affairs, State Department) on 20 June 1972.

[179] TNA PREM 15/1365, Prime Minister to Secretary of State for Foreign and Commonwealth Affairs, Prime Minister's Personal Minute No. M57/72, 27 June 1972.

[180] TNA PREM 15/1365, Alec Douglas-Home to Prime Minister, 13 July 1972.

was in no doubt about British views, certainly in light of the demarche in Washington prior to the Moscow summit. It would therefore be inadvisable for a joint demarche with the French at that stage. Further consultation with the US would be needed, and in any case substantive second phase talks with the Russians would not take place until early 1973, so there was plenty time to develop a UK view. Heath accepted this advice.[181] A further round of bilateral talks with the US was organised for September 1972.[182]

Officials in the GEN 1 Committee produced a more considered review of the first phase SALT and ABM Treaty outcomes in mid July, but this time looking in detail at the implications for Britain and at aspects requiring further attention.[183] This drew heavily on Macklen's assessment given to Lord Carrington on 14 June. Looking to the future, officials assumed that nothing should be done in SALT that might either jeopardise US support for the deterrent required for the foreseeable future, or contribute to a weakening of the US strategic commitment to European defence. The 1958 Mutual Defence Agreement and the Polaris Sales Agreement must not be affected retrospectively by any no-transfer or no-conflict clauses that might appear in a definitive agreement on offensive weapons, the expected outcome from SALT II. The 1958 agreement was important because it was essentially an enabling agreement that allowed the UK and US to agree on new obligations from time to time. Moreover, the US had assured the UK that it would continue to support Polaris. Whether the UK could buy Poseidon under the Polaris Sales Agreement, or obtain continuing support under the 1958 Agreement for Super Antelope would depend on a political decision by the Administration. Although Nixon might be willing to support UK efforts, he might well be constrained by Congress.

Officials doubted that UK direct participation in SALT II would put it in any stronger position to secure US support for sustaining Polaris' continuing effectiveness. Indeed there were serious political risks, or so officials thought; disparity in stockpile sizes would inevitably lead to Britain playing a minor role in any SALT II negotiation. The UK would have little power to influence the negotiations' course, and because of dependence on US support, the UK would have little choice but to accept the nature of limitations agreed by the superpowers. If the UK were formally involved its voice would inevitably have been drowned by the US–Soviet dialogue, presenting few chances to influence.

181 TNA PREM 15/1365, Bridges to J.A.N. Graham, FCO, 17 July 1972.

182 TNA FCO 46/827, C.M. Rose to Permanent Under Secretary, Private Secretary, 18 July 1972.

183 TNA CAB 130/547, GEN 1 (72) 4, 14 July, SALT: Implications for the United Kingdom of the Moscow Agreements; TNA DEFE 24/784, First Phase SALT Agreements: Report to Ministers, W.F. Mumford, Head of DS 22 to PS to Secretary of State, 18 July 1972.

By remaining outside and through bilateral contacts and consultations UK officials convinced themselves that this approach better enabled them to ensure British interests were taken into account. Moreover, this made them better placed to exploit the US desire that the UK should retain a viable 'independent' deterrent and the US formal commitments to help the UK do so.[184] These same arguments no doubt applied too to justify keeping out of SALT II lest such participation lead to direct limits on Polaris. For the immediate future, however, officials concluded that there were still too many uncertainties to see clearly what UK attitudes should be on these key questions. However, the first opportunity to ascertain US intentions would present itself with Trend's visit to Washington at the end of July.

Sir Burke Trend (Cabinet Secretary) saw Kissinger in Washington on 28 July.[185] Trend told him that the UK hoped to submit to Ministers during the autumn the basic choice between the two ways of ensuring that the UK retained an effective deterrent: the hardening of Polaris by the Super Antelope scheme or the substitution of Polaris with Poseidon. The choice depended on several factors – the costs, the implications of SALT I and SALT II and US willingness and ability to assist British efforts. If a clear case were put to Ministers, then officials would need to know how far the US would be prepared to go to help. Discussions to see what could be offered would be ideal, preferably very soon. Kissinger thought that there would be no difficulty with Super Antelope. Nor did he foresee any real difficulty about its subsequent stages; the current stage was with the President for approval. There was nothing in SALT so far that would be a barrier to Super Antelope's completion; and even if there were, the US would see that it was circumvented somehow. This assumed, however, that the UK was not talking about a MIRVed weapon; that would be different, although the President had made clear to Heath in Bermuda back in December that he wanted the UK to remain a nuclear power and would be prepared to go the extra mile to ensure that it did. Kissinger expected that there would no doubt be loud Soviet opposition if the US assisted the UK to modernise its deterrent with MIRVs, not to mention Congressional complaints. However, things would be clearer and easier after the election in November, especially if the President were re-elected with a larger majority and the Republicans controlled Congress. It would be easier too to get things done before SALT II progressed too far. Kissinger was also reassuring about no-transfer not being an issue in SALT II.

---

184 Personal Correspondence: these commitments were of course the 1958 Mutual Defence Agreement and the 1963 Polaris Sales Agreement.

185 TNA FCO 46/827, Discussions between the Soviet Union and United States on Limitations of Strategic Armaments 1972, Extracts from Sir Burke Trend's Discussion with Dr H. Kissinger in Washington on 28 July 1972.

Ahead of the autumn bilateral consultations in Washington the Americans made clear that they expected the British to be as constructive as possible at the meeting in their suggestions as to how the UK might best handle its chief concerns that the Americans had been able to evade in SALT I, but would find much more difficult to ignore in SALT II.[186] At that stage of the game the UK counted on obtaining a US assessment of Soviet intentions for allied nuclear forces and an interpretation of various hints that the Russians might seek to involve other nuclear powers in SALT II.[187] Washington wanted to reach a clear understanding of whether the UK was prepared for account to be taken of its nuclear forces in the follow-on agreement, and if so, how. However, the UK was not in a position to make the sort of commitment that the US sought, not least until Ministers had been consulted and for that more detail of US views were needed.[188] For the present, however, the UK would stick with its line that it could not accept that talks in which it did not take part could create binding obligations or inhibitions for the UK. The Embassy thought that the Americans hoped that the UK would accept this; the Americans believed that it would hardly be possible to leave the third country forces question hanging with another rebuttal of the kind used in SALT I. If explicit account were taken of UK forces, then the no-transfer issue would be much easier to deal with as the Russians could not then argue credibly that the US should be prevented from assisting Britain modernise its forces. London expected that the US would in fact share any draft texts that it might have in mind to deal with the no-transfer issue.[189] Two tentative ideas were soon provided unofficially by Ray Garthoff (ACDA) to the Embassy: either 'Each State Party undertakes not to circumvent the provisions or effectiveness of the treaty by the transfer to other States of strategic offensive arms' or 'Each State Party undertakes not to transfer to other States systems or components [but not technical descriptions, blueprints etc.] limited by the Treaty.'[190] Macklen commented forcibly that the

186 TNA DEFE 24/785, W.J.A. Wilberforce, Washington to A.E. Davidson, Defence Department FCO, 25 August 1972.

187 TNA DEFE 24/785, C.M. Rose, FCO to W.J.A. Wilberforce, Washington, 4 September 1972.

188 TNA DEFE 24/785, A.E. Davidson, Defence Department, FCO to A.P. Hockaday, Cabinet Office, 11 September 1972, Draft Brief, paragraph 7, see final version US/UK Talks on SALT – 28/29 September 1972, Allied Nuclear Forces in SALT II, FCO, 19 September 1972.

189 TNA DEFE 24/785, C.M. Rose, FCO to W.J.A. Wilberforce, Washington, 4 September 1972 and C.M. Rose, FCO to A.P. Hockaday, Cabinet Office, 11 September 1972.

190 TNA DEFE 24/785, W.J.A. Wilberforce, Washington to C.M. Rose, 11 September 1972.

second option would make it impossible for the UK to keep Polaris going. UK officials would therefore have to stress in Washington the very serious drawbacks of such a formula.[191] Consequently that part of the officials' brief on no-transfer for the Washington bilateral was toughened up.[192] London hoped that the US would tell the UK officially about any form of words that it was considering for a no-transfer clause in SALT II. UK officials, however, did not intend to do any more than elucidate US thinking; HMG's official position would be strictly reserved.[193]

Part of the problem in discussing the no-transfer issue with ACDA was that none of its officials, including Gerard Smith, were aware of the content of the separate technical UK–US exchanges on Super Antelope, or that they were even taking place; nor could they be mentioned by the British side in meetings on SALT II.[194] For this reason, no doubt, Cabinet Office officials – Arthur Hockaday and Bob Press – would have much preferred that the basic UK line was that there should be no provision on no-transfer at all in SALT II.[195] However, one official was fully aware – Ron Spiers – which meant that Clive Rose could speak quite frankly to him about UK concerns and the specifics of UK technical requirements.[196]

The talks themselves took place in Washington on 28 and 29 September 1972 and these concentrated on the US approach to SALT II. The UK aim was to elucidate US views and to ensure that nothing was done in SALT to make it more difficult to maintain Polaris' effectiveness or to undermine NATO's security requirements.[197] Nor could there be any question of the UK accepting limitations on its forces as a result of negotiations in which the UK did not take part.[198] This was the essence of the British approach in SALT II; there was no change in tack and London continued to chivvy the Americans if there was any sign that the no-transfer issue was raising its ugly head again, as in

191 TNA DEFE 24/785, S.S. Crew, DS 22, MOD to AUS (D Staff), 18 September 1972.

192 TNA DEFE 24/785, A.P. Hockaday to C.M. Rose, 18 and 19 September 1972.

193 TNA DEFE 24/785, Anglo/United States Bilateral Talks Note by Officials, 20 September 1972.

194 TNA FCO 46/828, A.P. Hockaday, Cabinet Office to C.M. Rose, 18 September 1972; and US/UK Talks on SALT – 28/29 September 1972, No Transfer/No Conflict in SALT II, FCO, 19 September 1972.

195 TNA FCO 46/824, A.P. Hockaday, Cabinet Office to C.M. Rose, 19 September 1972.

196 Personal Correspondence.

197 TNA PREM 15/1365, Burke Trend to Prime Minister, SALT, 20 September 1972.

198 TNA FCO 46/829, C.M. Rose to Permanent Under Secretary and Private Secretary, 25 October 1972.

the possible draft of a declaration of principles proposed by the Russians that would govern the conduct of SALT II negotiations and shape of an eventual agreement between the US and USSR.[199] Now whilst the ABM Treaty limited the problem that Polaris modernisation would have to deal with, there were still three threats to what Macklen described as the UK's relative peace of mind. First, there was always a chance that either side might denounce the treaty, but this was not a likely prospect. Second, exotic ABM systems might be developed – and Macklen and colleagues in the MOD and AWRE had considered two types (laser and particle beams), but doubted that these would pose a likely threat for some time. If such an ABM system were to emerge, no treaty could prevent its deployment if it turned out to have a large area capability.[200] Third, it was not clear how the yet to be established Consultative Commission could control ABM missile numbers – launcher re-loads. The US had made clear that the ABMT's future depended heavily on there being a follow-on agreement covering offensive weapons after the Interim Agreement expired in five years time. For this reason Britain wanted SALT II to succeed as no agreement, or an agreement unacceptable to Congress, might lead Washington to decide that its supreme interests were jeopardised and thus withdraw from the ABMT, with the possible result that Soviet ABM launcher and missile numbers would expand thereby threatening Polaris' viability, even if improved.[201]

Once again the Americans assured their anxious UK colleagues that they would consult fully with their allies before they put any formula on no-transfer to the Russians. London repeated the same point to the NSC: it would be essential to ensure that any no-transfer clause was fully compatible with the measures of support for Super Antelope that were being worked out separately. US SALT II objectives included deferral of discussion on this point until late in the negotiations when it would become clear what systems would be covered. If it proved necessary to table anything at an earlier stage they would base themselves on a general non-circumvention formula on lines proposed in 1970.

---

199 TNA PREM 15/1365, FCO telegram 998 to Washington, 1 May 1973; Extract – Record of Meeting Sir Burke Trend/Dr Kissinger, 1 May 1973, Strategic Arms Limitation Talks (SALT); TNA FCO 46/824, A.E. Davidson, Defence Department to A. Hockaday, Cabinet Office, SALT: Preparation for Bilateral Talks in Washington, SALT: Bilateral Discussions in Washington, September 28/29; TNA FCO 46/824, US/UK Talks on SALT – 28/29 September 1972, Allied Nuclear Forces in SALT II, 19 September 1972.

200 TNA FCO 46/828, V.H.B. Macklen to Dr R. Press, Cabinet Office, 8 September 1972.

201 TNA FCO 46/1001, M.E. Pellew, Defence Department to Mr Rose, 10 January 1973.

That suited London just fine.[202] In response to all of this British officials made three points. First, they hoped that the UK would be consulted in advance on the precise wording of any no-transfer clause. (Rose noted that the White House was aware of both the restricted exchanges on Polaris modernisation and SALT and would thus be able to watch this point.[203]) It was essential that any clauses did not prejudice the kind of support the UK was continuing to receive from the Americans.[204] The White House was reminded privately of the need to ensure that any no-transfer clause that was put forward was compatible with the offers of support that had been made to the UK in separate and restricted exchanges about future US support for the UK deterrent.[205] Second, officials would like to see the text of even any holding formula on non-circumvention of a future strategic agreement before it was tabled; and, third, London wanted to be assured that any no-conflict article would relate to future obligations and not have any retroactive effect, thus preserving the 1958 and 1963 Agreements. Farley confirmed that British interest in these points was fully understood, he must have been becoming fed up having to do so; and, when the Russians raised the no-transfer issue in the first round of SALT II, proposing an approach based on the ABMT's no-transfer provisions in its Article X, the Americans repeated their point in the unilateral 18 April 1972 statement that the ABM formula was no precedent for a treaty on future offensive arms.[206] So the UK could draw some comfort that the US would endeavour to hold the line. Despite these positive noises, when GEN 1 discussed the outcome of the Washington meeting in mid October, Rose argued that the US had in fact given little indication of its negotiating position in the second round on the various matters of concern to the UK. This was primarily because these positions had not been decided.[207]

---

202 TNA DEFE 24/786, Prime Minister's Visit to Washington, Brief No. A5: SALT II and Nuclear Matters; TNA FCO 46/828, Washington telegram 2890, 29 September to FCO; Prime Minister's Meeting with President Nixon, 10 January 1973, Draft Brief for Visit, Section (e): No-Transfer, M.E. Pellew, Defence Department to Mr Rose, 10 January 1973.

203 TNA FCO 46/829, C.M. Rose, Note for the Record, SALT, 27 September 1972.

204 TNA FCO 46/824, Record of UK/US Bilateral Talks on SALT, 28 September 1972.

205 TNA FCO 46/828, H.E. Pellew, Defence Department to Mr Rose, SALT: US/UK Bilateral Talks (see Attached Draft Report from GEN 1 to Ministers), September 1972.

206 TNA FCO 46/1001, W.J.A. Wilberforce, Washington to C.M. Rose, 4 January 1973.

207 TNA CAB 130/547, GEN 1 (72), 7th Meeting, 12 October 1972, and GEN 1 (72) 6, 23 October 1972, Anglo-United States Bilateral Talks, Note by the Secretary, 23 October 1972.

1973: Manning the Watchtower

In contrast to the position when SALT I opened, UK defence interests were very clearly understood when SALT II began in November 1973. Super Antelope, which evolved to become Chevaline and US assistance to it, was the core defence interest requiring protection. Officials thus continued to man the watchtowers to guard against any creeping encroachment on UK–US nuclear cooperation. The conclusion of SALT II was of course some years off (1979), but a careful eye on the negotiations had to be maintained nonetheless. As we saw in the discussion on UK exchanges with Kissinger on Super Antelope and other modernisation options, the only channel for this work was via his office and designated officials only. Others in the State Department or ACDA were totally unaware of these exchanges and did not appreciate exactly why the UK was so animated by the transfer issue. Nevertheless, the US promised to oppose further Soviet attempts to include unfavourable no-transfer and non-circumvention provisions, so when for example the Soviet delegation referred to its 17 May statement in December 1972, the Americans repeated their 24 May rebuttal.

The Russians proposed to the Americans that they should agree a declaration of principles that could serve as a basis for further negotiations and suggested that Nixon and Brezhnev should sign it; this emerged at the end of April 1973. The first Soviet draft appeared in April and Kissinger was quick to share a copy with the UK. (This issue was under very close hold in Washington, so British access was indeed a privilege.)[208] The offending passage on no-transfer and non-circumvention read:

> ... in order to ensure greater effectiveness and stability of agreements to be concluded between them on limitation of strategic arms, the Soviet Union and United States will not transfer to third countries the strategic arms which are covered by the Agreement to be concluded by the sides and also will not encourage and assist them in development and deployment of such types of arms and in general of new types of arms, the manufacturing or acquisition of which by third countries would increase the danger of an outbreak of war.

London did not like this and urged the Americans to change it. The Prime Minister had been quick to spot the text's unwelcome elements and suggested that it would be prudent to let Dr Kissinger know of the UK objections before he went to Moscow. While the Americans were no doubt aware of the difficulty

---

[208] TNA FCO 46/1002, Washington telegrams 1407 and 1408 to FCO, 27 April 1973.

that such a formulation would cause the UK, Heath thought that it would be sensible to remind them.[209] Instructions were sent to Washington accordingly.[210] And these made clear that it was essential that no doubt should be cast on the continued validity of the 1958 Mutual Defence Agreement and the Polaris Sales Agreement; London also needed to keep open the option for a further agreement which might be required in the context of Polaris Improvements; in other words if Poseidon could not be provided under the rubric of the 1963 Agreement. If the US had to accept some formulation in place of the Soviet version, then it should not go beyond a non-circumvention provision in as general and innocuous terms as possible.

Kissinger was grateful for UK comments and pointed out that the US alternative safeguarded UK interests. As an alternative the US proposed:

> in working out more complete measures on the limitation of strategic offensive arms the two sides agree to establish equitable limitations on ICBMs, SLBMs and heavy bombers: to resolve the question of those other nuclear delivery systems/ (types of strategic offensive arms) which one side considers as a strategic threat, they agree that neither side will circumvent a permanent agreement through systems not limited.[211]

This revised draft was seen as a great improvement as the offending Soviet text on no-transfer had been entirely eliminated.[212] A subsequent version merely noted that, 'In working out new arrangements between the USSR and the US on the questions of strategic offensive arms limitation the sides will proceed from the premise that these agreements should not be weakened or circumvented through a third state'. The UK hoped that this would be deleted and it was. Instead the Americans inserted, 'in working out a permanent agreement between themselves on more complete measures to limit strategic offensive arms, the two sides will proceed from the premise that such an agreement will not affect agreements with third countries to which either side subscribes'. The original Soviet draft passage on no-transfer was dropped from the June version as was a US proposal to send a separate letter to the Russians that would have stated that the US would not circumvent the agreement through third parties.

---

209 TNA FCO 46/1002, Tom Bridges, 10 Downing Street to M.O.D.B. Alexander, FCO, 30 April 1973.

210 TNA FCO 46/1002, FCO telegram 998 to Washington, 1 May 1972.

211 TNA FCO 46/1002, Washington telegram 1458 and 1459 to FCO, 3 May 1972.

212 TNA FCO 46/1002, Draft Brief for Sir Burke Trend's Talks with Dr Kissinger on 10 May 1973.

Further exchanges took place between the US and UK on revised versions, including further Russian comments, of the draft between mid May and early June 1973.[213] Heath, Carrington and Home were all consulted and the Prime Minister was particularly anxious that the Embassy in Washington should emphasise to the White House the strong hope that if the revised Soviet text was adopted, that an explicit statement should be made to the effect that the US government interpreted the phrase 'recognition of each others' equal security interests' as applying on the US side to the security interests of the NATO alliance. Nor indeed should there be any side letter from Nixon concerning an assurance to the Russians on no-transfer or non-circumvention.[214] A key phrase in this statement was 'equal security interests'. The US told the North Atlantic Council on the same day that this would be interpreted as including the concerns of their allies. Kissinger had made this clear to the Russians. The relevant section read: 'Both sides will be guided by recognition of each other's equal security interests and by the recognition that efforts to obtain unilateral advantage, directly or indirectly, would be inconsistent with the strengthening of peaceful relations between the US and USSR'. In the end, the SALT declaration of principles, which was signed by Nixon and Brezhnev on 21 June 1973, contained no reference to no-transfer or non-circumvention.[215] London could breathe more easily, especially since it was approaching crunch time on a final decision on Polaris modernisation. The decision to go ahead with Super Antelope and reject the other modernisation options was taken at the end of October 1973.[216] The Polaris modernisation Project Definition Study had already noted that even with ratification, the ABMT did not remove the need to improve Polaris to enable it to threaten Russian targets; the numerical limitation on missiles under the Treaty was not very much different from that previously assumed. Furthermore, assumption from the previous autumn still stood: the Russians would have an inventive to improve ABM radars in the Moscow area, as they were permitted to do so by the ABMT, so as to maximise the effectiveness of the missiles that they

---

213 TNA CAB 130/547, GEN 1 (73) 2, 14 June 1973, Joint US/Soviet Declaration on SALT, Note by the Secretary; TNA FCO 46/1002, Washington telegram 1767 to FCO, 5 June 1973; FCO telegram 1183 to Washington, 1 June 1973, Draft Soviet Declaration; FCO telegram 232 to UKDel NATO, 6 June 1972; FCO telegram 1203 to Washington, 6 June 1973, SALT Declaration; Washington telegram 1781 to FCO, 6 June 1973.

214 TNA FCO 46/1002, FCO telegram 1203 to Washington, 6 June 1973, and FCO telegram 1217 to Washington, 7 June 1973.

215 TNA FCO 46/1003, UKDel NATO telegram 455 to FCO and Cabinet Office and MOD, 21 June 1973.

216 TNA PREM 15/2038, M/17/2, 1 November 1973, Cabinet, Defence Expenditure, Minutes of a Meeting on 30 October 1973.

were allowed to deploy.[217] The British put on record its position on no-transfer and on the Soviet unilateral statement about allied nuclear forces. The point was made orally to White House staff that, 'it would be important to ensure that the text of any no-transfer clause was compatible with such measures of US support for the UK nuclear deterrent as were being worked out with us in separate and restricted exchanges'.

Whilst this was going on in October, the Soviet delegation tabled a stringent no-transfer provision. This included a prohibition on the transfer of technical descriptions or blueprints of nuclear weapons specially worked out for the development of strategic offensive armaments limited by the agreement, or action which would result in weakening or circumvention of the agreement's provisions through a third state or states. In response, the Americans told the Russians that such a provision failed to meet the agreed principles of 'essential equivalence in security'. After this the no-transfer and non-circumvention issues disappeared from the scene for the remainder of our period.

A continuing concern for the UK was just how the Americans would propose to rebut Soviet proposals and whether such rebuttals would be consistent throughout. Anxieties over US true intentions were never far away. One US official in thinking aloud had indicated, for example, that UK and French forces perhaps should be included under US totals as a solution to the no-transfer problems. This raised a good few eyebrows in London. Despite this the Americans initially suggested to the allies that a generalised non-circumvention provision, perhaps allied to a loose form of freeze on non-central systems, might be the best option.[218] UK officials thought that such a formula offered the least harmful way of dealing with Russian concerns on FBS while avoiding any significant impact on them. Despite US refusals to negotiate on third country forces and their resistance to any Soviet claims for compensation for any increase in force levels, there was still an air of concern in Whitehall. The abiding UK fear was that the US might at the end of the day in a bid to secure agreement meet the Soviet Union half way and in so doing undermine the UK deterrent. As we saw in Chapter 6, UK interests in the draft BTWC compliance provisions – effective machinery for the effective investigation of alleged BW use – in the spring of 1971 had been sacrificed on the altar of larger bilateral US relations

[217] TNA DEFE 19/207, Report of the KH 793 Project Review Board, 3 October 1973.

[218] This formula would serve to inhibit each side from making substantial changes on non-limited systems which could be viewed as circumventing the effectiveness of and jeopardising the viability of the agreement.

with the Soviet Union, so there were grounds for concern.[219] In particular, the main preoccupation was that although a future SALT II agreement might not contain clear numerical limits on Polaris, the agreement would render the US less willing to assist the UK to augment or improve its forces. Whitehall was bothered too by signs that the Americans would be inclined to recognise in a final agreement that factors might arise concerning third forces that could threaten to upset the strategic balance.

Conclusion

British policy during SALT mirrored that of the late 1950s and mid 1960s when US policies on a test ban treaty and a cut-off in the production of fissile material for weapons purposes also threatened the integrity of the emerging UK deterrent. UK enthusiasm for arms control was tempered by the then state of the weapons programme – warhead design and material availability issues. Immediate introduction of such arms control measures would prevent the UK from developing its nuclear programme as it would have wished. One of the paradoxes was that for the UK to support US arms control polices, the Americans would have to provide greater assistance, both in knowledge and materials, to ensure that Britain could meet its modest force requirements. SALT was no different. The timing of US arms control imperatives invariably seemed to clash with UK weapons programme timetables. Despite repeated US assurances, both in the context of the closely held exchanges on Polaris modernisation and on US SALT negotiating positions, there were sustained anxieties that US–Soviet bilateral requirements might just edge out British interests. At the end of the day these fears were misplaced. There was nothing in SALT I that prevented the US from providing the essential technical assistance required for the Polaris Improvement Programme. The Russians never, however, made no-transfer a sticking point in the negotiations, it may have been more of a rhetorical flourish rather than a substantive point. Britain of course could not be sure, but had the Russians made no-transfer a make or break issue it is not clear how the US would have responded. Whether the Americans would have been ready to trample on UK interests must remain an unknown. As in 1958 with the nuclear test moratorium and impending test ban negotiations, critical US technical assistance was forthcoming despite arms control pressures and this assistance made a decisive difference to the development and sustainability of the

[219] See TNA FCO 66/299, UK Policy on Chemical and Biological Warfare and Draft Biological Warfare Convention 1971; and Chapter 6.

UK nuclear weapons programme. Once again the UK was able to support arms control agreements, and in this case perhaps more grudgingly, whilst ensuring that its core nuclear defence interests were not constrained.

Chapter 11
Conclusions: Britain and Nuclear, Chemical and Biological Arms Control and Disarmament

So given this 20 year review of UK attitudes and role in nuclear, biological and chemical arms control, what have we learned? To what extent has the UK's need to protect its own programmes constrained its role in international negotiations? And what, if any, role has the UK's own programmes' experiences played in shaping its policies? Does the UK experience tell us anything more generally about the arms control process? The first thing to be said, and this is perhaps not that surprising, is that for both political reasons – domestic as well as international – and because of an intrinsic sense of the value of arms control and disarmament in its foreign and defence policy, the UK has both been a strong supporter of arms control, but it has also sought to moderate its impact where its provisions could damage its own programmes. For the UK arms control was about hard-headed calculations of national security rather than grand moral gestures – very much in the mould of Schelling and Bull. This is certainly true in the nuclear context as we have seen with both the fissile material cut-off and SALT/ABM issues. It does not apply in the biological area since the UK had long since ceased to have an offensive programme by the time disarmament became a possibility, although care was taken to ensure that any measures would not hamper protective programmes of the sort that were underway at the Microbiological Research Establishment. And of course it was the UK that took the initiative here in promoting the cause of biological arms control; its experience with both national offensive and defence programmes had possibly helped shape its approach to the problem given the clear awareness of the verification problems. One can be sure that if an offensive biological warfare programme had still existed in 1968, it is likely that UK enthusiasms would have probably been more muted. The chemical warfare case is slightly more nuanced. The 1925 Geneva Protocol certainly was a consideration in the 1960s when the UK was looking into acquiring incapacitating chemical agents for a range of scenarios, some of which could only entail retaliation if the UK were attacked first by chemical agents. By the end of the 1960s the UK has changed its interpretation of the scope of the Geneva Protocol to exclude CS. This was only achieved after internal disputes between the MOD and FCO were overcome at the highest levels of government, but on the rules for use of agents such as CR there was nothing that fundamentally split the then Labour government or its immediate successor.

To What Extent has the UK's Need to Protect its Own Programmes Constrained its Role in Arms Control?

In both the cut-off debates that surfaced periodically from the 1950s through to the mid 1960s and again during the SALT and ABM negotiations when the UK was not even a direct participant, the need to ensure that whatever was agreed did not adversely affect the UK nuclear weapons programme was a paramount policy objective. Ensuring too that the NPT did not interfere with the established patterns of nuclear cooperation with the US was also a core requirement in the mid 1960s. In all cases the UK was successful, although in the case of the cut-off it was never really put to the test. Anxieties were misplaced as a negotiation on a fissile material cut-off treaty never occurred let alone was a treaty ever agreed, or even close to being agreed. More than 40 years later at the time of writing a cut-off treaty is no nearer realisation than it was in the late 1950s or mid 1960s; in both periods, however, a cut-off treaty would have seriously constrained UK plans for its nuclear stockpile in terms of total numbers and its composition (the mix of strategic and tactical weapons). It may also have damaged its nuclear submarine plans – both for its hunter killers and ballistic missile boats. Despite this the UK did not want to be seen obstructing an arms control initiative, especially one promoted by the US, and wanted to be in a position to support it whenever it could. Provided UK supplies could be guaranteed to meet its requirements, then the UK stood ready to back any initiative. Contingency planning is thus a key feature in UK approaches to arms control, and this was especially true in the SALT/ABM case. This is much the same story that we saw in the case of the Comprehensive Nuclear Test Ban Treaty in the 1950s.[1]

In SALT/ABM the UK was a bystander to the superpower strategic arms control negotiations in Vienna and Helsinki, albeit with a prime ringside seat. British officials enjoyed privileged access to US thinking and negotiating strategies, both formally through ACDA and through the back-channel with Kissinger. Britain's stated public position was one of support for attempts to

---

1 See John R. Walker, *British Nuclear Weapons and the Test Ban 1954–1973* (Farnham, 2010).

constrain the strategic nuclear arms race, as greater stability and predictability were seen as desirable in their own right, but there were two central issues that could have significant implications for the UK nuclear deterrent's continued viability. Although the US fended off Soviet attempts to include 'third country' forces in the scope of the strategic arms limitation talks, a much more serious threat was through pressure for no-transfer and non-circumvention clauses that would prevent or constrict US nuclear weapons cooperation with the UK. Given the time line of the talks from 1969 onwards, this coincided directly with UK efforts to maintain the viability and effectiveness of its strategic nuclear force and for which US assistance was essential. If the well of US technical, material, infrastructure and technology aid dried up then the UK programme would have been in dire straits – dependence made the UK much more of a supplicant than was perhaps envisaged back in 1958 and 1963. There was, however, a very real prize to be had from the US–Soviet talks. The main driver for the Polaris Improvement Programme had been fears over the growing capabilities of Soviet ballistic missile defences that, if unchecked, could seriously undermine the UK's modest deterrent's effectiveness. Seeking maximum constraints and preferably no ABMs at all was clearly in the UK interest, though officials recognised that a complete ban was not going to materialise. Nevertheless, limitations on missile defence systems – numbers of sites, launchers, re-loads and radars – could help sustain the credibility of the deterrent as holding Moscow hostage was the core consideration in UK deterrent criteria. However, as Vic Macklen pointed out at the time, welcome though the Anti-Ballistic Treaty (ABMT) was, it did not obviate completely the need for the Polaris Improvement Programme. The Treaty did, however, put things on a more predictable footing and even more so with the later amendment in 1974 that reduced the number of permitted ABM launcher sites from two to one. The UK was successful in ensuring that the Interim Agreement signed by Nixon and Brezhnev in Moscow in May 1972 contained no provision that would have conflicted with either the 1958 Mutual Defence Agreement or the 1963 Polaris Sales Agreement. No sooner than SALT I had been agreed than negotiations began on SALT II, and it was clear that the UK would still need to be on its guard against the twin threats of no-transfer or non-circumvention clauses. Although not addressed in detail in this volume, UK attitudes were exactly the same throughout the SALT II negotiations for the remainder of the decade.[2]

---

[2] See TNA FCO 46/1001, Strategic Arms Limitation Talks (SALT) between the Soviet Union and USA, 1973 and TNA FCO 46/1002 to 1006, 1973; TNA FCO 46/1157 to 1162 for 1974 for instance.

What Role, if Any, has the UK's Own Programme Experiences Played in Shaping its Policies?

When we review the UK's role in the origins of the Biological and Toxin Weapons Convention, there does not appear to have been a direct linkage between the experiences from the offensive programme in the Second World War and in the immediate post-war years and the decision to seek a treaty banning BW. As we saw, the immediate precursor was the perceived need for the UK to be seen as active in the disarmament field after the conclusion of the NPT, but this was not meant to be a case of window-dressing. Ministers were quite prepared to push ahead with an initiative on BW even though the US was not fully convinced at the time and a good deal of senior effort, including Ministers, went into the framing of the UK's working paper for the ENDC that would make the case for the separation of BW from CW for disarmament purposes. MOD experts, including the Director of the MRE, were closely involved in this process so undoubtedly a biological defence perspective informed policy making. What such a treaty banning BW should look like and what would be required for an effective one were uppermost in UK thinking. We can see this clearly enough from the strenuous efforts that the UK made to have an investigation system for cases of alleged use of biological weapons as a central feature of the proposed treaty. Although the UK recognised at the time that 'verification' (as the term was understood in the mid to late 1960s) was not feasible in the case of BW, it still wanted to see the most effective and meaningful compliance procedures included in any final agreement. That the UK ultimately failed does not diminish the importance attached to this objective. As we have seen from the subsequent history of the BTWC, addressing this compliance machinery deficit has been a recurring preoccupation of some of its States Parties ever since the Convention's first Review Conference in 1980 right down to the present.[3] We see this in UK approaches too in the nuclear area, whether Comprehensive Nuclear Test Ban Treaty, or cut-off; there should be effective verification to deter cheating. Support for effective verification and compliance mechanisms is one of the distinguishing features of UK arms control policies; a simple agreement was never enough, although they may not always have succeeded as the BTWC case shows. And this legacy was to continue shaping British approaches to arms control and disarmament well into the twenty-first century. Often times it had to remain an aspiration thwarted by the wiles of other states that were much less fussy or probably had something to hide in the first place.

---

3 See Jez Littlewood, *The Biological Weapons Convention: A Failed Revolution* (Aldershot, 2005) and Nicholas A. Sims, *The Future of Biological Disarmament: Strengthening the Treaty Ban on Weapons* (London, 2009).

Britain's own military and civil nuclear programmes provided a test bed for development of approaches to fissile material cut-off as well as the emergence of safeguards for the civil nuclear industry to ensure that material was not diverted for explosive devices. Knowledge of the types of facilities required and those aspects of their design and operation that could compromise classified defence information if divulged to inspectors gave the UK a head start in working out verification requirements. Almost from the start of the nuclear programmes in the UK, when the key fuel cycle facilities had been only running for a few years, we see practical verification exercises run to inform policy and technical approaches to the problems. Further work was done in the mid and late 1960s too as the UK contributed to the development of the IAEA safeguards system for the NPT; and this is one area where there is an untapped source of primary material in The National Archives that deserves further study for those wishing to delve deeper into UK policies.[4] We saw comparable work in seismology as UK efforts to support the development of an effective verification regime for a CTBT. Pragmatic fieldwork on verification technologies and procedures would be a hallmark of UK approaches to arms control verification again in the 1980s and 1990s during work for the future Chemical Weapons Convention as well as the attempts to develop a verification protocol for the BTWC.[5] Precedents set in early work in the 1950s and 1960s underline a strong empirical trait in British approaches to the problems of arms control and effective verification.

Does the UK Experience Tell us Anything More Generally About the Arms Control Process?

Drawing general lessons from the UK experience in nuclear and biological arms control and disarmament policies and experiences is perhaps of limited value. It is clear that the UK experience was unique. As a medium-sized nuclear power, its capabilities were always at the lower end of a 'minimum' deterrent and maintaining its credibility in the face of both technological change, and

4 See for example TNA FCO 10/193 to 205, Safeguards and Non-proliferation Treaty, 1967.

5 John R. Walker, 'Chemical Weapons Verification: The UK's Practice Challenge Inspection Programme at Government Facilities', in J.B. Poole (ed.), *Verification Report 1991: Yearbook on Arms Control and Environmental Agreements* (London, 1991); John R. Walker, A.P. Phillips and Lorna Miller, 'Verification of the Biological and Toxin Weapons Convention and the UK's Practice Compliance Inspection Programme: Final Report', in J.B. Poole and R. Guthrie (eds.), *Verification 1996: Arms Control, Peacekeeping and the Environment* (London, 1996).

pressures real or imagined from arms control and disarmament proved a constant challenge. It is clear too from the record that the UK wanted to be seen as a supporter of arms control, not just for public presentational reasons at home and abroad, but because in some circumstances it saw strategic and security value in effective agreements. Such agreements could either be multilateral in character or bilateral – as long as UK core defence programme equities were protected, then it could support actively the negotiation and conclusion of arms control and disarmament agreements. States act in their security interests; and in this sense the UK experience applies to others. Probably no other state, apart from the US and USSR involved in arms control or affected by the process, had the same range of defence interests as the UK and had to balance the at times conflicting pressures of the need to safeguard national capabilities and be seen as an international good citizen supporting arms control and disarmament initiatives.

As we saw in the chapters discussing the BTWC negotiations, considerable efforts went into arguing over the meaning of words and phrases. What from the outside might appear to be pedantic nit-picking was extremely important since the future scope and obligations not just for the UK but for all other States Parties would hinge on the proper interpretation of the final language agreed in the Convention. Meanings had to be as clear as possible, except in those cases where constructive ambiguity, such as in the NPT's Article I, was a necessary negotiating artefact to enable all parties to come away convinced that their interests were protected and that it did not serve anyone's aims to probe too deeply into the meaning of the language agreed. The UK's attempt to convince itself that certain formulations and phrases on issues such as use and research were just about sufficient to enable it to claim that at least some of its original core BW negotiating objectives were met. It was also clear that without US support the UK was just not strong enough to push through its demands, or to stand out alone and withhold consensus on a disarmament agreement. That this should be the case is hardly surprising and is no doubt a general rule applicable to all international treaty negotiations. Failure to secure the UK's initial objectives in the BTWC came to have a decisive impact on the Convention's future; States Parties, or at least some of them, have been striving to address the compliance provisions deficit ever since. Getting it wrong in the negotiations can come back to haunt a treaty's subsequent operation and implementation.

The Last Word – For Now

The archives surveyed for this study present a clear picture of a key 20-year period in British arms control, disarmament and defence policy: for most of the time the UK successfully balanced its national defence requirements with support for, and active and pragmatic contributions to, arms control and disarmament. This applies especially in the nuclear area where strategic nuclear weapons were central to UK defence policy; support for biological disarmament could be far more extensive and forward leaning precisely because there were no weapon programmes to protect, but that is not to devalue the nature of the UK contribution, which was immense. Without the UK imagination, drive and persistence there might have been no BTWC at all. Despite aspirations to reacquire an offensive CW capability and to develop incapacitating agents for a range of military purposes, the UK only ended up with a single agent – CR – and even here it was so nervous about its perception that it made it almost impossible for itself to use this agent in domestic law and order situations for which it had been developed. This tells us that there does not just have to be arms control pressures in play – fears of public reaction – domestic and international – can place significant constraints on capabilities too, at least certainly in a Western parliamentary democracy. Conversely, when things suit core defence interests, as in the reinterpretation of the Geneva Protocol in 1969, Britain was ready to weather the storm of protest.

Much more remains to be written about the UK and disarmament in this period as there are still plenty of untapped papers in The National Archives covering the period of this book, such as the negotiations behind the NPT's Article III (the safeguards Article); the grandiose plans for general and complete disarmament in the late 1950s and early 1960s; the first years of the BTWC's negotiations in 1969 and 1970; and the early debates on chemical disarmament in the 1970s, including the UK's own draft Convention tabled in 1976. Looking in detail at US and Russian attitudes to the UK as revealed in archival sources in Washington and Moscow would also be well worth investigating.

Indeed there are many files covering arms control issues beyond it too, such as the 1977–80 Trilateral Nuclear Test Ban talks between the UK, US and USSR all of which are now open, certainly in the FCO class, as are papers on the UK and the SALT II negotiations through to 1979. Moreover, as papers become available and with the move to a 20-year rule, we shall soon see even more detail that will reveal how and whether the legacy, experiences, attitudes and approaches that emerged and shaped UK policy in the 1950s and 1960s lived on into the late 1970s, 1980s and 1990s.

Annex
Draft Texts of a Biological and Toxin Weapons Convention 1969–1971[1]

UK Draft, 10 July 1969, ENDC/255

Article I

Each of the Parties to the convention undertakes never in any circumstances, by making use for hostile purposes of microbial or other biological agents causing death or disease by infection or infestation in man, other animals, or crops, to engage in biological methods of warfare.

Article II

Each of the Parties to the Convention undertakes
(a) not to produce or otherwise acquire, or assist in or permit the production or acquisition of
(i) microbial or other biological agents of types and in quantities that have no independent peaceful justification for prophylactic or other purposes;
(ii) ancillary equipment or vectors the purpose of which is to facilitate the use of such agents for hostile purposes.
(b) not to conduct, assist or permit research aimed at production of the kind prohibited in sub-paragraph (a) of this Article; and
(c) to destroy, or divert to peaceful purposes, within three months after the Convention comes into force for that Party, any stocks in its possession of such agents or ancillary equipment or vectors as have been produced or otherwise acquired for hostile purposes.

[1] Main Operative Articles only included here.

Article III

1. Any Party to the Convention which believes that biological methods of warfare have been used against it may lodge a complaint with the Secretary-General of the United Nations, submitting all evidence at its disposal in support of the complaint, and request that the complaint be investigated and that a report on the result of the investigation be submitted to the Security Council.
2. Any Party to the Convention which believes that another Party has acted in breach of its undertakings under Article I and II of the Convention, but which is not entitled to lodge a complaint under paragraph 1 of this Article may similarly lodge a complaint with the Security Council and request that the complaint be investigated.
3. Each of the Parties to the Convention undertakes to co-operate fully with the Secretary-General and his authorised representatives in any investigation he may carry out, as a result of a complaint, in accordance with the Security Council Resolution No...

Article IV

Each of the Parties to the Convention affirms its intention to provide or support appropriate assistance, in accordance with the United Nations Charter, to any other Party to the Convention, if the Security Council concludes that methods of biological warfare have been used against that Party.

Article V

Each of the Parties to the Convention undertakes to pursue negotiation in good faith on effective measures to strengthen the existing constraints on the use of chemical methods of warfare.

Article VI

Nothing contained in the present Convention shall be construed as in any way limiting or derogating from obligations assumed by any State under the Protocol for the Prohibition of the Use in War of Asphyxiating, Poisonous or other Gases, and of Bacteriological methods of Warfare signed at Geneva on 17 June 1925.

USSR and Socialist Group Draft, 30 March 1971, CCD/325

Article I

Each State Party to this Convention undertakes not to develop, produce, stockpile or otherwise acquire:
(1) microbiological or other biological agents or toxins of such types and in such quantities as are not designed for the prevention of disease or for other peaceful purposes;
(2) auxiliary equipment or means of delivery designed to facilitate the use of such agents or toxins for hostile purposes.

Article II

Each State Party to this Convention undertakes to destroy within a period of three months after the entry into force of the Convention – observing all the necessary precautions – or to convert to peaceful uses all previously accumulated weapons in its possession as well as the equipment and means of delivery mentioned in Article I of the Convention.

Article III

Each State Party to the Convention undertakes not to assist, encourage or induce any particular State, group of States or international organisations to take action contrary to the provisions of this Convention.

Article IV

Each State Party to the Convention shall be internationally responsible for compliance with its provisions by legal or physical persons of that State.

Article V

Each State Party to the Convention undertakes to take as soon as possible, in accordance with its constitutional procedures, the necessary legislative and administrative measures for prohibiting the development, production and stockpiling of the weapons, equipment and means of delivery mentioned in Article I of the Convention, and for destroying them.

Article VI

The States Parties to the Convention undertake to consult one another and to cooperate in solving any problems which may arise in the application of the provisions of this Convention.

Article VII

1. Each State Party to the Convention which finds that action of any other State Party constitutes a breach of the obligations assumed under the provisions of this Convention may lodge a complaint with the Security Council of the United Nations. Such a complaint should include all possible evidence confirming its validity, as well as a request for its consideration by the Security Council. The Council shall inform the States Parties to the Convention of the result of the investigation.
2. Each State Party to the Convention undertakes to co-operate in carrying out any investigations which the Security Council may undertake, in accordance with the provisions of the United Nations Charter, on the basis of the complaint received by the Council.

Article VIII

Nothing in this Convention shall be interpreted as in any way limiting or detracting from the obligations assumed by any State under the Geneva Protocol of 17 June 1925 on the Prohibition of the Use in War of Asphyxiating, Poisonous or other Gases, and of Bacteriological methods of Warfare, an instrument which embodies generally recognised rules of international law.

Article IX

Each State Party to this Convention undertakes to conduct negotiations in good faith on effective measures for prohibiting the development, production and stockpiling of chemical weapons and for their destruction, and on appropriate measures concerning equipment and means of delivery specifically designed for the production or use of chemical weapons as means of warfare.

Article X

1. The States Parties to this Convention undertake to facilitate, and have the right to participate in, the fullest possible exchange of equipment, materials

and scientific and technological information for the use of bacteriological (biological) agents and toxins for peaceful purposes.
2. This Convention shall be implemented in a manner designed to avoid hampering the economic or technological development of States Parties to the Convention or international co-operation in the field of peaceful bacteriological (biological) activities, including the international exchange of bacteriological (biological) agents and toxins and equipment for the processing, use or production of bacteriological (biological) agents and toxins for peaceful purposes in accordance with the provisions of the Convention.

Identical US and USSR Drafts, 5 August 1971, CCD/337 and CCD/338

Article I

Each State Party to this Convention undertakes not to develop, produce, stockpile or otherwise acquire or retain:
(1) Microbial or other biological agents or toxins of such types and in quantities that have no justification for prophylactic or other peaceful purposes;
(2) Weapons, equipment or means of delivery designed to use such agents or toxins for hostile purposes or in armed conflict.

Article II

Each State Party to this convention undertakes to destroy or to convert to peaceful purposes, as soon as possible but not later than ... months after the entry into force of the Convention – all agents, toxins, weapons, equipment and means of delivery specified in Article I of the Convention, which are in its possession or under its jurisdiction or control. In implementing the provision of this Article all necessary safety precautions shall be observed to protect the population and the environment.

Article III

Each State Party to this Convention undertakes not to transfer to any recipient whatsoever, directly or indirectly, and not in any way to assist, encourage, or induce any State, group of States or international organisations to manufacture or otherwise acquire any agent, toxin, weapon, equipment or means of delivery specified in Article I of the Convention.

Article IV

Each State Party to this Convention shall, in accordance with its constitutional processes, take any necessary measures to prohibit and prevent development, production, stockpiling, acquisition or retention of the agents, toxins, weapons, equipment and means of delivery specified in Article I of the Convention, within the territory of such State, under its jurisdiction or under its control anywhere.

Article V

The States Parties to this Convention undertake to consult one another and to co-operate in solving any problems which may arise in relation to the application of the provisions of the Convention.

Article VI

1. Each State Party to this Convention which finds that actions of any other State Party constitute a breach of obligations assumed under the provisions of the Convention may lodge a complaint with the Security Council of the United Nations. Such a complaint should include all possible evidence confirming its validity, as well as a request for its consideration by the Security Council. The Security Council shall inform the States Parties to the Convention of the results of the investigation.
2. Each State Party to this Convention undertakes to co-operate in carrying out any investigation which the Security Council may undertake, in accordance with the provisions of the Charter of the United Nations, on the basis of the complaint received by the Council.

Article VII

Nothing in this Convention shall be interpreted as in any way limiting or detracting from the obligations assumed by any State under the Geneva Protocol of 17 June 1925 for the Prohibition of the Use in War of Asphyxiating, Poisonous or Other Gases, and of Bacteriological Methods of Warfare.

Article VIII

Each State Party to this Convention undertakes to conduct negotiations in good faith on effective measures for the prohibiting of their development, production and stockpiling and for their destruction, and on appropriate measures concerning equipment and means of delivery specifically designed for the production or use of chemical weapons for warfare.

Article IX

1. The States Parties to this Convention undertake to facilitate, and have the right to participate in, the fullest possible exchange of equipment, materials and scientific and technological information for the use of bacteriological (biological) agents and toxins for peaceful purposes.
2. This Convention shall be implemented in a manner designed to avoid hampering the economic or technological development of States Parties to the Convention or international co-operation in the field of peaceful bacteriological (biological) activities, including the international exchange of bacteriological (biological) agents and toxins and equipment for the processing, use or production of bacteriological (biological) agents and toxins for peaceful purposes in accordance with the provisions of the Convention.

Final Draft, 28 September 1971, in CCD/356[2]

Article I

Each State Party to this Convention undertakes never in any circumstances to develop, produce, stockpile or otherwise acquire or retain:
(1) microbial or other biological agents, or toxins whatever their origin or method of production, of types and in quantities that have no justification for prophylactic, protective or other peaceful purposes;
(2) weapons, equipment or means of delivery designed to use such agents or toxins for hostile purposes or in armed conflict.

Article II

Each State Party to this Convention undertakes to destroy, or to divert to peaceful purposes, as soon as possible but not later than nine months after the entry into force of the Convention, all agents, toxins, weapons, equipment and means of delivery specified in Article I of the Convention, which are in its possession or under its jurisdiction or control. In implementing the provisions of this Article

[2] This was effectively the final text that was opened for signature on 10 April 1972. This text was co-sponsored by Bulgaria, Canada, Czechoslovakia, Hungary, Italy, Mongolia, Netherlands, Poland, Romania, USSR, UK and USA.

all necessary safety precautions shall be observed to protect populations and the environment.

Article III

Each State Party to this Convention undertakes not to transfer to any recipient whatsoever, directly or indirectly, and not in any way to assist, encourage, or induce any State, group of States or international organisations to manufacture or otherwise acquire any of the agents, toxins, weapons, equipment or means of delivery specified in Article I of the Convention.

Article IV

Each State Party to this Convention shall, in accordance with its constitutional processes, take any necessary measures to prohibit and prevent the development, production, stockpiling, acquisition or retention of the agents, toxins, weapons, equipment and means of delivery specified in Article I of the Convention, within the territory of such State, under its jurisdiction or under its control anywhere.

Article V

The States Parties to this Convention undertake to consult one another and to co-operate in solving any problems which may arise in relation to the objective of, or in the application of the provisions of, this Convention. Consultation and co-operation pursuant to this Article may also be undertaken through appropriate international procedures within the framework of the United Nations and in accordance with its Charter.

Article VI

1. Any State Party to this Convention which finds that any other State Party is acting in breach of obligations deriving from the provisions of this Convention may lodge a complaint with the Security Council of the United Nations. Such a complaint should include all possible evidence confirming its validity, as well as a request for its consideration by the Security Council.
2. Each State Party to this Convention undertakes to co-operate in carrying out any investigation which the Security Council may initiate, in accordance with the provisions of the Charter of the United Nations, on the basis of the complaint received by the Council. The Security Council shall inform the States Parties to the Convention of the results of the investigation.

Article VII

Each State Party to this Convention undertakes to provide or support assistance, in accordance with the United Nations Charter, to any Party to the Convention which so requests, if the Security Council decides that such Party has been exposed to danger as a result of violation of the Convention.

Article VIII

Nothing in this Convention shall be interpreted as in any way limiting or detracting from the obligations assumed by any State under the Geneva Protocol of 17 June 1925 for the Prohibition of the Use in War of Asphyxiating, Poisonous or Other Gases, and of Bacteriological Methods of Warfare.

Article IX

Each State Party to this Convention affirms the recognised objective of effective prohibition of chemical weapons and, to this end, undertakes to continue negotiations in good faith with a view to reaching early agreement on effective measures for the prohibition of their development, production and stockpiling and for their destruction, and on appropriate measures concerning equipment and means of delivery specifically designed for the production or use of chemical agents for weapons purposes.

Article X

1. The States Parties to this Convention undertake to facilitate, and have the right to participate in, the fullest possible exchange of equipment, materials and scientific and technological information for the use of bacteriological (biological) agents and toxins for peaceful purposes. Parties to the Convention in a position to do so shall also co-operate in contributing individually or together with other States or international organisations to the further development and application of scientific discoveries in the field of bacteriology (biology) for the prevention of disease, or for other peaceful purposes.
2. This Convention shall be implemented in a manner designed to avoid hampering the economic or technological development of States Parties to the Convention or international co-operation in the field of peaceful bacteriological (biological) activities, including the international exchange of bacteriological (biological) agents and toxins and equipment for the processing,

use or production of bacteriological (biological) agents and toxins for peaceful purposes in accordance with the provisions of the Convention.

Bibliography

Primary Sources: The National Archives (TNA), Kew

Biological and Toxin Weapons Convention and Origins

AIR 20/8727 Air Staff Target No. O.R./1065 Biological Warfare Agents and Weapons, and the associated problems of their storage and handling under service conditions, Air Ministry, (D.O.R. (A)), 16 July 1954.

CAB 21/4505 Biological and chemical warfare, 1952–60.

CAB 148/71/10 Biological warfare (BW), Note by the Secretaries, 1970.

CAB 148/92/16 Biological Warfare, Memorandum by the Minister of State for Foreign and Commonwealth Affairs, 1969.

CAB 168/125 Disarmament proposals concerning chemical and biological warfare, 1967–69.

DEFE 11/672 Biological and chemical warfare, 1968–69.

FCO 10/179 Biological and chemical weapons, 1967–68.

FCO 10/181 Biological and chemical weapons, 1968.

FCO 10/182 Biological and chemical weapons, 1968.

FCO 10/183 Biological and chemical weapons, 1968.

FCO 10/184 Biological and chemical weapons, 1967.

FCO 66/133 British initiative on biological warfare, 1968–69.

FCO 66/134 British initiative on biological warfare, 1968–69.

FCO 66/136 Convention for the prohibition of methods of biological warfare, 1969.

FCO 66/137 Convention for the prohibition of methods of biological warfare, 1969.

FCO 66/138 Convention for the prohibition of methods of biological warfare, 1969.

FCO 66/139 Convention for the prohibition of methods of biological warfare, 1969.

FCO 66/223 British draft convention for prohibition of methods of biological warfare, 1970.

FCO 66/224 British draft convention for prohibition of methods of biological warfare, 1970.

FCO 66/225 British draft convention for prohibition of methods of biological warfare, 1970.
FCO 66/298 UK policy on chemical and biological warfare and draft Biological Warfare Convention, 1970–71.
FCO 66/299 UK policy on chemical and biological warfare and draft Biological Warfare Convention, 1971.
FCO 66/300 UK policy on chemical and biological warfare and draft Biological Warfare Convention, 1971.
FCO 66/301 UK policy on chemical and biological warfare and draft Biological Warfare Convention, 1971.
FCO 66/302 UK policy on chemical and biological warfare and draft Biological Warfare Convention, 1971.
FCO 66/303 UK policy on chemical and biological warfare and draft Biological Warfare Convention, 1971.
FCO 66/304 UK policy on chemical and biological warfare and draft Biological Warfare Convention, 1971.
FCO 66/309 Draft UK convention for prohibition of methods of chemical and biological warfare, 1971.
FCO 66/1520 Sverdlovsk (including bilateral with Americans), 1980.
FCO 66/1521 Sverdlovsk – follow up, 1980.
FO 366/3386, Creation of research unit for disarmament, 1964.
FO 371/181368 FO Disarmament Research Unit and Advisory Panel on Disarmament, 1965.
FO 371/181369 FO Disarmament Research Unit and Advisory Panel on Disarmament, 1965.
FO 371/181370 FO Disarmament Research Unit and Advisory Panel on Disarmament, 1965.
FO 371/187448 Chemical and Biological Warfare, 1966.
PREM 13/3464 Scientific Warfare, Biological and chemical warfare, 1965–70.

Chemical Weapons Issues (Including Incapacitants and Riot Control Agents)

AIR 8/2391 Chemical and Biological Warfare, 1963–67.
CAB 21/4505 Biological and chemical warfare, 1952–60.
CAB 128/44 Part 2 C (69) 168, The Geneva Protocol and the Use of Riot Control Agents in War, 16 December 1969.
CAB 129/146 Part 2 CC (69) 61st Conclusions Item 3, 18 December 1969.
CAB 130/389 Working Group on Chemical and Biological Warfare: Meetings: 1–2, 1968.
CAB 131/16 Cabinet Defence Committee, October 1955.

CAB 131/17 Cabinet Defence Committee, D.C. (56) 13 (Revise), 4 July 1956.
CAB 131/17 Cabinet Defence Committee, D (56) 6th Meeting, 10 July 1956.
CAB 131/28 Cabinet Defence Committee, D (63) 14, 16 April 1963.
CAB 131/28 Cabinet Defence Committee, D (63) 6th Meeting, 3 May 1963.
CAB 148/71 Chemical and Biological Warfare. Note by the Secretaries 1969; Chemical Warfare and the Disarmament Negotiations. Note by the Secretaries 1969; Chemical Warfare. Memorandum by the Ministry of Defence 1969; Chemical Warfare, 1969.
CAB 148/91/4 The Geneva Protocol and the use of Riot Control Agents in War. 5. Chemical Warfare and the Disarmament Negotiations, 1969.
CAB 148/93 Chemical Warfare and the Disarmament Negotiations. Memorandum by the Minister of State for Foreign and Commonwealth Affairs, 1969.
CAB 148/110 Possible Arms Control Measures in the Field of Chemical Warfare, 1970.
CAB 148/111 Possible Arms Control Measures in the Field of Chemical Warfare. Memorandum by the Secretary of State for Foreign and Commonwealth Affairs, 1970.
CAB 148/113 Chemical Warfare and the Disarmament Negotiations. Note by the Secretaries 1970; United Kingdom Tactics on Chemical and Biological Warfare. Note by the Secretaries, 1970.
CAB 151/148 CS gas and the Geneva Protocol, 1970.
CAB 168/126 Disarmament proposals concerning chemical and biological warfare: discussions on use of CS gas, 1969–70.
CAB 168/127 Disarmament proposals concerning chemical and biological warfare: discussions on use of CS gas, 1971.
CAB 168/129 Disarmament of chemical and biological weapons, 1970.
CJ 4/456 The Maze Prison: proposed and existing security arrangements. Use of CR gas for riot control and to prevent escapes. Medical and political consequences. Consideration by Cabinet, MOD and FCO, 1973.
CJ 4/1338 Use of SPAD/CR gas in prisons, 1973–75.
DEFE 5/131 Chiefs of Staff Committee: Memoranda Nos 401–40, Chemical and Biological Warfare COS (62) 432, 1962.
DEFE 11/672 CW/BW – Future Policy, 1968–69.
DEFE 13/1055 Chemical warfare: defence establishments and international protocols, 1968–71.
DEFE 24/405 Biological and chemical warfare: Geneva Protocol and use of riot control agents, 1969–70.
DEFE 24/966 Northern Ireland: use of troops and chemical control materials in Northern Ireland prisons, 1975–78.

DEFE 24/967 Northern Ireland: dibenzoxazepine (CR) gas; chemical agent for use in riot control, 1974–77.
DEFE 24/1739 Disarmament and arms control: CB warfare, 1977.
FCO 8/712 Egyptian use of poison gas in Yemen, 1967.
FCO 8/713 Egyptian use of poison gas in Yemen, 1967.
FCO 66/216 UK policy on use of non-lethal chemical agents in war, 1970.
FCO 66/217 UK policy on use of non-lethal chemical agents in war, 1970.
FCO 66/218 UK policy on use of non-lethal chemical agents in war, 1970.
FCO 66/219 UK policy on use of non-lethal chemical agents in war, 1970.
FCO 66/221 UK policy on use in war of CS gas, 1970.
FCO 66/226 CS gas and Geneva Protocol 1925: problem of constitution involved in re-opening issue of use in war of CS gas, 1970.
FCO 66/500 UK policy on the use of CS gas and CR gas, 1973.
FCO 66/513 Trials of CR gas in Hong Kong, 1973.
FCO 66/514 Trials of CR gas in Hong Kong, 1973.
PREM 11/3099 Policy on chemical warfare, 1952–60.
WO 32/20163 Development of defensive chemical warfare capability with incapacitating agents, 1962–66.
WO 188/2704 Incapacitating and anti-riot agents, 1965–70.
WO 188/2705 Incapacitating and anti-riot agents, 1967–70.
WO 188/2706 Incapacitating and anti-riot agents: CS (orthochlorobenzylidene malononitrile) and CR (dibenzoxazepine) gases, 1952–72.
WO 188/2744 General Staff Target (GST) 3562: non-lethal temporarily incapacitating system ('aids to capture'), 1971–75.
WO 189/156 Biological testing of incapacitating agents: part 2; test using drugs with known effects on man, 1961.
WO 189/357 Oripavine derivative (TL 2636) as a potential incapacitating agent, 1963.
WO 189/422 Biological testing of incapacitating agents part IV, 1964.
WO 189/1078 The biological testing of incapacitating agents: Part 1, 1961.
WO 189/1104 The biological testing of incapacitating agents: Part 3, 1961.
WO 189/1493 New sensory irritants: possibility of producing persistent incapacitating skin pain from a chemical agent, 1965.
WO 189/2861 A physiological approach to physical incapacitation, 1961.
WO 195/15019 Conclusions and recommendations of CW Agents Group at 15th Tripartite conference, 1960.
WO 195/15033 Tripartite toxicological warfare: screening methods for incapacitating agents (paper TC/15/16), 1960.
WO 195/15169 Biological testing of incapacitating agents: Pt 1 methods, 1961.
WO 195/15170 Biological testing of incapacitating agents: Pt 2 results, 1961.

WO 195/15273 Biological testing of incapacitating agents, 1961.
WO 195/15386 Conclusions of 16th tripartite conference (CW agents group), 1962.
WO 195/15445 Non parametric statistics in assessment of incapacitation trials, 1962.
WO 195/15503 Oripavine derivative as a potential incapacitating agent, 1963.
WO 195/15518 Incapacitation by central inhibition: minutes of a meeting, 1963.
WO 195/16213 Summary of work on lysergic acid diethylamide [LSD], 1966.
WO 195/16214 Proposals for future work on LSD 25, 1966.
WO 195/16216 Paper on assessment of effects of LSD, 1966.
WO 195/16228 Applied Biology Committee: the estimation of LSD, 1966.
WO 195/16269 Biology Committee: note on pharmacological studies with LSD 25, 1966.
WO 195/16312 Exercise RECOUNT: preliminary report on field experiment (trials involving use of LSD), 1966.
WO 195/16429 Review of concepts of incapacitation, 1967.
WO 195/16691 Applied Biology Committee: effects of small doses of TL.2636 [a CW drug causing incapacitation] (ABC.58), 1968.
WO 195/16694 Applied Biology Committee: SMALL CHANGE [field trials with an incapacitant] (ABC.54), 1968.
WO 286/77 Chemical Warfare Policy, including the future of Nancekuke, 1953–57.

Cut-off and Safeguards

AB 6/2267 Allocation of fissile material and 'cut-off', 1959–60.
AB 9/2078 Cut-off and Safeguards Working Party: papers 1–23B (68), 1–29 (69), 1–19 (70) volume 1, 1968–70.
AB 9/2430 Cut-Off and Safeguards Working Party Minutes-agenda, 1966–70.
AB 16/1950 Operation Circus, 1956–64.
AB 16/2172 Sir William Penney's Working Party on Fissile Production Control Scheme, 1957–58.
AB 16/3301 International control of fissile material production, 1959–63.
AB 16/3302 International control of fissile material production, (extract from AB 16/3301), 1959.
AB 16/3708 Geneva negotiations for nuclear tests treaty: correspondence, 1960.
AB 16/4189 Future of Capenhurst, 1961–64.

AB 16/4778 The Technical Possibility of International Control of Fissile Material Production: paper by Sir W. Penney for Conference on Disarmament, 1962.
AB 38/176 Production Board: correspondence, 1961–67.
AB 38/187 Inspection and control: reports, 1962–67.
AB 38/401 Safeguarding of fissile material, 1955.
AB 38/515 Operation Circus: a plutonium safeguards exercise, 1955–58.
AB 38/607 Future of Calder, Chapelcross; correspondence and papers, 1968–70.
AB 41/90 Committee on Safeguards for Nuclear Materials: Technical Co-ordinating Committee for Safeguards and Cut-Off; correspondence, 1966.
AB 41/240 Committee on Safeguards for Nuclear Materials: Technical Co-ordinating Committee for Safeguards and Cut-Off, 1966.
AB 48/195 NRDC Papers and briefs, 1962–63.
AB 48/196 International Control of Fissile Material Safeguards Working Party, 1963–66.
AB 48/197 Defence liabilities for the U-235 Barter, 1962.
AB 48/335 Nuclear Requirements for Defence Committee: Working Party on the Implications of Reducing and Cutting Off of Supplies of Fissile material, 1964.
AB 48/336 Nuclear Requirements for Defence Committee: Working Party on the Implications of Reducing and Cutting Off of Supplies of Fissile Material, 1964–65.
AB 48/642 Disarmament: cut-off of production of fissile material for weapons purposes, 1961.
AB 48/1073 EURATOM Safeguards in the UK, 1972–73.
AB 65/186 Cut-off correspondence, 1962–66.
AIR 2/13745 Warhead for the medium range ballistic missile (OR 1139 and 1142), 1955–56.
AIR 2/18209 Nuclear Weapons for the Royal Air Force, 1969.
AIR 8/2393, Nassau Agreement – Shape and Role of RAF deterrent up 1970s, 1963–64.
AIR 20/12080 Nuclear Weapons policy, 1967–69.
AIR 20/12198 Nuclear weapons policy: RAF units in Germany, 1969.
AVIA 65/877 Procedure: service orders for nuclear weapons setting up of NRDC and AWPC, 1958–61.
AVIA 65/878 Procedure: service orders for nuclear weapons setting up of NRDC and AWPC, 1958–61.
AVIA 65/1111 Maralinga tests minor trials, 1957–59.
CAB 21/5728 Defence Review, 1964–65.
CAB 21/6037 Nuclear Tests (General), 1962–65.

CAB 21/6038 Nuclear Requirements for Defence Committee General, 1960–64.

CAB 130/130 Working Group on Disarmament: Meetings 1–18; Papers 1–23, 1957–60.

CAB 130/197 Supplies of Fissile Material: Meeting 1, 1964.

CAB 131/27 Cabinet Defence Committee, Meetings and Papers, 1962.

CAB 134/756 Cabinet Official Committee on Atomic Energy A.E. (O) (55) 127 Fissile Material for Civil Purposes. Note by the Atomic Energy Authority, 9 November 1955.

CAB 134/1330 Official Committee on Atomic Energy: Meetings 1–3; Papers 1–18, 1958.

CAB 134/1331 Official Committee on Atomic Energy: Meeting 1 (1959); Papers 1–10 (1959); Meetings 1–5 (1960); Papers 1–22 (1960).

CAB 134/2241 Nuclear Requirements for Defence Committee, 1964.

CAB 134/2296 Official Committee on Disarmament OD (63) 28, 19 December 1963.

CAB 134/2296 Official Committee on Disarmament OD (64) 10, 18 March 1964.

CAB 134/3014 Committee on Safeguards for Nuclear Materials: Meetings 1–6; Papers 1–17, 1967.

CAB 134/3015 Committee on Safeguards for Nuclear Materials: Meetings 1–8; Papers 1–19, 1968.

CAB 134/3016 Committee on Safeguards for Nuclear Materials: Meetings 1–4; Papers 1–14, 1968.

CAB 134/3017 Committee on Safeguards for Nuclear Materials: Meetings 1–3; Papers 1–11, 1970.

CAB 134/3020 Official Committee on Nuclear Policy General Sub-Committee, 1970.

CAB 134/3420 Official Committee on Nuclear Policy General Sub-Committee, 1971.

CAB 148/4 Defence and Oversea Policy (Official) Committee, Minute and papers, 1964.

CAB 148/4/7 The 'cut-off' as a measure to be taken in advance of stage 1 of a disarmament treaty. Note by the Ministry of Defence, 1964.

CAB 148/4/10 The 'cut-off' as a measure to be taken in advance of stage 1 of a disarmament treaty. Note by the Foreign Office, 1964.

CAB 148/4/19 The 'cut-off' as a measure to be taken in advance of stage 1 of a disarmament, 1964.

CAB 148/18/29 1 Nuclear Safeguards, 1965.

CAB 148/20/40 Defence and Oversea Policy (Official) Committee, Inspection by the International Atomic Energy Agency of the Central Electricity Generating Board Nuclear Power Station at Bradwell, Memorandum by the Minister of State for Foreign Affairs, 1965.
CAB 148/23/16 Nuclear Safeguards and the new UK/US Agreement, Note by the Minister of Technology, 1965.
CAB 148/50 Defence and Oversea Policy (Official) Committee Sub-Committee on Disarmament, 1964.
CAB 148/71 Defence and Oversea Policy (O) Committee Sub-committee on Disarmament OPDO, 1966–69.
CAB 148/71/2 The present position in the disarmament negotiation and the outlook. 2. Forthcoming bilateral discussions with the Americans: (i) Text of a non-proliferation treaty. (ii) Suspension of underground tests. (iii) Forthcoming bilateral discussions with the Americans: The question of the reduction of nuclear delivery vehicles, 1966.
CAB 165/745 Committee on safeguards for nuclear materials; composition and terms of reference, 1966–70.
CAB 165/746 Committee on safeguards for nuclear materials: chairman's brief, 1966–67.
CAB 165/747 Committee on safeguards for nuclear materials: arrangements of committee business, 1966–70.
DEFE 7/1888 Sub-Committee on Strategic Aspects of Atomic Energy, 1958–62.
DEFE 7/2217 Possible future nuclear deterrent weapons system, 1959–64.
DEFE 7/2379 Nuclear Tests Policy Committee, 1959.
DEFE 19/60 Safeguards for the non-proliferation treaty, 1968.
DEFE 19/61 Safeguards for the non-proliferation treaty, 1967–68.
DEFE 19/125 Production of weapons and fissile material, 1964–69.
DEFE 24/291 British Nuclear Tests, 1963–65.
DEFE 24/304 Storage, transportation and safety arrangements for nuclear weapons: special storage in Germany, 1969–70.
EG 1/115 Safeguarding of fissile material, 1955–57.
EG 1/495 Safeguards aspects of President Johnson's Disarmament proposals, 1964–65.
EG 8/84 Implementation of the UK/US offer on NPT safeguards, 1970–73.
FCO 10/162 Cut-off of fissile material for nuclear weapons, 1967–68.
FCO 10/188 Supply of fissile material, 1967–68.
FCO 10/193 Safeguards and non-proliferation treaty, 1967.
FCO 10/194 Safeguards and non-proliferation treaty, 1967.
FCO 10/195 Safeguards and non-proliferation treaty, 1967.

FCO 10/196 Safeguards and non-proliferation treaty, 1967.
FCO 10/197 Safeguards and non-proliferation treaty, 1967.
FCO 10/198 Safeguards and non-proliferation treaty, 1967.
FCO 10/199 Safeguards and non-proliferation treaty, 1967.
FCO 10/200 Safeguards and non-proliferation treaty, 1967.
FCO 10/201 Safeguards and non-proliferation treaty, 1967.
FCO 10/202 Safeguards and non-proliferation treaty, 1967.
FCO 10/203 Safeguards and non-proliferation treaty, 1967.
FCO 10/204 Safeguards and non-proliferation treaty, 1967.
FCO 10/205 Safeguards and non-proliferation treaty, 1967.
FCO 46/304 Missiles and Nuclear Matters; cooperation with the USA – 1958 Nuclear Agreement, 1968–69.
FCO 66/116 Cut-off of fissile material, 1969.
FO 371/123702 UN disarmament sub-committee, 1956.
FO 371/129818 Disarmament: interpretation of US proposal for cut-off date for production of fissile material for weapons, 1957.
FO 371/129820 Discussion on disarmament at UN, 1957.
FO 371/129825 Cut-off date for the production of fissionable material for military purposes: note from Minister of Defence to Prime Minister, 12 June 1957.
FO 371/129836 Disarmament treaty: Ministry of Defence comment on non-inclusion in treaty of any provision which will lead to cutting off fissile material production before a satisfactory control is in operation, 1957.
FO 371/129839 Revised US nuclear policy: letter from Harold Stassen to David Ormsby Gore, UK Minister for Foreign Affairs, commenting on 'cut-off' arrangements and resumption of nuclear testing 23 August 1957.
FO 371/132686 Geneva Negotiations on the suspension of nuclear tests, 1958.
FO 371/140520 Discussions between UK and US on cut-off of production of fissile material for weapon purposes, 1959.
FO 371/149355 Five Power Talks: US hesitation over talks on cut-off, 1960.
FO 371/149396 Cut-off, reconversion and missiles: extract from record of meeting held 6 May 1960.
FO 371/149437 Mr Cabot Lodge's statement on the cut-off and transfer of fissile material 16 August 1960.
FO 371/149438 Comments on the US proposal on cut-off and transfer of fissile material, 1960.
FO 371/171132 Proposed control and cut-off in production of fissile material, 1963.
FO 371/171133 Proposed control and cut-off in production of fissile material, 1963.

FO 371/176370 Control of fissile material production for defence purposes, 1964.
FO 371/176371 Control of fissile material production for defence purposes, 1964.
FO 371/176372 Control of fissile material production for defence purposes, 1964.
FO 371/176373 Control of fissile material production for defence purposes, 1964.
FO 371/181374 Cut-off in production of fissile material, 1965.
FO 371/181375 Cut-off in production of fissile material, 1965.
FO 371/187445 Cut-off in production of fissile material, 1966.
PREM 11/2168 Lord Cherwell's views on proceedings at UN Disarmament Committee on nuclear testing, 1957.
PREM 11/2841 Increased production of fissile material, 1953–59.
PREM 11/5172 Nuclear weapons programme: requirements for fissile material, 1963–64.
PREM 11/5181 Continuation of disarmament negotiations: test ban treaty; part 33, 1963–64.
PREM 11/5182 Negotiations on disarmament and suspension of nuclear tests, 1964.
PREM 13/2004 Supplies of tritium from US after 1968: production of lithium, tritium and uranium in UK; memorandum by Lord Penney; project Acarus; co-operation with European countries on nuclear energy, 1965–68.
T225/2500 Disarmament – Cut-off in production and transfer to peaceful uses of fissionable materials used for weapons, 1963–64.

ANF and NPT

CAB 21/6047 Atlantic nuclear force, 1964–65.
CAB 21/6048 Atlantic nuclear force, 1965.
CAB 130/212 Atlantic nuclear force, meeting 1964.
CAB 130/213 Atlantic nuclear force: meetings: 1–8; papers 1–14, 1964–65.
CAB 148/19/5 Atlantic nuclear force. Note by the Chairman of the Defence and Overseas Policy (Official) Committee, 1965.
CAB 148/35/19 Non-Proliferation Treaty, 1968.
CAB 148/41/3 and 4 Atlantic nuclear force, 1965.
CAB 148/41/11 Defence and Oversea Policy (Official) Committee, Atlantic Nuclear Force, Brief for Multilateral Discussion, 1965.
CAB 148/48 Sub-committee on Atlantic nuclear force, 1965–67.
CAB 148/51 Sub-committee on Disarmament, 1965.

CAB 151/138 Non-Proliferation Treaty, 1968.
CAB 168/16 Non-Proliferation Treaty and International Atomic Energy Agency (IAEA) safeguards, 1966–69.
DEFE 13/871 Disarmament, 1964–68.
DEFE 19/169 Non-Proliferation Treaty; working papers, 1967–73.
DEFE 25/33 Atlantic nuclear force, 1964–65.
DEFE 25/34 Atlantic nuclear force, 1965.
FO 115/4612 Non-dissemination and non-acquisition of nuclear weapons, 1965.
FO 115/4618 Non-dissemination and non-acquisition of nuclear weapons, 1966.
FO 371/179080 Atlantic nuclear force, 1964.
FO 371/181386 Non-dissemination of nuclear weapons, 1965.
FO 371/181388 Non-dissemination of nuclear weapons, 1965.
FO 371/181389 Non-dissemination of nuclear weapons, 1965.
FO 371/181390 Non-dissemination of nuclear weapons, 1965.
FO 371/181391 Non-dissemination of nuclear weapons, 1965.
FO 371/181392 Non-dissemination of nuclear weapons, 1965.
FO 371/181393 Non-dissemination of nuclear weapons, 1965.
FO 371/184406 Atlantic nuclear force, 1965.
FO 371/184407 Atlantic nuclear force, 1965.
FO 371/181411 Atlantic nuclear force, 1965.
FO 371/184412 Atlantic nuclear force, 1965.
FO 371/184413 Atlantic nuclear force, 1965.
FO 371/184414 Atlantic nuclear force, 1965.
FO 371/190665 Sharing of nuclear knowledge and non-proliferation, 1966.
FO 371/190666 Sharing of nuclear knowledge and non-proliferation, 1966.
FO 800/951 Secretary of State comments on the Atlantic Nuclear Force (ANF), 1964–65.
PREM 13/220 Atlantic nuclear force as alternative to multilateral force: part 4, 1965.
PREM 13/652 United Nations: disarmament negotiations, 1964–65.
PREM 13/2442 Disarmament negotiations: signing of non-proliferation treaty; part 6, 1968.

SALT/ABM

CAB 130/430 Strategic Arms Limitation Talks (SALT): Meeting: 1, 1969.
CAB 130/440 Strategic Arms Limitation Talks (SALT): Paper: 1, 1970.
CAB 130/446 Strategic Arms Limitation Talks (SALT): Meeting: 1, 1970.

CAB 130/547 International Aspects of Nuclear Defence Policy, 1972–73.
CAB 163/107 Soviet policy on disarmament: Russian position at the Strategic Arms Limitation Talks (SALT), 1969.
CAB 164/394 International aspects of nuclear defence policy: discussions concerning (Strategic Arms Limitation Talks) SALT, 1969.
CAB 164/936 Ministerial and official correspondence and briefs discussing various aspects of nuclear defence policy, positions of the US and USSR at the Strategic Arms Limitation Talks (SALT), 1970–71.
CAB 164/937 Ministerial and official correspondence and briefs discussing various aspects of nuclear defence policy, positions of the US and USSR at the Strategic Arms Limitation Talks (SALT), 1971–71.
CAB 168/277 Strategic Arms Limitation Talks (SALT), 1969–71.
DEFE 11/792 UK strategic nuclear deterrent, 1972–73.
DEFE 13/872 Disarmament, 1969–70.
DEFE 13/1032 Disarmament: Strategic Arms Limitation Talks (SALT), 1970–71.
DEFE 19/207 Ad hoc committees on Polaris improvements, 1972–73.
DEFE 24/770 Strategic Arms Limitation Talks (SALT), 1970.
DEFE 24/771 Strategic Arms Limitation Talks (SALT), 1970.
DEFE 24/772 Strategic Arms Limitation Talks (SALT), 1970–71.
DEFE 24/774 Strategic Arms Limitation Talks (SALT), 1971.
DEFE 24/775 Strategic Arms Limitation Talks (SALT), 1971.
DEFE 24/776 Strategic Arms Limitation Talks (SALT), 1971.
DEFE 24/777 Strategic Arms Limitation Talks (SALT), 1971.
DEFE 24/778 Strategic Arms Limitation Talks (SALT), 1972.
DEFE 24/779 Strategic Arms Limitation Talks (SALT), 1971–72.
DEFE 24/780 Strategic Arms Limitation Talks (SALT), 1972.
DEFE 24/781, Strategic Arms Limitation Talks (SALT), 1972.
DEFE 24/782 Strategic Arms Limitation Talks (SALT), 1972.
DEFE 24/783 Strategic Arms Limitation Talks (SALT), 1972.
DEFE 24/784 Strategic Arms Limitation Talks (SALT), 1972.
DEFE 24/785 Strategic Arms Limitation Talks (SALT), 1972.
DEFE 24/786 Strategic Arms Limitation Talks (SALT), 1972–73.
DEFE 24/888 Disarmament and arms control: Strategic Arms Limitation Treaty, 1970.
DEFE 24/889 Disarmament and arms control: Strategic Arms Limitation Treaty, 1970.
FCO 46/291 European talks: Strategic Arms Limitation (SALT), 1969.
FCO 46/292 European talks: Strategic Arms Limitation (SALT), 1969.
FCO 46/293 European talks: Strategic Arms Limitation (SALT), 1969.

FCO 46/294 European talks: Strategic Arms Limitation (SALT), 1969.
FCO 46/295 European talks: Strategic Arms Limitation (SALT), 1969.
FCO 46/296 European talks: Strategic Arms Limitation (SALT), 1969.
FCO 46/297 European talks: Strategic Arms Limitation (SALT), 1969.
FCO 46/298 European talks: Strategic Arms Limitation (SALT), 1969.
FCO 46/299 European talks: Strategic Arms Limitation (SALT), 1969.
FCO 46/300 European talks: Strategic Arms Limitation (SALT), 1969.
FCO 46/301 European talks: Strategic Arms Limitation (SALT), 1969.
FCO 46/302 European talks: Strategic Arms Limitation (SALT), 1969.
FCO 46/590 Talks between the Soviet Union and USA on Strategic Arms Limitation, 1970.
FCO 46/591 Talks between the Soviet Union and USA on Strategic Arms Limitation, 1970.
FCO 46/592 Talks between the Soviet Union and USA on Strategic Arms Limitation, 1970.
FCO 46/593 Talks between the Soviet Union and USA on Strategic Arms Limitation, 1970.
FCO 46/594 Talks between the Soviet Union and USA on Strategic Arms Limitation, 1970.
FCO 46/595 Talks between the Soviet Union and USA on Strategic Arms Limitation, 1970.
FCO 46/596 Talks between the Soviet Union and USA on Strategic Arms Limitation, 1970.
FCO 46/597 Talks between the Soviet Union and USA on Strategic Arms Limitation, 1970.
FCO 46/598 Talks between the Soviet Union and USA on Strategic Arms Limitation, 1970.
FCO 46/696 Strategic Arms Limitations Talks (SALT) between Soviet Union and USA, 1971.
FCO 46/697 Strategic Arms Limitations Talks (SALT) between Soviet Union and USA, 1971.
FCO 46/698 Strategic Arms Limitations Talks (SALT) between Soviet Union and USA, 1971.
FCO 46/699 Strategic Arms Limitations Talks (SALT) between Soviet Union and USA, 1971.
FCO 46/700 Strategic Arms Limitations Talks (SALT) between Soviet Union and USA, 1971.
FCO 46/701 Strategic Arms Limitations Talks (SALT) between Soviet Union and USA, 1971.

FCO 46/702 Strategic Arms Limitations Talks (SALT) between Soviet Union and USA, 1971.
FCO 46/703 Strategic Arms Limitations Talks (SALT) between Soviet Union and USA, 1971.
FCO 46/704 Strategic Arms Limitations Talks (SALT) between Soviet Union and USA, 1971.
FCO 46/822 Strategic Arms Limitations Talks (SALT) between Soviet Union and USA, 1972.
FCO 46/823 Strategic Arms Limitations Talks (SALT) between Soviet Union and USA, 1972.
FCO 46/824 Strategic Arms Limitations Talks (SALT) between Soviet Union and USA, 1972.
FCO 46/825 Strategic Arms Limitations Talks (SALT) between Soviet Union and USA, 1972.
FCO 46/826 Strategic Arms Limitations Talks (SALT) between Soviet Union and USA, 1972.
FCO 46/827 Strategic Arms Limitations Talks (SALT) between Soviet Union and USA, 1972.
FCO 46/828 Strategic Arms Limitations Talks (SALT) between Soviet Union and USA, 1972.
FCO 46/829 Strategic Arms Limitations Talks (SALT) between Soviet Union and USA, 1972.
FCO 46/830 Strategic Arms Limitations Talks (SALT) between Soviet Union and USA, 1972.
FCO 46/1001 Strategic Arms Limitation Talks between the Soviet Union and USA, 1973.
FCO 46/1002 Strategic Arms Limitation Talks between the Soviet Union and USA, 1973.
FCO 46/1003 Strategic Arms Limitation Talks between the Soviet Union and USA, 1973.
FCO 46/1004 Strategic Arms Limitation Talks between the Soviet Union and USA, 1973.
FCO 46/1005 Strategic Arms Limitation Talks between the Soviet Union and USA, 1973.
FCO 46/1006 Strategic Arms Limitation Talks between the Soviet Union and USA, 1973.
FCO 46/1157 Strategic Arms Limitation Talks (SALT) between Soviet Union and USA, 1974.
FCO 46/1158 Strategic Arms Limitation Talks (SALT) between Soviet Union and USA, 1974.

FCO 46/1159 Strategic Arms Limitation Talks (SALT) between Soviet Union and USA, 1974.
FCO 46/1160 Strategic Arms Limitation Talks (SALT) between Soviet Union and USA, 1974.
FCO 46/1161 Strategic Arms Limitation Talks (SALT) between Soviet Union and USA, 1974.
FCO 46/1162 Strategic Arms Limitation Talks (SALT) between Soviet Union and USA, 1974.
FCO 66/109 Strategic Arms Limitation Talks (SALT), 1968.
FCO 66/110 Strategic Arms Limitation Talks (SALT), 1968.
FCO 66/111 Strategic Arms Limitation Talks (SALT), 1968.
FCO 66/112 Strategic Arms Limitation Talks (SALT), 1968.
FCO 66/113 Strategic Arms Limitation Talks (SALT), 1968.
FCO 66/114 Strategic Arms Limitation Talks (SALT), 1968.
FCO 66/115 Strategic Arms Limitation Talks (SALT), 1968.
FCO 66/210 Strategic Arms Limitation Talks (SALT), 1970.
FCO 66/485 Strategic Arms Limitation Talks (SALT), 1973.
PREM 13/2569 Strategic Arms Limitation Talks (SALT), 1969.
PREM 13/3131 Strategic Arms Limitation Talks (SALT): part 2, 1969–70.
PREM 15/289 Talks on strategic arms limitation (SALT), 1970–71.
PREM 15/676 Soviet Ambassador, Mr Smirnovsky, called on Prime Minister to deliver Soviet Government statement on nuclear matters, including SALT and disarmament conference, 1971.
PREM 15/1365 Strategic Arms Limitation Talks (SALT): Operation Hullabaloo; part 2, 1972–73.
PREM 15/2038 Polaris improvement: consideration of Super Antelope/ Option M; future of UK strategic nuclear deterrent; part 3, 1973–74.

Secondary Sources

Alibek, Ken, with Handelman, Stephen, *BioHazard* (London: Hutchinson, 1999).
Arnold, Lorna, *Windscale 1957: Anatomy of a Nuclear Accident*, Third Edition (London: Palgrave, 2007).
Balmer, Brian, *Britain and Biological Warfare Expert Advice and Science Policy, 1930–65* (London: Palgrave, 2001).
Balmer, Brian, 'The UK Biological Weapons Program', in Mark Wheelis, Lajos Rozsa and Malcolm Dando (eds.), *Deadly Cultures: Biological Weapons Since 1945* (Cambridge, MA: Harvard University Press, 2006).

Balmer, Brian, 'Keeping Nothing Secret: United Kingdom Chemical Warfare Policy in the 1960s', *The Journal of Strategic Studies* vol. 33, no. 6 (2010), pp. 871–93.

Baylis, John and Stoddart, Kristan, 'Britain and the Chevaline Project: The Hidden Nuclear Programme, 1967–1982', paper prepared for Mountbatten Centre Fifth Seminar on the History of the UK Nuclear Weapon Programme, Thursday 9 October 2003.

Brennan, Donald G. (ed.), *Arms Control and Disarmament and National Security* (New York: George Braziller, 1961).

Bull, Hedley, *The Control of the Arms Race* (New York: Frederick A. Praeger, 1961).

Buteux, Paul, *The Politics of Nuclear Consultation in NATO 1965–1980* (Cambridge: Cambridge University Press, 1983).

Carter, Gradon B., *Chemical and Biological Defence at Porton Down 1916–2000* (London: The Stationery Office, 2000).

Carter, Gradon B., and Pearson, Graham S., 'Past British Chemical Warfare Capabilities', *RUSI Journal* vol. 141, no. 1 (February 1996), pp. 59–68.

Chevrier, Marie Isabelle, 'The Politics of Biological Disarmament', in Mark Wheelis, Lajos Rozsa, and Malcolm Dando (eds.), *Deadly Cultures: Biological Weapons Since 1945* (Cambridge, MA: Harvard University Press, 2006).

Cmnd 4725, *Further Documents on Disarmament: The Disarmament Negotiations 1970* (London: HMSO, August 1971).

Corrigan, Gordon, *Mud, Blood and Poppycock* (London: Cassell, 2003).

Davison, Neil, *'Non-Lethal' Weapons* (London: Palgrave, 2009).

Documents on Disarmament 1945–1959, Volume I: 1945–1956, Department of State, Historical Office Bureau of Public Affairs (August 1960).

Domaradskij, Igor V., and Orent, Wendy, *BioWarrior: Inside the Soviet/Russian Biological War Machine* (New York: Prometheus Books, 2003).

English, Richard, *Irish Freedom The History of Nationalism in Ireland* (London: Macmillan, 2006).

Foster, Robert Fitzroy, *Luck and the Irish: A Brief History of Change 1970–2000* (London: Allen Lane, 2007).

Garthoff, Raymond L., *Détente and Confrontation: American–Soviet Relations from Nixon to Reagan* (Washington, DC: The Brookings Institution, 1985).

Gowing, Margaret, *Independence and Deterrence: Britain and Atomic Energy 1945–1952*, Vol. 1: *Policy Making* and Vol. 2: *Policy Execution* (London: Macmillan, 1974).

Hammond, Peter M., and Carter, Gradon B., *From Biological Warfare to Healthcare: Porton Down, 1940–2000* (London: Palgrave, 2002).

Hansen, Chuck, *U.S. Nuclear Weapons: The Secret History* (Arlington, TX: Aerofax, 1988).

Hoffman, David E., *The Dead Hand: Reagan, Gorbachev and the Untold Story of the Cold War Arms Race* (London: Icon Books, 2011).

Howlett, Darryl A., *EURATOM and Nuclear Safeguards* (Basingstoke: Macmillan, 1990).

International Atomic Energy Agency, INFCIRC/86, *The Texts of the Agreements for the Application of Agency Safeguards to the Bradwell Reactor Facilty in the United Kingdom*, 23 September 1966.

International Atomic Energy Agency, INFCIRC/263, *The text of the Agreement of 6 September 1976 between the United Kingdom of Great Britain and Northern Ireland, the European Atomic Energy Community and the Agency in connection with the Treaty on the Non-Proliferation of Nuclear Weapons*, October 1978.

Kelle, Alexander, Nixdorff, Kathryn, and Dando, Malcolm, *Controlling Biochemical Weapons* (Basingstoke: Palgrave, 2006).

Littlewood, Jez, *The Biological Weapons Convention: A Failed Revolution* (Aldershot: Ashgate, 2005).

Macmillan, Harold, *Riding the Storm 1956–1959* (London: Macmillan, 1971).

McCamley, N.J., *Secret History of Chemical Warfare* (Barnsley: Pen and Sword, 2006).

McIntyre, Donald, *UK Nuclear History Working Paper Number 5: Project Crystal: Lithium 6 for ThermonuclearWeapons*, at http://www.mcis.soton.ac.uk/Site_Files/pdf/nuclear_history/Working_Paper_No_5.pdf, accessed 31 March 2011.

Meselson, Mathew, 'Tear Gas in Vietnam and the Return of Poison Gas', *Bulletin of Atomic Scientists* vol. 27, no. 3 (March 1971), pp. 17–19.

Moore, Richard, *Nuclear Illusion, Nuclear Reality: Britain, the United States and Nuclear Weapons, 1958–64* (London: Palgrave Macmillan, 2010).

Palazzo, Albert, *Seeking Victory on the Western Front: The British Army and Chemical Warfare in World War I* (Lincoln, NE and London: University of Nebraska Press, 2000).

Pearson, Graham, and Carter, Gradon, 'North Atlantic Chemical and Biological Research Collaboration: 1916–1995', *The Journal of Strategic Studies* vol. 19, no. 1 (March 1996), pp. 74–103.

Programme for Promoting Nuclear Non-Proliferation, *NPT Briefing Book Volume 1* (Southampton, April 2004).

Robinson, J.P., *Disabling Chemical Weapons: A Documented Chronology of Events 1945–2003* (Harvard-Sussex Program, 2003).

Schelling, Thomas C., and Halperin, Morton H., *Strategy and Arms Control* (Washington, DC: Pergamon-Brassey's, McLean, 1985).

Seaborg, Glenn T., with Loeb, Benjamin S., *Stemming the Tide: Arms Control in the Johnson Years* (Lexington, MA: Lexington Books, 1987).

Shaker, Mohamed I., *The Nuclear Non-Proliferation Treaty: Origin and Implementation 1959–1979 Volume 1* (London: Oceana Publications, 1980).

Sims, Nicholas A., *The Diplomacy of Biological Disarmament: Vicissitudes of a Treaty in Force, 1975–85* (London: Macmillan, 1988).

Sims, Nicholas A., *The Evolution of Biological Disarmament, SIPRI Chemical and Biological Warfare Studies 19* (Oxford: Oxford University Press, 2001).

Sims, Nicholas A., *The Future of Biological Disarmament: Strengthening the Treaty Ban on Weapons* (London: Routledge, 2009).

SIPRI, *The Problem of Chemical and Biological Warfare, Volume I: The Rise of CB Weapons* (Stockholm: Almqvist & Wiksell, 1971).

SIPRI, *Ecological Consequences of the Second Indochina War* (Stockholm: Almqvist & Wiksell, 1976).

Smith, Gerard, *Double Talk: The Story of SALT 1* (London: University Press of America, 1985).

Stocker, Jerry, *Britain and Ballistic Missile Defence 1942–2002* (London: Frank Cass, 2004).

Stoddart, Kristan, *The Wilson Government and British Responses to Anti-Ballistic Missiles, 1964–1970*, paper prepared for British Rocket Oral History Programme Charterhouse Conference, 2–4 April 2005.

Strengthening the Biological and Toxin Weapons Convention: Countering the Threat from Biological Weapons, Cm 5484 Presented to Parliament by the Secretary of State for Foreign and Commonwealth Affairs by Command of Her Majesty (London: The Stationery Office, April 2002).

UKAEA, *Annual Eleventh Annual Report 1964–65* (London: HMSO, 1965).

United Nations, *Chemical and Bacteriological (Biological) Weapons and the Effects of their Possible Use, Report of the Secretary-General* (New York, 1969).

Walker, John R., 'Chemical Weapons Verification: The UK's Practice Challenge Inspection Programme at Government Facilities', in J.B. Poole (ed.), *Verification Report 1991: Yearbook on Arms Control and Environmental Agreements* (London: VERTIC, 1991), pp. 97–104.

Walker, John R., *British Nuclear Weapons and the Test Ban 1954–1973 Britain, the United States, Weapons Policies and Nuclear Testing: Tensions and Contradictions* (Farnham: Ashgate, 2010).

Walker, John R., Phillips, A.P., and Miller, Lorna, 'Verification of the Biological and Toxin Weapons Convention and the UK's Practice Compliance

Inspection Programme: Final Report', in J.B. Poole and R. Guthrie (eds.), *Verification 1996: Arms Control, Peacekeeping and the Environment* (London: VERTIC and Westview Press, 1996), pp. 171–91.

Wheelis, Mark, Rozsa, Lajos, and Dando, Malcolm (eds.), *Deadly Cultures: Biological Weapons since 1945* (Cambridge, MA: Harvard University Press, 2006).

World Health Organisation, *Public Health Response to Biological and Chemical Weapons: WHO Guidance, Second Edition of Health Aspects of Chemical and Biological Weapons: Report of a WHO Group of Consultants, Geneva World Health Organization, 1970* (Geneva: WHO, 2004).

Wright, Susan, 'Geopolitical Origins', in Susan Wright (ed.), *Biological Warfare and Disarmament: New Problems/New Perspectives* (Lanham, MD: Rowman & Littlefield, 2002), pp. 313–42.

Wynn, Humphrey, *RAF Nuclear Deterrent Forces* (London: HMSO, 1994).

Index

For Product Safety Concerns and Information please contact our EU representative GPSR@taylorandfrancis.com Taylor & Francis Verlag GmbH, Kaufingerstraße 24, 80331 München, Germany